Methods in Molecular Biology

Series Editor
John M. Walker
School of Life and Medical Sciences
University of Hertfordshire
Hatfield, Hertfordshire AL10 9AB, UK

For further volumes:
http://www.springer.com/series/7651

Avian and Reptilian Developmental Biology

Methods and Protocols

Edited by

Guojun Sheng

IRCMS, Kumamoto University
Kumamoto, Kumamoto, Japan

Humana Press

Editor
Guojun Sheng
IRCMS
Kumamoto University
Kumamoto, Kumamoto, Japan

ISSN 1064-3745 ISSN 1940-6029 (electronic)
Methods in Molecular Biology
ISBN 978-1-4939-7215-9 ISBN 978-1-4939-7216-6 (eBook)
DOI 10.1007/978-1-4939-7216-6

Library of Congress Control Number: 2017943683

Printed on acid-free paper

This Humana Press imprint is published by Springer Nature
The registered company is Springer Science+Business Media LLC
The registered company address is: 233 Spring Street, New York, NY 10013, U.S.A.

Preface

Birds and nonavian reptiles are collectively referred to as sauropsid amniotes. The chick model, representing both the birds and the saurospids, made significant contributions to the field of developmental biology in the twentieth century. Its popularity has been waning due to the lack of tractable genetic tools, in contrast to rapid breakthroughs made in mammalian model systems. In the age of omics and stem cell biology, however, reliance on classical genetics is no longer a necessity. We are currently witnessing the re-emergence of avian and reptilian models as powerful tools in the developmental, evolutionary, eco-devo, learning, and behavioral research fields.

The aim of this book is to present readers with recent advances in avian and reptilian biology that are making such re-emergence possible. In addition to chapters focusing on the chick model, more than a half of the book describes techniques used in other avian and reptilian models. Based on the topics they cover, 23 chapters are grouped into four parts: (1) *Genomics and Transcriptomics*; (2) *Genetic Manipulation*; (3) *Stem Cells*; and (4) *New Model Systems*. The first part includes chapters detailing how to perform genomic and transcriptomic analyses in birds and reptiles. This part is especially timely given that high-throughput sequencing has become a routine in most labs. The second part highlights recent technological advancement in avian genetic manipulation, including utilization of the CRISPR/Cas9 system for gene editing and methods for germline and transient transgenesis. The third part focuses on techniques concerning the handling of pluripotent cells, and the last part covers emerging models in avian and reptilian developmental biology. Fitting for a forward-looking methods book, the opening commentary reminds us that as we embrace new technologies we should not lose sight of the embryo. By bringing together researchers taking diverse experimental approaches and specializing in different sauropsid models, this MIMB book on *Avian and Reptilian Developmental Biology* is expected to facilitate cross-talk and collaboration among its readers and to push this field forward to a new level.

I would like to thank Prof. John Walker for his continuous support, Ms. Yuka Endo for her secretarial assistance, and all the authors for their time and patience in the course of the preparation of this book.

Kumamoto, Japan ***Guojun Sheng***

Contents

Part III Stem Cells

Part IV New Model Systems

Contributors

ALAA ABDELHAMID • *Qassim College of Dentistry, Qassim University, Buraydah, Saudi Arabia*

KATIE L. AYERS • *Murdoch Children's Research Institute, Royal Children's Hospital, The University of Melbourne, Melbourne, VIC, Australia; Department of Paediatrics, Royal Children's Hospital, The University of Melbourne, Melbourne, VIC, Australia*

FEDERICA BERTOCCHINI • *Instituto de Biomedicina y Biotecnologia de Cantabria-CSIC-Universidad de Cantabria-Sodercan, Santander, Spain*

YINGGUANG FRANK CHAN • *Friedrich Miescher Laboratory of the Max Planck Society, Tübingen, Germany*

CHENG-MING CHUONG • *Integrative Stem Cell Center, China Medical University Hospital, China Medical University, Taichung, Taiwan; Department of Pathology, Keck School of Medicine, University of Southern California, Los Angeles, CA, USA; Research Center for Developmental Biology and Regenerative Medicine, National Taiwan University, Taipei, Taiwan*

RUSLAN DEVIATIIAROV • *Institute of Fundamental Medicine and Biology, Kazan Federal University, Kazan, Russia*

RAUL E. DIAZ JR. • *Department of Biology, La Sierra University, Riverside, CA, USA; Natural History Museum of Los Angeles County, Los Angeles, CA, USA; Department of Biology, University of California, Riverside, CA, USA*

SCOTT V. EDWARDS • *Department of Organismic and Evolutionary Biology, Harvard University, Cambridge, MA, USA; Museum of Comparative Zoology, Harvard University, Cambridge, MA, USA*

AMR ELKARARGY • *Qassim College of Dentistry, Qassim University, Buraydah, Saudi Arabia*

AURELIE FUET • *Univ Lyon, Université Lyon 1, INSERM, INRA, Stem Cell and Brain Research Institute, U1208, USC1361, Bron, France*

PHIL GRAYSON • *Department of Organismic and Evolutionary Biology, Harvard University, Cambridge, MA, USA; Museum of Comparative Zoology, Harvard University, Cambridge, MA, USA*

MARIANA B. GRIZANTE • *School of Life Sciences, Arizona State University, Tempe, AZ, USA*

OLEG GUSEV • *Institute of Fundamental Medicine and Biology, Kazan Federal University, Kazan, Russia; Division of Genomic Technologies, Center for Life Science Technologies (CLST), RIKEN, Yokohama, Japan; Preventive Medicine and Diagnosis Innovation Program, RIKEN, Yokohama, Japan; RIKEN Innovation Center, Wako, Japan*

JAE YONG HAN • *Department of Agricultural Biotechnology, College of Agriculture and Life Sciences, Seoul National University, Seoul, South Korea*

CARINA HANASHIMA • *Laboratory for Neocortical Development, RIKEN Center for Developmental Biology, Kobe, Japan; Department of Biology, Graduate School of Science, Kobe University, Kobe, Japan*

HIDEKIYO HARADA • *Genetics and Development Division, Krembil Research Institute, University of Toronto, Toronto, ON, Canada; Department of Physiology, Faculty of Medicine, University of Toronto, Toronto, ON, Canada*

CLAIRE E. HIRST • *Department of Anatomy and Development Biology, Monash University, Clayton, VIC, Australia*

PEI-SHAN HOU • *Laboratory for Neocortical Development, RIKEN Center for Developmental Biology, Kobe, Japan*
MARIAN HRUSKA-PLOCHAN • *Institute of Molecular Life Sciences, University of Zurich, Zurich, Switzerland*
DAVID HUSS • *Department of Radiology and Developmental Neuroscience Program, Saban Research Institute, Children's Hospital Los Angeles, Los Angeles, CA, USA; Translational Imaging Center, University of Southern California, Los Angeles, CA, USA*
MAKIKO IWAFUCHI-DOI • *Smilow Center for Translational Research, University of Pennsylvania Perelman School of Medicine, Philadelphia, PA, USA*
M. KHALIL KHAN • *Qassim College of Dentistry, Qassim University, Buraydah, Saudi Arabia*
BONNIE K. KIRCHER • *Department of Biology, University of Florida, Gainesville, FL, USA*
HISATO KONDOH • *Faculty of Life Sciences, Kyoto Sangyo University, Kita-ku, Kyoto, Japan*
TAKUMA KUMAMOTO • *Laboratory for Neocortical Development, RIKEN Center for Developmental Biology, Kobe, Japan; Sorbonne Universités, UPMC Univ Paris 06, INSERM U968, CNRS UMR 7210, Institut de la Vision, Paris, France*
KENRO KUSUMI • *School of Life Sciences, Arizona State University, Tempe, AZ, USA*
YUNG-CHIH LAI • *Integrative Stem Cell Center, China Medical University Hospital, China Medical University, Taichung, Taiwan; Department of Pathology, Keck School of Medicine, University of Southern California, Los Angeles, CA, USA; Research Center for Developmental Biology and Regenerative Medicine, National Taiwan University, Taipei, Taiwan*
RUSTY LANSFORD • *Department of Radiology and Developmental Neuroscience Program, Saban Research Institute, Children's Hospital Los Angeles, Los Angeles, CA, USA; Keck School of Medicine, University of Southern California, Los Angeles, CA, USA*
BO RAM LEE • *Department of Agricultural Biotechnology, College of Agriculture and Life Sciences, Seoul National University, Seoul, South Korea*
JING LI • *Life Sciences Institute, The Key Laboratory of Conservation Biology for Endangered Wildlife of the Ministry of Education, Zhejiang University, Hangzhou, Zhejiang, China*
WAN-CHUN LIU • *Department of Psychology and Neuroscience Program, Colgate University, Hamilton, NY, USA*
MARINA LIZIO • *Division of Genomic Technologies, Center for Life Science Technologies (CLST), RIKEN, Yokohama, Japan; Omics Science Center (OSC), RIKEN, Yokohama, Japan*
CHRISTOPHE MARCELLE • *EMBL Australia, Australian Regenerative Medicine Institute, Monash University, Clayton, VIC, Australia; Institut NeuroMyoGène, INMG, Faculty of Medicine Laënnec, University Lyon1, Lyon, France*
MEI MATSUZAKI • *Faculty of Agriculture, Shizuoka University, Shizuoka, Japan; United Graduate School of Agricultural Science, Gifu University, Gifu, Japan*
ATSUSHI MIYANOHARA • *Vector Development Core Laboratory, Department of Anesthesiology, UCSD School of Medicine, La Jolla, CA, USA*
SHUSEI MIZUSHIMA • *Department of Biological Sciences, Faculty of Science, Hokkaido University, Kita-ku, Sapporo, Hokkaido, Japan*
VALÉRIE MORIN • *Institut NeuroMyoGène, INMG, Faculty of Medicine Laënnec, University Lyon1, Lyon, France*
HARUKAZU NAKAMURA • *Frontier Research Institute for Interdisciplinary Science (FRIS), Tohoku University, Aoba-ku, Sendai, Japan*
NAOKO NISHIMURA • *Shizuoka Prefecture Special Education Schools, Yahatano, Ito, Shizuoka, Japan*

TADASHI NOMURA • *Developmental Neurobiology, Kyoto Prefectural University of Medicine, Sakyoku, Kyoto, Japan*
MINORU OMI • *Department of Anatomy I, School of Medicine, Fujita Health University, Toyoake, Aichi, Japan*
BERTRAND PAIN • *Univ Lyon, Université Lyon 1, INSERM, INRA, Stem Cell and Brain Research Institute, U1208, USC1361, Bron, France*
TOBIAS PIEPER • *Anatomy and Embryology, University Medical Centre, Göttingen, Germany*
KELLY N. ROESZLER • *Murdoch Children's Research Institute, Royal Children's Hospital, The University of Melbourne, Melbourne, VIC, Australia; Department of Paediatrics, Royal Children's Hospital, The University of Melbourne, Melbourne, VIC, Australia*
TIMOTHY B. SACKTON • *Informatics Group, Faculty of Arts and Sciences, Harvard University, Cambridge, MA, USA*
THOMAS J. SANGER • *Department of Biology, Loyola University in Chicago, Chicago, IL, USA*
TOMOHIRO SASANAMI • *Faculty of Agriculture, Shizuoka University, Shizuoka, Japan; United Graduate School of Agricultural Science, Gifu University, Gifu, Japan*
OLIVIER SERRALBO • *Australian Regenerative Medicine Institute (ARMI), Monash University, Clayton, VIC, Australia*
TOYO SHIMIZU • *Developmental Neurobiology, Kyoto Prefectural University of Medicine, Sakyoku, Kyoto, Japan*
SIMON Y. W. SIN • *Department of Organismic and Evolutionary Biology, Harvard University, Cambridge, MA, USA; Museum of Comparative Zoology, Harvard University, Cambridge, MA, USA*
CRAIG A. SMITH • *Department of Anatomy and Development Biology, Monash University, Clayton, VIC, Australia*
CLAUDIO D. STERN • *Department of Cell & Developmental Biology, University College London, London, UK*
TAKAYUKI SUZUKI • *Division of Biological Science, Graduate School of Science, Nagoya University, Chikusa-ku, Nagoya, Japan*
HANS-GEORG SYDOW • *Anatomy and Embryology, University Medical Centre, Göttingen, Germany*
MIKIKO TANAKA • *Department of Life Science and Technology, Tokyo Institute of Technology, Yokohama, Japan*
PAUL A. TRAINOR • *Stowers Institute for Medical Research, Kansas City, MI, USA; Department of Anatomy and Cell Biology, University of Kansas Medical Center, Kansas City, KS, USA*
NIKOLOZ TSIKOLIA • *Anatomy and Embryology, University Medical Centre, Göttingen, Germany*
ATHANASIA C. TZIKA • *Laboratory of Artificial & Natural Evolution (LANE), Department of Genetics & Evolution, University of Geneva, Geneva, Switzerland; SIB Swiss Institute of Bioinformatics, Geneva, Switzerland; Institute of Genetics and Genomics of Geneva (iGE3), University of Geneva, Geneva, Switzerland*
MASANORI UCHIKAWA • *Graduate School of Frontier Biosciences, Osaka University, Suita, Osaka, Japan*
SHOGO UEDA • *Department of Life Science and Technology, Tokyo Institute of Technology, Yokohama, Japan*

ASIER ULLATE-AGOTE • *Laboratory of Artificial & Natural Evolution (LANE), Department of Genetics & Evolution, University of Geneva, Geneva, Switzerland; SIB Swiss Institute of Bioinformatics, Geneva, Switzerland; Institute of Genetics and Genomics of Geneva (iGE3), University of Geneva, Geneva, Switzerland*

NADÈGE VÉRON • *EMBL Australia, Australian Regenerative Medicine Institute, Monash University, Clayton, VIC, Australia*

CHRISTOPH VIEBAHN • *Anatomy and Embryology, University Medical Centre, Göttingen, Germany*

RANDALL B. WIDELITZ • *Department of Pathology, Keck School of Medicine, University of Southern California, Los Angeles, CA, USA*

PING WU • *Department of Pathology, Keck School of Medicine, University of Southern California, Los Angeles, CA, USA*

BENHUA XU • *Department of Radiation Oncology, The Union Hospital Affiliated with Fujian Medical University, Fuzhou, Fujian, China*

CINDY XU • *School of Life Sciences, Arizona State University, Tempe, AZ, USA*

WATARU YAMASHITA • *Developmental Neurobiology, Kyoto Prefectural University of Medicine, Sakyoku, Kyoto, Japan*

ZHICAO YUE • *Institute of Life Sciences, Fuzhou University, Fuzhou, Fujian, China*

JILIN ZHANG • *Division of Functional Genomics, Department of Medical Biochemistry and Biophysics, Karolinska Institutet, Solna, Sweden*

QI ZHOU • *Life Sciences Institute, The Key Laboratory of Conservation Biology for Endangered Wildlife of the Ministry of Education, Zhejiang University, Hangzhou, Zhejiang, China*

Chapter 1

Some Thoughts on Experimental Design

Claudio D. Stern

Abstract

Perhaps even more important than the techniques themselves are the quality of the biological questions asked and the design of the experiments devised to answer them. This chapter summarizes some of the key issues and also touches on how the same principles affect scholarly use of the scientific literature and good peer-reviewing practices.

Key words RNA-seq, Hypothesis-driven research, Rigor, Control experiments, Peer review, Citations

1 Introduction

Over the last few years, I have become increasingly aware that designing good experiments appears to be a dying art. The increasing popularity of next-generation sequencing techniques (such as RNA-seq), because of their accessibility and decreasing cost, is accompanied by increasing prevalence of other data collecting methods (such as measuring a number of parameters, "imaging", etc.). PhD students and postdocs seem to find it increasingly difficult to design hypothesis-driven experiments or any approach that does not rely entirely on collecting data. But the problem seems to be much more widespread—reviewing grant applications for a number of bodies over the last few years has made me aware that this is a rapidly growing trend even among more established biologists. It isn't that data collecting is intrinsically bad—just that it often seems to be the only approach and people appear to be having increasing difficulty not only in designing hypothesis-driven experiments but also in formulating a clear, testable hypothesis. In fact, next-generation sequencing and other data gathering approaches can be part of a well-designed, hypothesis-driven project, and it is probably impossible to ask a good, new question without starting with some basic observations. So, although this short contribution is not specific to doing experiments on avian and reptilian embryos, its main purpose is to guide young scientists entering a new area

Guojun Sheng (ed.), *Avian and Reptilian Developmental Biology: Methods and Protocols*, Methods in Molecular Biology, vol. 1650, DOI 10.1007/978-1-4939-7216-6_1, © Springer Science+Business Media LLC 2017

about some general principles that could lead to better questions and better experiments, or at least to encourage them to think about these issues.

2 Picking a Question and Hypothesis-Driven Research Versus Data Collection

The first challenge is of course to select a question to address. In my opinion, a prerequisite is that the experimenter must be truly curious to discover the answer to the question that will be posed. This simple premise should usually be sufficient to color the type of approaches that will be followed—are we likely to be truly curious to have a list of genes or a set of measurements of sizes or shapes? To make such observations fit into a more "interesting" question, they must refer to a clear biological principle. In developmental biology, questions like "how does the XXX develop?" are likely to be more intrinsically interesting than the examples above, but how does one tackle such an open-ended question? When entering a new field for the first time, there is probably no alternative to just observing the system to be studied and noting its salient features. My UCL colleague Lewis Wolpert often uses the example "what developmental mechanisms are responsible for ensuring that our left and right limbs are the same length as each other?" Observing that the left and right arms of an individual are always the same length (but can differ greatly between individuals), despite the fact that they develop from buds that are not adjacent to each other during development, suggests the existence of a "global" mechanism that defines the extent of growth and also that this must apply differently to legs and arms of the same individual, despite the fact that the fore- and hind limb buds develop at approximately the same time. The initial observation can be, as in this example, examination of its anatomical or cellular features or other behaviors, but it can also be derived from other methods including next-generation sequencing, or measuring parameters, or acquiring a time-lapse movie. The key is that this data gathering cannot be the end of the process—it is the beginning. A good question starts from the observation and then leads to inquiries about causality or other aspects.

Most grant-giving bodies and their review panels tend to require applications for funding to be "hypothesis driven". Likewise, some journals in developmental biology state that they prefer to publish papers that are "mechanistic, rather than descriptive". Both of these terms refer to the same thing—studies that draw connections and ideally establish causality, rather than just a list of observations or measurements. But the term "hypothesis driven" is often misunderstood and misused. It should not refer to a proposal of a single mechanism put forward to explain observations, which the researchers then set out to prove. Rather, it should involve two or more alternative explanations and experiments proposed designed to distinguish between them. The most interesting and compelling hypothesis-driven research will result from situations where the alternative explanations proposed are equally likely (and equally interesting). The narrative explaining such an approach will generally be much more engaging to the reader than one describing a data-collection exercise—this is true both for a proposal for future work (as in a grant application) and for a description of work done (as in a scientific paper).

When using this approach, the experimenter should think about all the possible results that could be obtained and what would be concluded in each case. Then, for each case, the experimenter would ask whether there could be any alternative interpretations—this will guide the choice of controls (see below) or prompt for revising the experimental strategy. This sort of planning makes the experiments not only more informative but also much more fun to design and should stimulate curiosity for the results much more effectively than data-collecting approaches.

3 Designing Next-Generation Sequencing (e.g., RNA-seq) and Other Data-Gathering Experiments

It follows from the above that there will be two situations in which data-collecting approaches can be used. At the start of a study, the researcher may wish to describe the system in order to notice salient features that will guide the driving question for the subsequent research. RNA-seq, time-lapse video movies, or any other type of observation can lead to questions about causality. Generally it is easiest to do this when comparing two or more situations. As a general rule, the more similar the situations being compared, the more likely it is that the differences found will relate directly to the nature of those differences and ideally also to the reasons for those differences. Therefore, comparing the transcriptomes of two different species is unlikely to lead to a good question, whereas comparing the fore- and hind limb buds of the same animal embryo, at the same stage in their development, might.

But the same techniques can also be used as part of an experimental design to distinguish between alternative mechanisms, rather than just to observe. The general rule here is the same as above: ideally, when a data gathering method is used to distinguish between two or more alternative mechanisms, the best experiments will be those where the situations being analyzed differ *only* by the variables being assessed. This usually requires very careful thought.

Investigators often forget to take into account the influence of *timing* in the selection of samples to compare. For example, studies designed to find targets of a particular gene for which they have a mutation frequently compare a mutant and a wild-type sample from stages very much later than when the mutation is likely to act (or even without reference to when the gene is first expressed). This is likely to generate a very long list of genes that are differently expressed in the two situations, and many of them will not be (direct) targets. The longer the time interval between when the mutation acts and the time of sampling, the more irrelevant the gene list obtained, and this relationship could increase exponentially. Following the same line of reasoning, the *tissue* selected for analysis should also be the same tissue that is directly affected by the mutation, and it should be as pure as possible. The choice of tissue may be different according to whether the mutation is in a gene likely to act cell autonomously (for example if it encodes a transcription factor) or not (if it encodes a secreted protein)—but including more than one of these "in case" is likely to muddle the list even further. Different, separate experiments should be designed with these possibilities in mind, aiming to separate the variables from each other.

In principle, a well-designed next-generation sequencing experiment should generate a relatively short list of differences between the situations being compared. If the list is long, it could be an indication that the experimental design is not optimal, and more time and effort will be required to select candidates worthy of further study.

4 Designing Good Controls

The idea that experiments (by any method) should be designed to distinguish between two or more equally likely alternative possibilities leads to the critical concept of *control experiments*. The purpose of a control experiment is to eliminate other (spurious) interpretations than the one(s) being tested. Each question and experimental situation needs its own specific control—in some cases, a sham, unoperated control may not be the most appropriate control for a microsurgical manipulation (for example a transplant). Depending on the question, it may be important, for example, to transplant a different tissue, or to cause a particular type of wound, to discount particular problems that could influence the result. To come up with the appropriate controls, the experimenter needs to think of possible confounding situations that could provide alternative explanations for the results, other than those being tested. Therefore the same experiment could require completely different controls according to the question being addressed.

This is probably the most important task assigned to people acting as reviewers of a manuscript being considered for publication. The reviewer should ask: is there an alternative (likely!) interpretation of the results obtained other than the ones being tested by the experimental design? If so, what is the best (and feasible!) control experiment that will test whether this new alternative has influenced the results?

5 Quantitative Versus Qualitative Approaches

In the last few years, there has also been increasing obsession with "quantitative" approaches. But there is nothing wrong with "qualitative" ones, and one could argue that the greater the accuracy needed to reveal an experimental result, the more subtle the effect and therefore potentially it will be less interesting than one that gives a qualitative, binary, "yes or no" answer. Again this should affect reviewers—it is not appropriate for a reviewer to demand more accurate quantification of a result unless the reviewer can clearly articulate the reasons for this: what specific (and interesting) effect could be revealed by greater accuracy that would be masked by a qualitative effect? I am not referring to statistical analysis here, but rather to quantitative (scalar) measurements. Imagine a situation where the results of an experiment are being assessed using in situ hybridization—this is not a quantitative method but a comparative one, allowing comparison of relative expression between different cells or regions of a tissue. It is not unusual for reviewers to ask for a quantitative method of analysis such as quantitative RT-PCR (qPCR) or RNA-seq. In many cases such a request is not appropriate. Consider, for example, a tissue that responds in a heterogeneous manner to a stimulus because it contains different cell types or some other reason. qPCR or RNA-seq experiments cannot distinguish many cells expressing a very low level of a transcript from a few cells expressing very high levels. In situ hybridization will however reveal this situation, as well as allow individual cells to be observed separately. As with the design of controls, whether quantitative measurements should be part of the experimental design or not depends on the question being addressed.

6 Elegance of Experimental Design

What is a "good" experiment? One answer to this question is obviously a set of experimental manipulations, or observations, that lead to a conclusion, accompanied by appropriate, rigorous controls designed to eliminate or minimize alternative interpretations. But there is also an element of *elegance* in experimental design—mathematicians often describe demonstrations of a principle as "beautiful proof" and in my opinion the same applies to other areas of Science. An experiment (or set of experiments) can be described as *elegant* when it is both rigorous and simple—ideally, when the result is not necessarily expected but it very clearly helps to distinguish between different alternatives.

One of my favorite examples of such an experiment in avian embryology is the demonstration of the principle of "non-equivalence" in limb development [1] based on experiments by Saunders, Gasseling, and Cairns [2]. The experiment consisted of a transplant of mesenchyme from the proximal leg bud (prospective thigh) into the distal part of the wing bud. The graft developed into distal leg, including toes, scales, and claws, showing that limb identity (wing versus leg) is determined prior to proximal-distal positional information and that the grafted leg mesenchyme "carries with it its 'leg' label, even though it can respond to signals within the wing bud affecting its proximal-distal character" [1]. I find this experiment and its interpretation particularly beautiful because it is typical of experimental embryology: it reveals rules that govern the behavior of a system even before the precise nature of the molecular constituents is known.

7 What Is More Important, Results or Conclusions?

Teaching developmental biology to advanced undergraduates and postgraduate students has also made me realize that many students tend to remember only the conclusions of published studies, rather than the results. I think that this is very dangerous. The conclusions of a study correspond to the interpretation of the results given by the authors, to the best of their ability. This is critically dependent on the experimental design including the controls performed, both of which are dependent on the knowledge available at the time the experiment was done. The same experiment done at a different time will produce identical results, but the conclusion could be quite different. In my opinion, we should advance our field by taking all results into account (in the context of how each of the experiments was done) because all results are potentially informative. As the field advances, the growing set of results, coupled with fluid, changing interpretations, will lead effectively to much greater understanding. Importantly, if we consider that it is the results, rather than the conclusions, of a study that are important and will never cease to be valid, then it is virtually impossible to be wrong. Conclusions are subjective by nature. Results are entirely objective, as long as the experimental situation is described in sufficient detail and with sufficiently clarity to allow reproducibility.

This also leads to the issue of how to cite the previous literature. Another relatively new trend is to cite a review in support of statements. This is not scientific. A declarative statement should be supported by reference(s) to where the evidence can be found that supports that statement directly. It is often said that some fields have grown so much that it is impossible to cite all the evidence. In those cases, I propose a simple rule of thumb: pick two references—the first and the best. The first time someone made the proposal of what is being said is much more important than the most recent; the finding was original, and the author deserves more credit and citations than the last ones to have done the same thing. "The best" refers to the paper(s) that contain the most compelling or clear evidence on which the statement is based, irrespective of when it was done. In large fields, this is often neither the first nor the last. Reviews can of course be cited, in an appropriate context, but never to support a statement of fact. But unless we are careful to cite the literature appropriately, reviews can actually be quite damaging to the progress of a field.

8 Good Peer Reviewing

Many of the pointers above give some clear indications of what good peer review should be. In my view (now also increasingly stated in their guidelines by some of the leading journals, such as *Development*), a good peer reviewer should assess the following, and *only* the following:

- *Originality.* Does the paper contain new information that was not available before? If not, the reviewer *must* give specific references that demonstrate that the specific finding was made previously.
- *Significance.* To what extent does the paper provide a significant advance over what was known before? This is a value judgment by the reviewer and the editor, as well as the authors, should be allowed to disagree. If the reviewer does raise this, then again specific references to what was known before must be included in the review.
- *Rigor.* Are the conclusions justified by the results presented, or are there some possible alternative interpretations? If the latter, the alternatives must be *likely.* And if this situation is suggested by the reviewer, the reviewer must specify what the alternative interpretations are, and what experiment(s) should be done to determine whether these confound the study. Importantly, the proposed experiments must be *feasible*, in the authors' laboratory. In my view authors should be allowed a certain amount of flexibility to speculate in the discussion of a paper—they may favor one interpretation over another, and the discussion section provides an explicit place where this can be done without effect on the rigor of the study as a whole. However, if this is done, it is important to ensure that the speculations are presented as such and do not color either the abstract or the title of

the paper, because this could make presentation of the study misleading.

9 Conclusions

I hope that this brief set of thoughts will encourage some to think about these important issues and influence them when designing experiments for grant applications, presenting their findings in published papers and evaluating the work of their colleagues in a fair and open way. With these principles, the techniques presented in this volume should become even more powerful.

References

1. Lewis JH, Wolpert L (1976) The principle of non-equivalence in development. J Theor Biol 62:479–490
2. Saunders JW Jr, Gasseling MT, Cairns JM (1959) The differentiation of prospective thigh mesoderm grafted beneath the apical ectodermal ridge of the wing bud in the chick embryo. Dev Biol 1:281–301

Part I

Genomics and Transcriptomics

Chapter 2

Comparative Genomics as a Foundation for Evo-Devo Studies in Birds

Phil Grayson, Simon Y.W. Sin, Timothy B. Sackton, and Scott V. Edwards

Abstract

Developmental genomics is a rapidly growing field, and high-quality genomes are a useful foundation for comparative developmental studies. A high-quality genome forms an essential reference onto which the data from numerous assays and experiments, including ChIP-seq, ATAC-seq, and RNA-seq, can be mapped. A genome also streamlines and simplifies the development of primers used to amplify putative regulatory regions for enhancer screens, cDNA probes for in situ hybridization, microRNAs (miRNAs) or short hairpin RNAs (shRNA) for RNA interference (RNAi) knockdowns, mRNAs for misexpression studies, and even guide RNAs (gRNAs) for CRISPR knockouts. Finally, much can be gleaned from comparative genomics alone, including the identification of highly conserved putative regulatory regions. This chapter provides an overview of laboratory and bioinformatics protocols for DNA extraction, library preparation, library quantification, and genome assembly, from fresh or frozen tissue to a draft avian genome. Generating a high-quality draft genome can provide a developmental research group with excellent resources for their study organism, opening the doors to many additional assays and experiments.

Key words Genome assembly, Library preparation, Avian, Bird, Developmental genomics, ALLPATHS-LG, Comparative genomics

1 Introduction

1.1 Avian Developmental and Genomic Resources

A long history of study has made the chicken into an important model for analyzing developmental patterns and processes across diverse time scales and cell types [1]. Avian developmental biology is not limited to chicken: developmental staging systems for quail, zebra finch, and emu are also available [2–4]. Arguably one of the most important advancements for developmental or molecular avian research was the sequencing of the chicken (*Gallus gallus*) genome in 2004 [5]. In the 10 years that followed, more than 50 additional avian genomes have been published, most of them with the December 2014 release of the *Science* Special Issue, "A Flock of Genomes," examining data from 48 bird species (*see* [6, 7]). The last 2 years have seen an increasing rate of avian genome

Guojun Sheng (ed.), *Avian and Reptilian Developmental Biology: Methods and Protocols*, Methods in Molecular Biology, vol. 1650, DOI 10.1007/978-1-4939-7216-6_2, © Springer Science+Business Media LLC 2017

sequencing; at the time of writing, the total number of draft bird genomes available in public databases has soared to over 70. With an eventual goal of generating draft genomes of all 10,500 species of living birds, the Bird 10,000 Genomes (B10K) project has promised to increase this total even more rapidly [8].

1.2 Avian Genomes

Avian genomes are relatively small, around 1–1.26 Gb, and low in repetitive elements compared to other amniotes [6, 9]. These two factors make achieving high-quality reference genomes relatively straightforward for research groups with the necessary protocols and infrastructure, including a few key instruments for library preparation and quantification, and access to Illumina sequencing machines. In the last several years, our laboratory has produced a number of high-quality draft genomes of birds. Each genome has required approximately two full lanes of Illumina HiSeq 2500 v4 high-output sequencing to achieve reasonable coverage and quality. The ALLPATHS-LG assembly approach [10] used by our group made a strong showing during the Assemblathon 2 comparative study of assembly methods and has since been applied frequently across diverse vertebrate taxa [11]. Sequencing technology is rapidly changing, as discussed in Subheading 1.4 below, and while the strategy outlined here has proved to be robust and cost-effective in our hands, new approaches will likely surpass the ALLPATHS-LG/ Illumina strategy in both cost and quality in the near future.

1.3 ALLPATHS-LG and Basic Sequencing Strategies

The ALLPATHS-LG assembly algorithm requires at minimum a single overlapping fragment library (e.g., 220 bp library sequenced using 2 × 125 bp paired-end reads), as well as a jumping library (e.g., 3 kb Nextera Mate Pair Library, also sequenced using 2 × 125 bp reads). The jumping library allows the assembler to "jump" over stretches of repetitive DNA because each end of a mate pair read provides linkage information about bases that appear approximately 3 kb from one another in the genome. In the preparation of the jumping library, DNA is "tagmented" (fragmented and tagged with junction adapters in one step using transposase) and then circularized (junction adapter to junction adapter), such that DNA on either side of the junction (the mate pairs) originated approximately 3 kb away on the same strand in the native genomic DNA. The circularized DNA is then fragmented a second time and only junction adapter-containing fragments are collected and used to produce the library. For genomes that are larger and more complex than those of birds, a larger jumping library (e.g., 6 kb) or a fosmid jumping library (e.g., 40 kb) can be incorporated to span long stretches of repeats [10].

Here, we report basic assembly statistics from 17 unpublished draft avian genomes generated in our laboratory to serve as a guide when planning the sequencing depth per library and total coverage for a new avian ALLPATH-LG project (Table 1). Genomes from

Table 1
Assembly statistics for 17 unpublished avian genomes assembled using ALLPATHS-LG

Common species name	Latin species name	Estimated genome size (Gb)	Total coverage	Contig N50 (kb)	Scaffold N50 (Mb)	Fragment reads (*E* +08)	Jump reads (*E*+08)	Total reads (*E* +08)
Emu	*Dromaius novaehollandiae*	1.31	64.8	139	3.32	5.01	3.98	8.99
Southern cassowary	*Casuarius casuarius*	1.37	54.1	133	3.70	4.19	4.60	8.79
Great spotted kiwi	*Apteryx haastii*	1.35	44.4	106	1.36	2.95	2.77	5.72
Little spotted kiwi	*Apteryx owenii*	1.36	60.5	128	1.62	4.48	4.52	8.99
Okarito brown kiwi	*Apteryx rowi*	1.40	54.3	120	1.67	3.58	4.36	7.94
Greater rhea	*Rhea americana*	1.24	59.9	68.7	4.08	3.90	3.76	7.66
Lesser rhea	*Rhea pennata*	1.27	42.4	55.9	3.85	3.13	3.31	6.44
Thicket tinamou	*Crypturellus cinnamomeus*	1.20	47.5	50.3	2.43	2.70	3.22	5.92
Elegant crested tinamou	*Eudromia elegans*	0.96	68.0	98.8	3.28	3.43	3.29	6.72
Chilean tinamou	*Nothoprocta perdicaria*	1.01	76.2	75.8	3.35	4.29	3.36	7.66
Leach's storm petrel	*Oceanodroma leucorhoa*	1.24	79.5	165	8.58	4.40	5.83	10.2
Wandering albatross	*Diomedea exulans*	1.22	28.3	42.8	1.10	1.50	2.06	3.56
Waved albatross	*Phoebastria irrorata*	1.24	41.1	63.0	1.51	2.24	2.76	5.00
Short-tailed albatross	*Phoebastria albatrus*	1.23	41.5	56.7	1.52	2.27	2.65	4.92
Laysan albatross	*Phoebastria immutabilis*	1.25	59.4	81.4	1.62	3.97	2.96	6.93
Black-footed albatross	*Phoebastria nigripes*	1.25	67.1	81.9	1.53	4.40	3.57	7.98
Northern cardinal	*Cardinalis cardinalis*	1.10	58.8	114	3.66	3.36	3.68	7.04

our laboratory were sequenced exclusively on Illumina HiSeq 2500s using 2 × 125 bp reads from Illumina HiSeq SBS V4 high-output sequencing runs alone, or in combination with a small proportion of reads (generated during test runs) using rapid run mode at 2 × 150 bp. Multiplexing and sequencing were carried out with the aim of providing between 500 million and 1 billion paired-end reads (between 1 and 2 lanes of a v4 flow cell) per genome. Because sequencing comprises the largest portion of the total cost in generating a draft genome, the genome of the wandering albatross (*Diomedea exulans*), at only 356 million reads and 28.3× coverage, cost approximately one half of the higher coverage genomes (e.g., the emu (*Dromaius novaehollandiae*)).

Scaffold and contig N50 scores are common metrics of genome quality, defined as the length of the element (scaffold or contig) above which 50% of the genome exists in elements of that length or longer. Following assembly, each contig contains mostly contiguous DNA sequence. Contigs are linked together with stretches of unknown bases in between to generate scaffolds. Reads from jumping libraries span regions that are difficult to sequence or assemble, aiding in the assembly of both contigs and scaffolds. Some trends in genome quality that emerge from analyzing the 17 avian genome assemblies include:

1. Contig N50 correlates with the total number of raw reads (both jumping and fragment) (Fig. 1).
2. Scaffold N50 also correlates with the total raw read number, but the correlation is less strong than for contig N50 (Fig. 2). Two groups emerge in this graph, one that appears to benefit from deeper sequencing and one that does not (assemblies from the latter group have scaffold N50 values below 2 Mb). Despite high sequencing depth in the two kiwi and two albatrosses near the bottom right of the figure, the scaffold N50 is low. The quality of the input DNA, properties of the genomes (e.g., repeat content or types), or other factors could explain this pattern.
3. We note several caveats regarding increasing sequencing depth as a means to achieve increased quality of assembly. Our laboratory has experimented with increasing sequencing depth without creating additional libraries in two instances, once for a very low coverage assembly of the waved albatross (*Phoebastria irrorata*) and once as an attempt to improve the 67× coverage assembly of the black-footed albatross (*Phoebastria nigripes*) (Table 1). For the waved albatross, the first assembly was produced with approximately 21× coverage, with contig and scaffold N50 values of 3.06×10^4 and 7.87×10^5 bp, respectively. By increasing the coverage to over 40×, this assembly was greatly improved (*see* Table 1). However, for the black-footed albatross, nearly doubling the coverage to

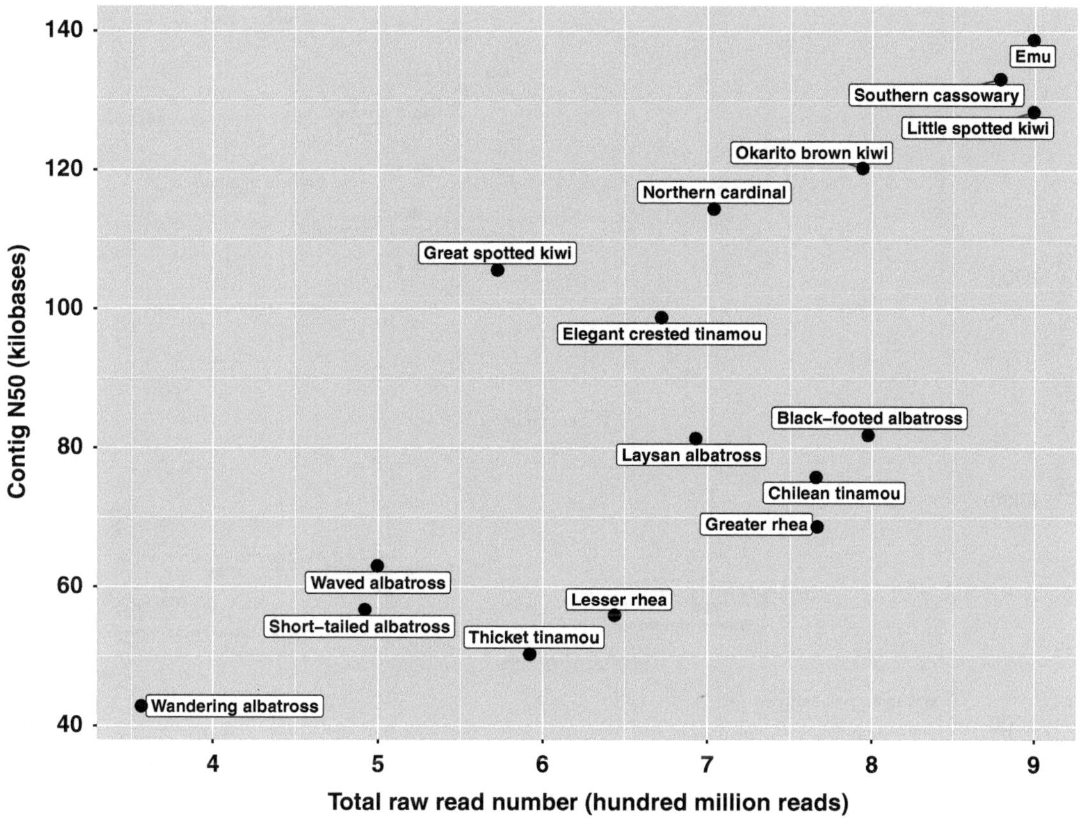

Fig. 1 Contig N50 in kilobases versus total number of sequencing reads (3 kb jumping and fragment libraries combined) for 16 draft avian genomes assembled using ALLPATHS-LG

approximately 110×, apparently did little to improve the assembly (contig and scaffold N50 values of 8.64×10^4 and 1.51×10^6 bp, respectively). This result suggests that there may be diminishing returns to higher coverage sequencing, at least in some cases. For example, if the input DNA is not of very high quality, the diversity of the resulting libraries may be low. Mapping raw sequencing reads back to the assembled genome could provide information pertaining to library diversity to determine if additional sequencing of the current libraries could improve an assembly [12].

4. Both our experience and the theory behind genome assembly suggest that the best way to improve scaffold size and N50 in an assembly is to generate an additional jumping library. Only one genome in our laboratory, that of the Leach's storm petrel (*Oceanodroma leucorhoa*), has been sequenced with both a 3 and 6 kb jumping library, and it has a scaffold N50 approximately twice as long as any of our other genomes, suggesting that an additional jumping library can greatly improve assembly quality (Table 1). The petrel assembly also has a slightly higher contig N50 compared to the other 16 genomes.

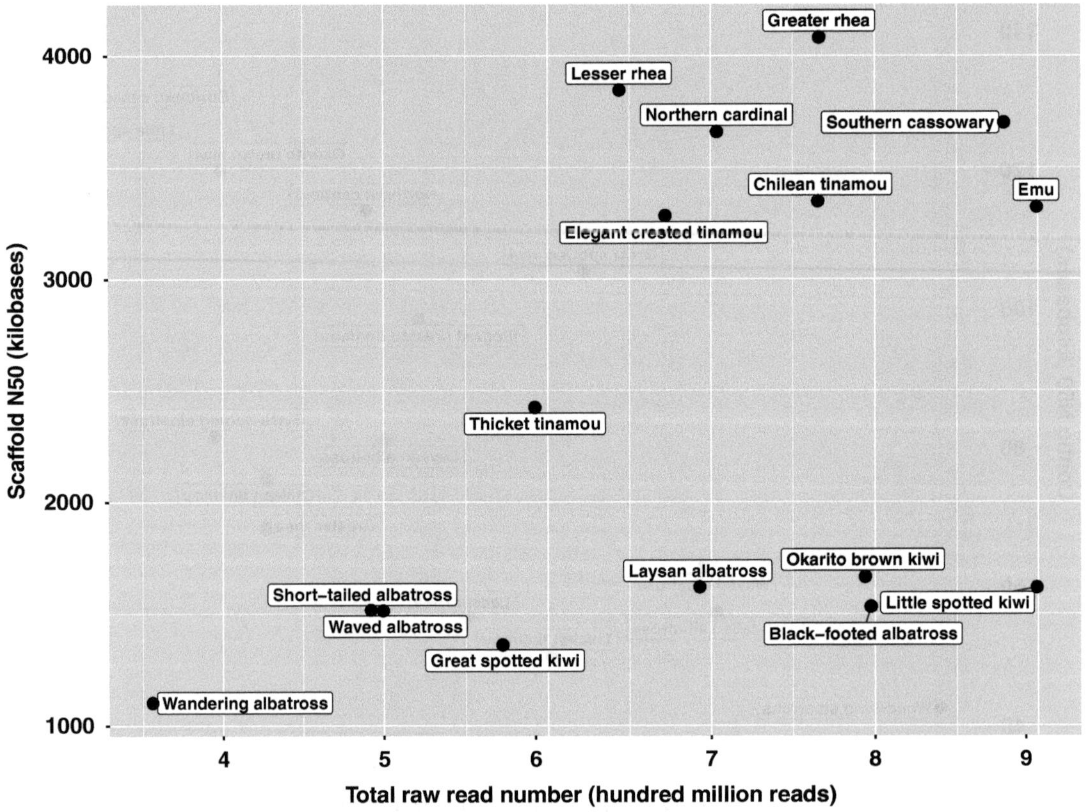

Fig. 2 Scaffold N50 in kilobases versus total number of sequencing reads (3 kb jumping and fragment libraries combined) for 16 draft avian genomes assembled using ALLPATHS-LG

1.4 Alternative Approaches to Sequencing Avian Genomes

At the time of writing, the Illumina/ALLPATHS-LG strategy we describe here is likely the most widely available and cost-effective method for producing high-quality draft avian genomes (Fig. 3). However, new and emerging sequencing technologies have the potential to produce better, more contiguous assemblies for a similar, or cheaper, cost. Oxford Nanopore and Pacific Biosciences both have the advantages of single-molecule long reads but still suffer from higher costs and error rates compared to the shorter Illumina reads [13, 14]. There is little doubt that deep sequencing on either of these platforms would provide the best possible assembly but might cost up to ten times more than the Illumina/ALLPATHS-LG approach described here [15]. Costs and error rate continue to rapidly decline on these platforms, and they may achieve parity with Illumina in the near future.

Another recent alternative is the Chromium library preparation system from 10× Genomics, which uses microfluidics to partition genomic DNA and label individual strands with over 1 million unique barcodes prior to sequencing on the Illumina platform [16]. By referring back to the original barcodes on the

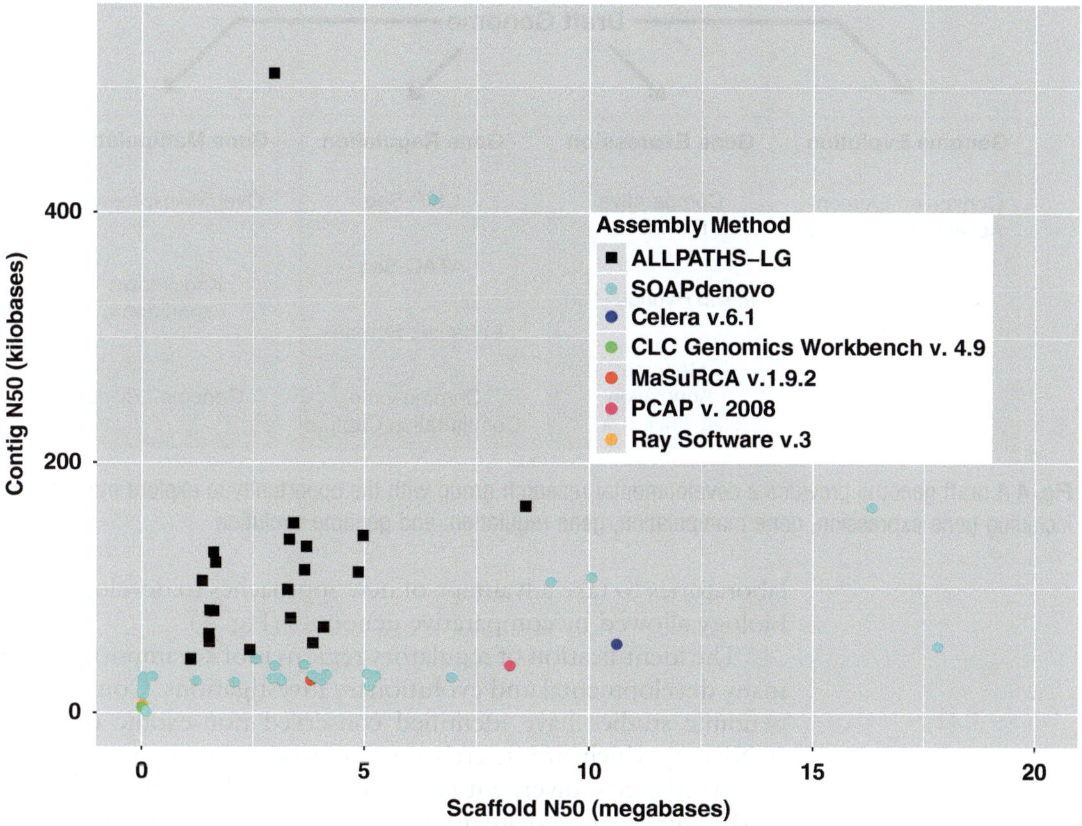

Fig. 3 Contig N50 in kilobases versus scaffold N50 in megabases for 17 draft avian genomes assembled in our laboratory alongside published avian genomes available from NCBI categorized by assembly method (total $n = 72$). NCBI avian genomes were included if the metadata contained assembly method, contig N50 and scaffold N50 values. This figure does not display the chicken or great tit (*Parus major*) since their contig N50 (chicken) and scaffold N50 (great tit) values are five and four times greater than all the other species, respectively. ALLPATHS-LG provides excellent contig N50 values and good scaffold N50 values across avian samples. Since gene annotation (and mapping of short reads for epigenomic assays) relies more heavily on large contigs than large scaffolds, the ALLPATHS-LG approach is appropriate for developmental genomics

unfragmented DNA, the assembly algorithm can identify the original native DNA molecule from which a particular read is derived, thus generating haplotype-phased genomes with high scaffold and contig N50. The additional costs of the library preparation for this approach compared to that for standard Illumina are relatively modest, but the efficiency and performance of the algorithms and methods in nonhuman contexts are not yet clear [17].

1.5 Draft Genomes as Foundations for Developmental Genomics

The first steps in sequencing a genome are the most critical. The high cost of errors during library preparation or quantification often places genome projects out of reach for many research laboratories. Our goal for this chapter was to provide an overview of an accessible method for generating draft avian genomes, allowing

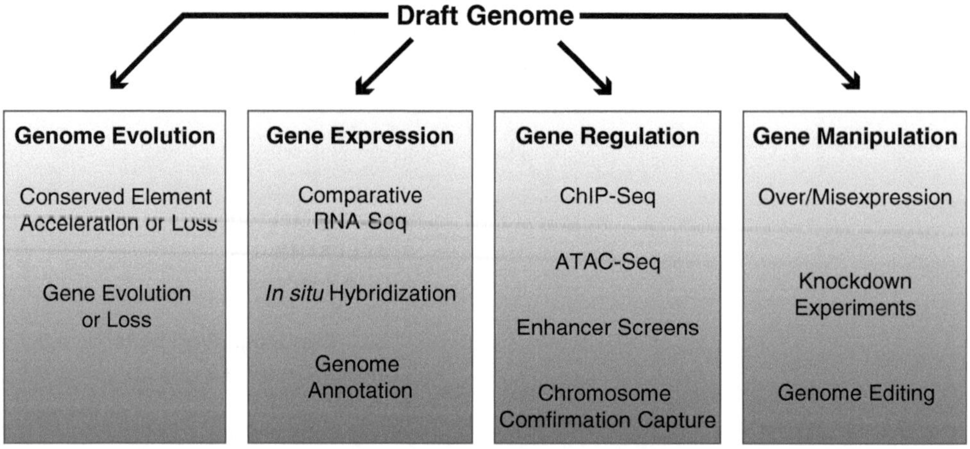

Fig. 4 A draft genome provides a developmental research group with the opportunity to explore many areas, including gene expression, gene manipulation, gene regulation, and genome evolution

laboratories to take advantage of new approaches to developmental biology allowed by comparative genomics (Fig. 4).

The identification of regulatory regions is of key importance for many developmental and evolutionary investigations. Comparative genomic studies have identified conserved non-exonic elements (CNEEs), which are stretches of noncoding sequence that are conserved across divergent taxa, evolve more slowly than a benchmark neutral rate, and are usually comprised of fourfold degenerate sites of protein-coding regions. These CNEEs, which are often identified through the phastCons program following whole-genome alignment [18], appear to serve developmental regulatory roles. When tested in a transgenic mouse enhancer assay, 115 out of 231 of the highly conserved human genomic regions drove reporter gene expression in various murine tissues, suggesting conserved enhancer activity [19].

Because of the important role of CNEEs in development, many computational methods have been used to predict their origins, functions, and the effect of repeated losses of orthologous elements across diverse taxa [18, 20–23]. Recent studies in birds suggest that CNEEs play an important role in evolutionary diversification and in the origin of morphological novelties, such as feathers [21, 24]. CNEE discovery techniques have proven to be most powerful when coupled with genomic assays for tissue-specific regulatory activity. Using a 60-way vertebrate alignment for CNEE identification, alongside chromatin immunoprecipitation (ChIP)-seq data for active enhancer marks (histone 3 lysine 27 acetylation or H3K27ac) from developing mouse limbs, Booker et al. were able to identify and experimentally validate novel regulatory regions (referred to as bat accelerated regions or BARs due to their rapid evolutionary rate in bat lineages) implicated in the development of the bat wing and conduct comparative in situ hybridization experiments on genes neighboring BARs of interest [25]. Since CNEEs are often found clustered near key developmental genes [20], interesting patterns of CNEE evolution can also aid in the generation of candidate gene lists for traits of interest.

Another recent paper on development and evolution of bat limbs used a similar comparative genomics approach to identify additional BARs [26]. These authors then examined comparative RNA-seq, in situ hybridization and both H3K27me3 (repressive mark) and H3K27ac (active mark) ChIP-seq data across three developmental stages to identify regulatory changes and expression differences for RNAs (both coding and long noncoding) implicated in shaping bat forelimb development through differential mesenchymal condensation.

In pigeons, comparative genomics, cross-strain breeding, and developmental experiments alongside ChIP-seq from developing limbs allowed researchers to suggest that genes implicated in the origins of feathered feet in multiple pigeon breeds drive an alteration of limb identity, with the developing feathered hindlimbs displaying a localized shift toward forelimb-specific expression patterns of *Tbx5* and *Pitx1* through cis-regulatory changes [27]. Furthermore, the authors identified similar *Tbx5* expression patterns in the developing hindlimbs of convergently feathered chicken breeds [27]. In other *Tbx5* research, an assay that identifies stretches of open chromatin from live intact cells (ATAC-seq) uncovered a CNEE present in humans, bony fish, and cartilaginous fish that appears to serve an essential role in the evolution of paired appendages across jawed vertebrates [28]. Injection of this enhancer region alongside *tbx5a* into finless zebrafish (*Danio rerio*) embryos rescued the finned phenotype, while injecting *tbx5a* alone was ineffective [28]. All of the developmental studies described above used comparative genomics alongside cutting-edge genomic and epigenomic assays for their model and trait of interest, illustrating the utility of one or more draft genomes for driving developmental research.

1.6 Overview

The methods section begins with avian tissue and ends with draft de novo genome assemblies (Fig. 5). Each section follows one or more published protocol(s) from the kits and instruments recommended; DNA extraction (Subheading 3.1) follows the protocol from the E. Z.N.A. Tissue DNA Kit. Fragment Library Preparation (Subheading 3.2) follows protocols from the Qubit dsDNA HS Assay Kit, PrepX ILM 32i DNA Library Kit, IntegenX PCR Cleanup

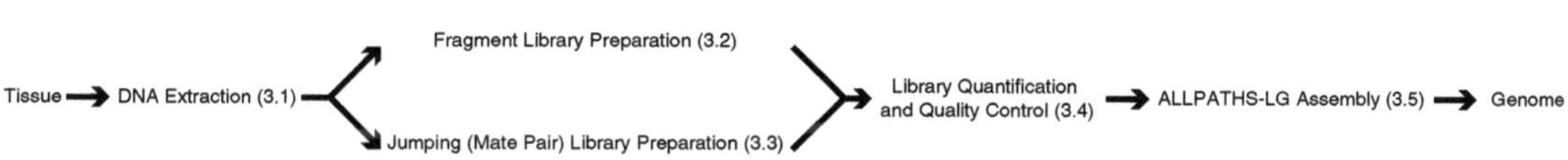

Fig. 5 Overview of methods presented in this chapter to produce ALLPATHS-LG genome assembly starting from high-quality avian tissue

8 Protocol, and the Covaris S220 instrument. Jumping (Mate Pair) Library Preparation (Subheading 3.3) follows protocols from the Qubit, Covaris, Nextera Mate Pair Sample Preparation Kit, and Pippin Prep instruments. Library Quantification and Quality Control (Subheading 3.4) follows protocols from Qubit, the High Sensitivity DNA Kit for the Agilent Bioanalyzer, and the KAPA Library Quantification Kit for next-generation sequencing. ALLPATHS-LG assembly (Subheading 3.5) utilizes commands from Fastq, cutadapt, Trim Galore!, and ALLPATHS-LG manuals. Notes (Subheading 4) interspersed throughout the Materials and Methods provide tips and best practices that have benefitted our laboratory throughout these protocols.

This chapter serves as a general guide to achieving high-quality draft genomes of birds, but we recommend referring to the protocols included in the kits themselves or updated online because companies often modify their protocols as they change reagents or optimize key steps.

2 Materials

2.1 DNA Extraction and Quality Control

1. Tissue sample (*see* **Note 1**).
2. DNA extraction kit such as E.Z.N.A. Tissue DNA Kit (cat. No. D3396, Omega bio-tek).
3. Standard 1.7 mL microcentrifuge tubes.
4. RNase A.
5. 100% ethanol.
6. 100% isopropanol.
7. Sterile scalpel blades.
8. Agarose.
9. 10× TAE buffer (cat. No. 15558042, Thermo Fisher).
10. 1 kb DNA Ladder and 6× loading dye (cat. No. N3232, NEB).
11. Agarose gel electrophoresis system.
12. Gel imager.
13. SYBR Safe DNA Gel Stain (cat. No. S33102, Thermo Fisher).

2.2 Fragment Library Preparation

1. DNA from Subheading 2.1.
2. Qubit Fluorometer.
3. Qubit dsDNA HS Assay Kit (cat. No. Q32854, Thermo Fisher Scientific).
4. Qubit assay tubes (cat. No. Q32856, Life Technologies) or Axygen PCR-05-C tubes (cat. No. 10011-830, VWR).
5. Apollo 324 NGS Library Preparation System (*see* **Note 2**).

6. Apollo consumables (*see* **Note 3**): PCR strip-tubes, microtiter plates, filter tips, piercing tips, PCR strip-caps, reservoirs, and 1.1 mL tubes (cat. No. 300019, 300026–300029, 300031, and 300033, respectively).
7. PrepX ILM 32i DNA Library Kit (*see* **Note 2**) (cat. No. 400076, Wafergen).
8. PrepX Complete ILMN DNA Barcodes (cat. No. 400077-1 to 400077-5, Wafergen).
9. Covaris S220 shearing device and microTUBE holder.
10. Covaris microTUBE (6 × 16 mm) glass tubes (cat. No. 520045, Covaris).
11. 70% EtOH.
12. 2.5 M NaCl.
13. UltraPure DNAse/RNAse-free distilled water (cat. No. 10977023, Thermo Fisher Scientific). This will be referred to as water throughout.

2.3 Jumping [Mate Pair] Library Preparation

1. Nextera Mate Pair Sample Preparation Kit (cat. No. FC-132-1001, Illumina).
2. Qubit Fluorometer.
3. Qubit dsDNA HS Assay Kit (cat. No. Q32854, Thermo Fisher Scientific).
4. Qubit assay tubes (cat. No. Q32856, Life Technologies) or Axygen PCR-05-C tubes (cat. No. 10011-830, VWR).
5. Covaris S220 shearing device and microTUBE holder.
6. Covaris T6 (6 × 32 mm) glass tubes (cat. No. 520031, Covaris).
7. Covaris Snap-Cap (8 mm) Teflon Silicone Septa (cat. No. 520042, Covaris).
8. Pippin Prep.
9. Pippin Prep 0.75% agarose gel cassettes with ethidium bromide (cat. No. CSD7510, Sage Science).
10. Maxymum Recovery 1.7 mL microcentrifuge tubes (cat. No. MCT-175-L-C, Axygen).
11. Genomic DNA Clean and Concentrator Kit (cat. No. D4010 or D4011, Zymo Research)
12. AMPure XP beads (cat. No. A63880, Beckman Coulter).
13. Dynabeads M-280 Streptavidin magnetic beads (cat. No. 112-05D, Invitrogen).
14. Magnetic rack for 1.7 mL microcentrifuge tubes; MagnaRack, for example (cat. No. CS15000, Invitrogen).

2.4 Library Quantification and Quality Control

1. Agilent 2100 Bioanalyzer.
2. High Sensitivity DNA Kit for Bioanalyzer (cat. No. 5067-4626, Agilent Technologies).
3. KAPA Library Quantification Kit for next-generation sequencing (cat. No. KK4824, Kapa Biosystems). Catalog number above is for the universal kit.
4. qPCR thermocycler.
5. Preferred diluent (10 mM Tris–HCl, pH 8.0 and optional 0.05% Tween 20).
6. qPCR plates or tubes with clear sealing film or lids.

2.5 ALLPATHS-LG Assembly

1. Computer with FastQC installed. Available here: http://www.bioinformatics.babraham.ac.uk/projects/fastqc/.
2. Computer with cutadapt installed. Available here: http://cutadapt.readthedocs.io/en/stable/installation.html.
3. Computer with Trim Galore! installed. Available here: http://www.bioinformatics.babraham.ac.uk/projects/trim_galore/.
4. High memory (>256 Gb) server running ALLPATHS-LG (*see* **Note 4**). Available here: http://software.broadinstitute.org/allpaths-lg/blog/.

3 Methods

3.1 DNA Extraction

1. This section follows the protocol from the E.Z.N.A. Tissue DNA Kit (*see* **Note 5**).
2. Preheat blocks or water baths to 55 and 70 °C, place elution buffer at 70 °C, and add appropriate volume of 100% ethanol and 100% isopropanol to DNA wash buffer and HBC buffer, respectively (varies with kit size), before beginning.
3. Weigh and mince 30 mg of tissue. Place in 1.7 mL microcentrifuge tube.
4. Add 220 μL TL Buffer and flick-mix by hand. Do not vortex at any point (*see* **Note 6**).
5. Add 25 μL OB Protease solution and mix.
6. Incubate at 55 °C with agitation, either constant shaking or frequent flick-mixing every 20–30 min (*see* **Note 7**).
7. Once tissue is digested, add 4 μL RNase A (100 mg/mL) per 30 mg tissue and incubate for 2 min at room temperature. This will remove RNA from your sample and allow you to better determine the quality of your DNA sample on an agarose gel.

8. Centrifuge at maximum speed (use max speed for all DNA extraction spins) for 5 min and transfer supernatant to sterile 1.7 mL microcentrifuge tube.
9. Add 220 μL BL Buffer and mix thoroughly.
10. Incubate at 70 °C for 10 min.
11. Add 220 μL of 100% ethanol and mix thoroughly.
12. Transfer full volume to HiBind DNA Mini Column placed in a collection tube.
13. Centrifuge at max speed for 1 min.
14. Discard filtrate and replace collection tube.
15. Add 500 μL HBC buffer (ensure that 100% isopropanol has been added).
16. Centrifuge at max speed for 30 s.
17. Discard collection tube and filtrate. Apply new collection tube to column.
18. Add 700 μL DNA wash buffer (ensure that 100% ethanol has been added).
19. Centrifuge at max speed for 30 s.
20. Discard filtrate and replace collection tube.
21. Repeat 700 μL DNA wash buffer steps one additional time.
22. Centrifuge empty column and collection tube for 2 min to remove excess wash buffer.
23. Transfer column to new collection tube and centrifuge for 1 additional minute to remove excess wash buffer.
24. Transfer HiBind DNA Mini Column into sterile nuclease-free 1.7 mL microcentrifuge tube.
25. Add 100–200 μL heated 70 °C elution buffer to column. Incubate columns at 70 °C for 2 min.
26. Centrifuge for 1 min. Second elution can be carried out with additional elution buffer.
27. Make 1× TAE stock from 10× TAE buffer using dH_2O.
28. Based on total volume of 1× TAE used, weigh and mix in appropriate amount of agarose to create 0.5–1% agarose gel. Microwave and mix solution until homogeneous and cool until comfortably warm to the touch. Add SYBR Safe (approximately 1 μL per 10 mL of gel solution). Pour, harden, load, and run gel at 120 V.
29. When the gel is imaged, samples should show single band with little to no streaking underneath (Fig. 6).

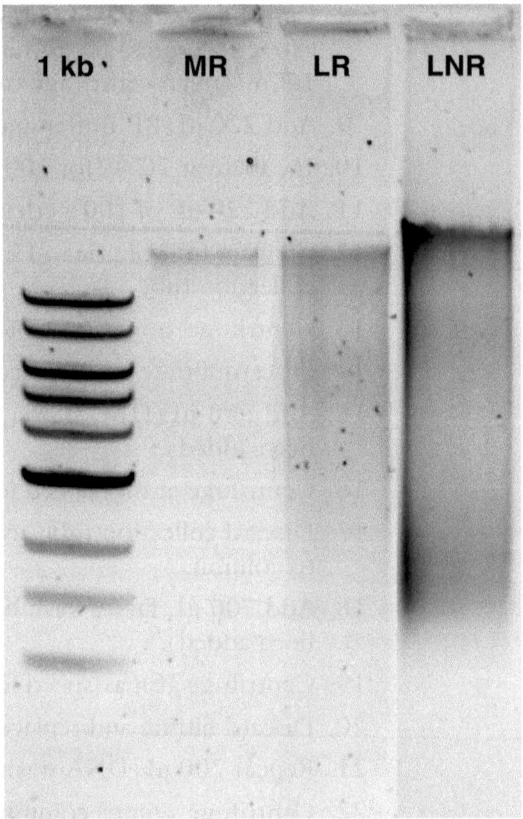

Fig. 6 Composite image of a 1 kb ladder (1 kb) and three DNA extractions using slightly different methods on tissues stored under identical conditions (frozen in RNA later at −80 °C) to show differences in DNA quality. The first sample (MR) was muscle tissue that was lysed for only 3 h and treated with RNAse A before extraction, the second (LR) is liver tissue treated in the same way, and the last sample (LNR–far right) is a piece from the same liver sample after being lysed overnight (18 h) without the addition of RNAse A. RNAse A (or the shorter lysis step) appears to have cleaned up some of the extended smear seen in LNR compared to LR. We utilized MR for genome sequencing since it appeared to have the least degradation

3.2 Fragment Library Preparation

1. This section follows protocols from Qubit dsDNA HS Assay Kit, PrepX ILM 32i DNA Library Kit, IntegenX PCR Cleanup 8 Protocol, and the Covaris S220 instrument.
2. Following the Qubit dsDNA HS Assay Kit protocol, quantify DNA prior to library preparation. The full protocol is included below.
3. Label tops of assay tubes for samples and two standards.
4. Calculate and mix the appropriate volume for Qubit working solution for all samples and standards in a plastic microcentrifuge or Falcon tube. The working solution is composed of

Qubit dsDNA HS reagent diluted 1:200 in Qubit dsDNA HS Buffer. Each sample requires 199 μL of working solution (if the readings are too low using 199 μL, the Qubit protocol allows volumes as low as 180 μL for samples), and the two standards both require 190 μL.

5. Each tube will be brought to 200 μL with standard or sample, vortexed, and incubated at room temperature for 2 min before reading (*see* **Note 8**).
6. On the Qubit, select DNA, then dsDNA High Sensitivity.
7. Read standard 1 and then standard 2.
8. Read sample and calculate concentration on the Qubit screen based on volume of sample used (e.g., 1 μL sample with 199 μL of working solution).
9. Following the Wafergen Biosystems PrepX ILM 32i DNA Library Kit protocol, the DNA concentration working range is 0.067–6.67 ng/μL. All our libraries were generated near the top end of this range.
10. Turn on Covaris and water bath at least 30 min prior to use.
11. Fill water tank with distilled water between the 10 and 15 on the fill scale, which should result in a reading between 10 and 15 on the run scale once the apparatus is lowered into the tank and water is displaced.
12. Turn the machine on and open the SonoLab software.
13. Allow the water to degas and reach optimal temperature (e.g., 6 °C).
14. To fragment DNA, pipette 50–130 μL of sample through the pre-slit cap on a Covaris microTUBE.
15. Place microTUBE in microTUBE holder and place holder in Covaris (*see* **Note 9**).
16. For 220 bp insert size, set the following protocol on the SonoLab software: duty cycle/duty factor, 10%; intensity, 4 (peak power intensity on the S220 is set to 140); cycles/bursts, 200; time/cycle, 9 s; number of cycles, 8; and total process time, 72 s.
17. Press run. Following fragmentation, remove sample and place on ice.
18. Prior to Apollo setup, anneal universal Wafergen adapters and barcodes in their plate using a two-step thermocycler protocol (95 °C for 5 min, 70 °C for 15 min).
19. Remove plate from thermocycler and cool on bench. Add 6 μL of water to each well and mix. Reseal plate after use and store at −20 °C. This is only done once (not each time a library preparation is carried out).

20. To utilize adapters, mix 2 μL of adapter (from the plate above) with 13 μL of water for each Apollo run (this is the adapter mix). Keep on ice.
21. To make Apollo ligation mix (per four reactions), combine 52.8 μL of PrepX Ligase Buffer, 4.4 μL of PrepX Ligase Enzyme, and 8.8 μL of water. Keep on ice.
22. Turn on Apollo and select Library Prep, DNA, ILM, and 220 bp.
23. Follow the on-screen instructions for setup of Apollo consumables.
24. Once the "Cooling" notification disappears, load the following in PCR strip-tubes as instructed on screen: 15 μL of Covaris-sheared sample, 15 μL of adapter mix, and 64 μL of ligation mix (per four reactions).
25. Load 225 μL of 2.5 M NaCl (supplied by Wafergen) to the Apollo microtiter plate as instructed on screen (per four reactions).
26. Add water and 70% ethanol to reservoirs 1 and 3 as instructed on screen.
27. Load the enzyme strip(s) and place retention plates on blocks 3 and 4 as instructed on screen.
28. Close the instrument door and press "Run." The run will end in 3 h, 20 min.
29. Retrieve products from block 3, rows 9–12. The libraries should be ~15 μL.
30. Qubit libraries as above and prepare the PrepX PCR reaction mixture (25 μL of PrepX PCR Master Mix, 2 μL of PrepX PCR Primers, 5 ng total of Apollo library, and water to make up the total volume to 50 μL).
31. Run the following PCR protocol: 98 °C for 30 s followed by five cycles of 98 °C for 10 s, 60 °C for 30 s, and 72 °C for 30 s. Following the five cycles, hold at 72 °C for 300 s and hold at 4 °C until removed from the thermocycler.
32. Proceed to PCR Cleanup 8 Protocol on the Apollo. Warm AMPure Beads to room temperature and vortex thoroughly just prior to running protocol.
33. Turn on the Apollo. Press "Utility Apps" and "PCR Cleanup 8."
34. Place empty consumables, 6 mL of 100% ethanol, and 6 mL of water on the Apollo deck as instructed on screen.
35. Once "Cooling" has disappeared, place strips containing 50 μL of sample and those containing 50 μL of freshly vortexed AMPure beads on Apollo deck as instructed on screen.

36. Press "Run."
37. After 16 min, collect clean products from Block 3, Row 4, and place on ice.
38. Bring 10 μL of clean library up to 15 μL with water.
39. Proceed to library quantification and quality control.

3.3 Jumping [Mate Pair] Library Preparation

1. This section follows protocols from the Nextera Mate Pair Sample Preparation Kit and the Qubit, Covaris, and Pippin Prep instruments.
2. For the Gel-Plus protocol (using the Pippin Prep), the Nextera Mate Pair Library Prep Reference Guide recommends 4 μg of input DNA.
3. Quantify DNA with Qubit as described in Subheading 3.2.
4. Thaw the Tagment Buffer Mate Pair on ice.
5. Set a heat block to 55 °C.
6. Add 4 μg of gDNA to 1.7 mL microcentrifuge tube (max volume 308 μL).
7. Make up remainder of 308 μL with water.
8. Add 80 μL of Tagment Buffer Mate Pair.
9. Add 12 μL of Mate Pair Tagment Enzyme.
10. Flick-mix (do not vortex) and collect volume at the bottom of the tube using minicentrifuge.
11. Incubate for 30 min at 55 °C.
12. Purify the tagmentation reaction using the Zymo Genomic DNA Clean & Concentrator kit. Begin by adding two volumes ($2 \times 400 = 800$ μL) of Zymo ChIP DNA Binding Buffer to the tagmentation reaction. Pipette to mix.
13. Transfer up to 800 μL of mixture to Zymo-Spin IC-XL column in a collection tube.
14. Centrifuge at top speed for 30 s. Discard flow-through.
15. Add remaining reaction to column and repeat previous step.
16. If this is the first time using the Zymo kit, add 24 mL 100% ethanol to the 6 mL (or 96 mL 100% ethanol to the 24 mL) DNA Wash Buffer concentrate.
17. Wash two times by adding 200 μL of Zymo DNA Wash Buffer to the column, spinning at top speed for 1 min and discarding the flow-through.
18. Centrifuge the empty column at top speed with the lid open for 1 min to remove excess wash buffer. Discard the collection tube and transfer the column to a new Axygen Maxymum Recovery 1.7 mL microcentrifuge tube (*see* **Note 10**).

19. Add 30 μL of Resuspension Buffer (RSB from Nextera Kit) and incubate at room temperature for 1 min before centrifuging for 1 min at full speed to elute purified DNA (*see* **Note 11**).
20. DNA can be stored at −20 °C for up to 24 h at this step, but to complete the protocol in 2 days, immediately proceed to the next step (*see* **Note 12**).
21. Thaw 10× Strand Displacement Buffer and dNTPs at room temperate and place on ice when completely thawed.
22. Set heat block to 20 °C.
23. To the 30 μL of purified tagmented DNA sample, add 132 μL water, 20 μL of 10× Strand Displacement Buffer, 8 μL of dNTPs, and 10 μL of strand displacement polymerase.
24. Flick-mix and collect volume at the bottom of the tube using minicentrifuge.
25. Incubate for 30 min at 20 °C. During this time, remove AMPure XP beads from 4 °C and allow them to reach room temperature.
26. Purify strand displacement mix using AMPure XP beads. Begin by vortexing beads thoroughly (vortex beads often to ensure they are evenly dispersed).
27. Prepare 800–900 μL of 70% ethanol per library preparation.
28. To the 200 μL of strand displaced DNA, add 100 μL of freshly vortexed AMPure XP beads.
29. Flick-mix and collect volume at bottom of the tube using minicentrifuge.
30. Incubate at room temperature for 15 min repeating the previous step every 2 min.
31. Place on magnetic rack for 5–10 min.
32. Leave tube on rack to remove and discard all supernatant without disturbing beads.
33. Leave tube on rack for two washes with 400 μL of 70% ethanol, removing and discarding supernatant after 30 s (*see* **Note 13**).
34. Once all supernatant has been removed, allow the beads to dry on the magnetic rack for 10–15 min. Small volumes of supernatant can be collected manually with a 10 μL pipette tip during the drying step.
35. Add 30 μL of RSB and remove tube from magnetic rack.
36. Flick-mix and collect volume at the bottom of the tube using minicentrifuge.
37. Incubate at room temperature for 5 min. Your sample is now in the supernatant.
38. Place on magnetic rack for 5 min.

39. Carefully collect supernatant and pipette to new 1.7 mL microcentrifuge tube.
40. DNA can be stored at −20 °C for up to 7 days at this step, but to complete the protocol in 2 days, proceed to the next step.
41. Size selection can be carried out using the Sage Science 0.75% cassette in the Pippin Prep for jumping libraries up to 8 kb. This will be described below. For jumping libraries 8–10 kb in size (which our lab has not utilized), size selection can be carried out with agarose gel electrophoresis, followed by a gel cleanup kit, or the BluePippin from Sage Science.
42. Turn on Pippin Prep. Click "Protocol Editor."
43. Select "NEW."
44. Select appropriate cassette (0.75%).
45. Select "End Run when Elution is Completed."
46. Under "Reference Lane," select a lane number for the marker D, then "apply reference to all lanes."
47. Select "Range" and supply values (e.g., 2–5 or 4–8 kb). If the Pippin Prep only allows tight cuts with the 0.75% cassette, supply your target size (e.g., 3 or 6 kb). Save the protocol.
48. Fill the rinse cassette with distilled water and place into the optical nest. Close the lid and wait 20 s.
49. Press CALIBRATE button on screen. This will prompt the message, "Calibration not done."
50. Place the calibration fixture dark side down over the five LED detectors on the rinse cassette.
51. Close the lid and ensure that 0.80 is in target setting.
52. Click CALIBRATE. If calibration is OK, press EXIT and proceed to samples.
53. The cassette is constructed as follows from left to right: Negative Buffer Chambers, Sample Wells, Elution Modules, and Positive Buffer Chambers.
54. Inspect the 0.75% cassette for damage to the gel and ensure that buffer reservoirs are a minimum of 50% full (add buffer if necessary). Do not use lanes that appear to be damaged.
55. Look at underside of cassette to ensure that there are no bubbles under the gel (between the gel and the plastic). Do not use lanes where gel bubbles appear to be present.
56. Tilt Positive Buffer Chambers up to gather bubbles near far right wall. Tap sides and bottom of cassette gently to dislodge any small bubbles between the Elution Modules and the Positive Buffer Chambers.

57. Place the cassette into the optical nest of the Pippin Prep while maintaining this position to ensure that bubbles do not get trapped between the Elution Modules and the Positive Buffer Chambers.
58. Place one hand on cassette and slowly remove adhesive strips by the white tabs provided.
59. Remove old buffer from Elution Modules and replace with 40 μL of fresh electrophoresis buffer (*see* **Note 14**).
60. Seal the Elution Modules with adhesive tape strips provided.
61. Fill Sample Wells to maximum with buffer if any are not full.
62. Test continuity by closing the lid and pressing the TEST button on screen.
63. Once PASS appears on the screen, hit RETURN (*see* **Note 15**).
64. Combine entire 30 μL of supernatant from AMPure bead purification with 10 μL of room temperature loading solution.
65. Flick-mix and collect volume at the bottom of the tube using minicentrifuge.
66. Remove 40 μL of buffer from Sample Well in target lane. Do not puncture agarose.
67. Add 40 μL of sample to Sample Well.
68. Repeat for additional samples and for marker D.
69. Begin run.
70. Once run is complete, carefully remove the adhesive strip over the Elution Modules one at a time and transfer size-selected product into new 1.7 mL microcentrifuge tube (*see* **Note 14**).
71. Purify the DNA from the Pippin Prep by once again using the Zymo Genomic DNA Clean & Concentration Kit, but note that the protocol different from above (*see* **Note 16**).
72. Pipette to mix five volumes of ChIP Binding Buffer into the Pippin Prep elution.
73. Transfer the mixture to a Zymo-Spin IC-XL column in a collection tube.
74. Centrifuge at top speed for 30 s. Discard flow-through.
75. Wash two times by adding 200 μL of Zymo DNA Wash Buffer to the column, spinning at top speed for 1 min and discarding the flow-through.
76. Discard the collection tube and transfer the column to a new 1.7 mL microcentrifuge tube.
77. Add 10 μL of RSB directly to the column matrix and incubate at room temperature for 1 min before centrifuging for 30 s at full speed to elute purified DNA.

78. DNA can be stored at −20 °C for up to 24 h at this step, but to complete the protocol in 2 days, proceed to the circularization step.
79. Run Qubit dsDNA HS Assay on samples as described in Subheading 3.2.
80. Utilize up to 600 ng of purified DNA for the circularization reaction (*see* **Note 17**).
81. Thaw Circularization Buffer at room temperature and then place on ice.
82. Set heat block to 30 °C.
83. Using up to 600 ng of DNA from the previous step, add enough water to make the total volume 268 μL, then 30 μL of Circularization Buffer 10×, and 2 μL of Circularization Ligase.
84. Flick-mix and collect volume at the bottom of the tube using minicentrifuge.
85. Incubate overnight (12–16 h) at 30 °C. This is a good stopping point for day 1.
86. Set two heat blocks to 37 and 70 °C.
87. Thaw Stop Ligation Buffer and place on ice.
88. Add 9 μL of exonuclease to overnight circularization reaction.
89. Flick-mix and collect volume at the bottom of the tube using minicentrifuge.
90. Incubate for 30 min at 37 °C.
91. During this incubation, turn on and prepare the Covaris S220.
92. Fill the reservoir with distilled water and allow machine to degas.
93. Chill Covaris S220 water bath to 6 °C.
94. Incubate samples from above for 30 min at 70 °C.
95. During this incubation, set up Covaris S220 protocol.
96. Set intensity to 8 (or peak power intensity to 240 on the S220), duty cycle/duty factor to 20%, cycles per burst to 200, and time to 40 s.
97. Flick-mix and add 12 μL of Stop Ligation Buffer to samples.
98. Flick-mix and collect volume at the bottom of the tube using minicentrifuge.
99. Transfer the entire volume of sample (~320 μL) to a Covaris T6 tube. Fill tube with water if space remains after the addition of sample. Carefully place the snap cap onto tube and press firmly (*see* **Note 18**).

100. Snap the T6 tube into the tube holder and place it into the Covaris S220.
101. Shear the DNA and transfer to a new 1.7 mL microcentrifuge tube on ice.
102. Vortex and shake M-280 Streptavidin magnetic beads to resuspend.
103. Transfer 20 μL of M-280 Streptavidin magnetic beads per library preparation to a new 1.7 mL microcentrifuge tube.
104. Place on magnetic rack for 1–2 min.
105. Remove and discard supernatant.
106. Wash the beads two times by adding 40 μL of Bead Bind Buffer for every 20 μL of original beads, incubating for 1–2 min, and removing all supernatant.
107. Remove the beads from the magnetic rack and add 300 μL of Bead Bind Buffer per 20 μL of original beads. Mix thoroughly and spin down.
108. Add 300 μL of beads in Bead Bind Buffer to 300 μL of sheared DNA.
109. Incubate for 15 min at 20 °C. Every 2 min during this incubation, flick-mix and collect volume at the bottom of the tube using minicentrifuge. Do not spin for more than a second to avoid pelleting the beads.
110. Following incubation, spin briefly in a minicentrifuge (5–10 s).
111. Place tube on the magnetic rack for 1–2 min. During this incubation, remove the End Repair Mix from the freezer and thaw at room temperature. Once thawed, move the End Repair Mix to ice.
112. Remove and discard supernatant from microcentrifuge tube.
113. Wash four times with Bead Wash Buffer by adding 200 μL of Bead Wash Buffer to the beads, flick-mixing and spinning down for 1–2 s, placing the mixture on the magnetic rack for 30 s, and removing all supernatant.
114. Wash two times with RSB by adding 200 μL of RSB to the beads, flick-mixing and spinning down for 5–10 s, placing the mixture on the magnetic rack for 30 s, and removing and discarding all supernatant. The supernatant from the second wash should not be removed until the End Repair Mix is thawed (from above) and the End Repair Mixture (in next step) has been made. Set a heat block to 30 °C during this step.
115. For each library preparation, mix 40 μL of End Repair Mix and 60 μL of water to create the End Repair Mixture.

116. Remove all supernatant from tubes.
117. Spin briefly (approximately 5 s) on the minicentrifuge to collect additional supernatant.
118. Remove all excess supernatant with a 10 μL pipette.
119. Add 100 μL of End Repair Mixture to beads.
120. Flick-mix and collect volume at the bottom of the tube using minicentrifuge.
121. Incubate for 30 min at 30 °C. During this incubation, place A-Tailing Mix on surface of ice in ice bucket (it will not thaw if placed into the ice).
122. Following incubation, spin in a minicentrifuge for 5–10 s.
123. Place the tube on magnetic rack for 1 min. If the A-Tailing Mix has not thawed on ice, begin thawing at room temperature now.
124. Remove and discard supernatant.
125. Wash four times with Bead Wash Buffer by adding 200 μL of Bead Wash Buffer to the beads, flick-mixing and spinning down for 5–10 s, placing the mixture on the magnetic rack for 30 s, and removing all supernatant.
126. Wash two times with RSB by adding 200 μL of RSB to the beads, flick-mixing and spinning down for 5–10 s, placing the mixture on the magnetic rack for 30 s, and removing and discarding all supernatant. The supernatant from the second wash should not be removed until the A-Tailing Mix is thawed (from above) and the A-tailing reaction mixture (in next step) has been made. Set a heat block to 37 °C during this step.
127. For each reaction, mix 12.5 μL A-Tailing Mix and 17.5 μL of water to create the A-tailing reaction mixture.
128. As above, remove all supernatant from the tube, spin for 5–10 s, and then remove excess using a 10 μL pipette.
129. Add 30 μL of A-tailing reaction mixture.
130. Flick-mix and collect volume at the bottom of the tube using minicentrifuge.
131. Incubate for 30 min at 37 °C. During this incubation, thaw DNA Adapter Indexes and Stop Ligation Buffer at room temperature. Once thawed, place on ice. Select individual DNA Adapters for each library (*see* **Note 19**).
132. To the 30 μL A-tailing reaction mix, add 2.5 Ligation Mix, 4 μL water, and 1 μL of unique DNA Adapter Index. Record which adapter has been used for each library.
133. Flick-mix and collect volume at the bottom of the tube using minicentrifuge.

134. Incubate for 10 min at 30 °C. During this incubation, place the Enhanced PCR Master Mix and PCR Primer Cocktail tubes at room temperature until thawed. Once thawed, place on ice.
135. Add 5 μL of Stop Ligation Buffer.
136. Spin for 5–10 s in a minicentrifuge.
137. Wash four times with Bead Wash Buffer by adding 200 μL of Bead Wash Buffer to the beads, flick-mixing and spinning down for 5–10 s, placing the mixture on the magnetic rack for 30 s, and removing all supernatant.
138. Wash two times with RSB by adding 200 μL of RSB to the beads, flick-mixing and spinning down for 5–10 s, placing the mixture on the magnetic rack for 30 s, and removing and discarding all supernatant. The supernatant from the second wash should not be removed until the Enhanced PCR Master Mix and PCR Primer Cocktail have thawed and the PCR reaction mixture (in next step) has been made.
139. For each library preparation, mix 20 μL of Enhanced PCR Master Mix, 5 μL of PCR Primer Cocktail, and 25 μL of water to create the PCR reaction mixture.
140. Program a thermocycler with the following PCR program: 98 °C for 30 s, followed by 10–15 cycles of 98 °C for 10 s; 60 °C for 30 s; and 72 °C for 30 s (*see* **Note 20**). Following these cycles, hold at 72 °C for 5 min and hold at 4 °C.
141. As above, remove and discard supernatant, spin tube to collect excess supernatant, replace the tube on the magnetic rack, and remove excess with a 10 μL pipette.
142. Add 50 μL of PCR reaction mixture to the beads and pipette to mix.
143. Transfer the mixture to PCR tubes and run PCR program.
144. Following PCR, the library can be kept at −20 °C for up to 7 days. This is a good stopping point for day 2.
145. Bring AMPure XP beads to room temperature.
146. Prepare 400 μL of fresh 70–80% ethanol per library.
147. Place PCR tubes on magnetic rack for 1 min. If necessary, pipette product from PCR tube into new 1.7 mL microcentrifuge tube to use the same magnetic rack from previous steps.
148. Transfer 45 μL of bead-separated solution (containing library) to a new 1.7 mL microcentrifuge tube.
149. Add 30 μL of freshly vortexed AMPure XP beads to this volume.

150. Flick-mix and collect volume at the bottom of the tube using minicentrifuge.
151. Incubate at room temperature for 5 min.
152. Place on magnetic rack for 5 min.
153. Remove and discard supernatant.
154. Wash two times with 70%–80% ethanol by adding 200 μL of ethanol to the beads, incubating for 30 s, and removing the supernatant. If beads are disturbed, allow them to settle once more before removing supernatant.
155. Dry the beads for 10–15 min. Remove excess ethanol during this time with a 10 μL pipette tip.
156. Remove the tube from the magnetic rack and resuspend beads in 20 μL of RSB. Flick-mix and collect volume at the bottom of the tube using minicentrifuge.
157. Incubate at room temperature for 5 min.
158. Incubate on the magnetic rack for 5 min. The library is now in the supernatant.
159. Transfer the supernatant to a new 1.7 mL microcentrifuge tube and proceed to library quantification and quality control.

3.4 Library Quantification and Quality Control

1. This section follows protocols from Qubit, the High Sensitivity DNA Kit for the Agilent Bioanalyzer, and the KAPA Library Quantification Kit for next-generation sequencing.
2. Run Qubit dsDNA HS Assay as described in Subheading 3.2 for all libraries.
3. To gain a better measure of library concentration and size distribution, run samples on Agilent 2100 Bioanalyzer HS DNA chip (*see* **Note 21**).
4. Up to 11 samples and 1 ladder can be run on an Agilent 2100 Bioanalyzer High Sensitivity DNA chip. The assay requires 1 μL of sample; dilute a small volume of each original library to approximately 1 ng/μL if necessary, and run from low to high concentration in well order on the Bioanalyzer chip.
5. As described in the Agilent High Sensitivity DNA Kit guide, bring High Sensitivity DNA dye concentrate and High Sensitivity DNA gel matrix to room temperature for around 30 min.
6. Vortex these tubes and combine 15 μL of dye concentrate with 300 μL of gel matrix. Vortex to mix and transfer to spin filter (provided with kit).
7. Centrifuge for 10 min at 2300 × *g*. Discard the filter and label the gel-dye mix with the date (it is stable when stored at 4 °C in the dark for 6 weeks).
8. Ensure that the electrode is clean (*see* **Note 22**).

9. Place the electrode in the electrode cartridge, and insert into bioanalyzer.
10. Examine back of the HS DNA chip to ensure that it does not appear broken.
11. Pipette 9 μL of gel-dye mix into the well labeled with the G in the black circle.
12. Place chip in priming station and close lid until click.
13. Set timer for 60 s and steadily press plunger on syringe until it rests under the silver clip.
14. After 60 s, release the silver clip. Watch for the plunger to rebound to at least 0.3 mL on the syringe (more likely 0.7 mL or above) and wait for it to stop moving.
15. Remove the chip and inspect the capillaries on the back of the chip for bubbles. Do not proceed if bubbles are present. Clean the priming station and begin again with a new chip.
16. Pipette 9 μL of gel-dye mix into the remaining 3 G wells.
17. Load 5 μL of High Sensitivity DNA marker into all remaining wells.
18. Warm samples to 37 °C for 1–2 min (while proceeding to the next step).
19. Place chip into the chip vortexer. Vortex starting from 0 rpm and ramping up to 2000–2200 rpm over a few seconds. Leave at 2200 rpm for 60 s.
20. Load 1 μL of sample (or water) into all Sample Wells and 1 μL of ladder into ladder well.
21. Vortex as above. Use chip within 5 min to minimize evaporation.
22. Open Expert software and select High Sensitivity DNA under Assays, DNA.
23. Once the run is complete, calculate size and concentration of libraries.
24. Following the bioanalyzer run, fragment libraries should be normally distributed with an average size of ~400 bp and jumping libraries often appear negative skewed with an average size of ~750 bp (Fig. 7). If additional peaks appear on the Bioanalyzer run, these can potentially be size-selected out using an additional cleaning step with AMPure XP beads. The ratio of beads used will be based on the size of the fragment being removed and the fragment that is to be kept.
25. Following Qubit and Bioanalyzer quantification, libraries should undergo a final quantification step using qPCR. The KAPA kits provide standards of known size with Illumina adapters to allow the most accurate quantification of the number of adapter-ligated molecules (those that will be sequenced) in a given library.

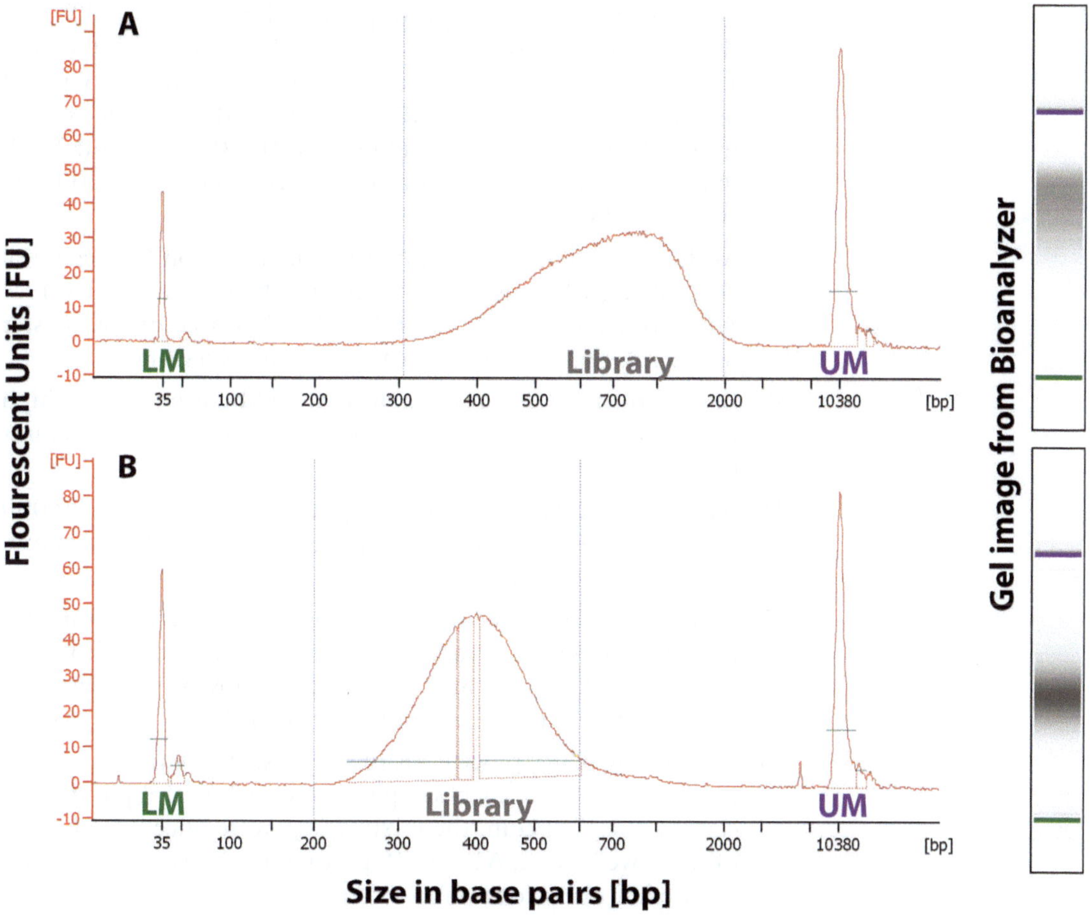

Fig. 7 Example electropherograms from an HS DNA Assay run on the Agilent 2100 Bioanalyzer for a Nextera 3 kb jumping library (**a**) and Apollo 220 bp fragment library (**b**). In the traces (*left* panel A and B), peaks represent the lower marker (green LM at left), the library (*grey* Library at *center*), and the upper marker (*purple* UM at *right*). In the Gel Image (*right* panel), the same color scheme is used, and the library is present between the green LM (at *bottom*) and the purple UM (at *top*). Average sizes are approximately 750 and 400 bp for the jumping and fragment libraries, respectively

26. Thaw all KAPA kit components on ice. Ensure that they are well mixed at each step.
27. Given that all libraries are at approximately 1–10 ng/μL, begin a serial dilution as described below (*see* **Note 23**).
28. Starting from the concentration described above, samples will be diluted to a final concentration of 1:200,000. This final concentration will be run alongside a 1:20,000 dilution. One possible set of volumes for this serial dilution is described here, where each successive step uses a volume from the previous step and the current concentration is listed for each step: (a) Add 1 μL library to 9 μL of water – 1:10. (b) Add 5 μL of previous dilution to 45 μL of preferred diluent – 1:100. (c) Add 5 μL of

previous dilution to 45 μL of preferred diluent – 1:1000. (d) Add 5 μL of previous dilution to 45 μL of preferred diluent – 1:10,000. (e) Add 25 μL of previous dilution to 25 μL of preferred diluent – 1:20,000. (f) Add 25 μL of previous dilution to 100 μL of preferred diluent – 1:100,000. (g) Add 25 μL of previous dilution to 25 μL of preferred diluent – 1:200,000 (*see* **Note 24**).

29. As described in the KAPA Library Quantification Kit Technical Data Sheet, when using the kit for the first time, add 1 mL of 10× Primer Premix to 2× KAPA SYBR FAST qPCR Master Mix 5 mL bottle. Vortex reaction mixture very briefly.
30. Based on the qPCR instrument that will be used, low or high ROX might be recommended. Check manufacturer instructions.
31. Each standard (1–6), library sample (1:20,000 and 1:200,000 dilutions), and a single no template control should be run in triplicate at 20 μL per replicate.
32. Calculate the number of samples needed and create a master mix of 12 μL reaction mixture with 4 μL of water (or 3.6 μL of water with 0.4 μL of 50× ROX high or low if necessary) for each sample (plus extra volumes for error).
33. In a qPCR plate, pipette 48 μL of master mix into each first replicate well. For each sample, three technical replicates are carried out. To begin, the mixture is pipette once into the first of three wells (e.g., A1 gets 48 μL of master mix, while A2 and A3 are left blank for now).
34. Pipette 12 μL of standard, sample, or water for no template control into each first replicate well and mix thoroughly.
35. Pipette 20 μL from first replicate well into second and third replicate wells. Using our example from above, A1 now has 48 μL of master mix and 12 μL of standard, sample, or water (60 μL total), while A2 and A3 are currently empty. Distribute this volume evenly across the three replicate wells by pipetting 20 μL from A1 to A2 and from A1 to A3.
36. Seal plate with clear film, spin down briefly, and place in instrument.
37. Set up instrument to read SYBR green and label the six tenfold dilution standards (from 20 to 0.0002 pM) if possible in software.
38. Run qPCR with 95 °C denaturation for 5 min, followed by 35 cycles of 95 °C for 30 s and 60 °C for 45 s with plate read (increase to 90 s if jumping libraries are included). Following amplification, add melt curve analysis at 65–95 °C.
39. Following qPCR run, check qPCR for adapter dimers (a double peak following the main peak in the melt curve). Additional

bead cleanups can remove adapter dimers, and qPCR should be run again to properly quantify libraries following this step.

40. If melt curve analysis looks clean, proceed to the calculation of efficiency. Remove any major outliers and plot the average Cq of the six standards against the log of their pM concentrations.
41. Calculate the slope and R^2 (which should be above 0.99). Ensure that efficiency is between 90 and 110% using a website such as https://www.thermofisher.com/us/en/home/brands/thermo-scientific/molecular-biology/molecular-biology-learning-center/molecular-biology-resource-library/thermo-scientific-web-tools/qpcr-efficiency-calculator.html
42. If all above values are within acceptable ranges, the Cq and quantity scores given for each sample can be utilized.
43. Multiply each quantity by the original dilution (20,000 or 200,000) and correct for library size with the following formula: (452/library size) × average quantity. This formula corrects for the difference in size between the library and the standards. Divide the output of this formula by 1000 to obtain nM (*see* **Note 25**).
44. Multiplex libraries to concentrations requested by sequencing center to fit individual budget and experimental needs.

3.5 ALLPATHS-LG Assembly

1. This section utilizes commands from the Fastq, cutadapt, TrimGalore!, and ALLPATHS-LG manuals.
2. All commands contain curly braces, {}, to indicate names (files, folders, etc.,) that need to be changed by the user to reflect the specific data being processed. Remove the curly braces and input user-specific information.
3. To run FastQC on one fastq file, use the command:

```
fastqc -outdir {outdir_name} {fastq_file.fq}
```

4. Repeat above command line for each fastq file.
5. Run Trim Galore! on each individual library to remove adapters with commands in steps below (*see* **Note 26**).
6. For fragment libraries, use the command:

```
trim_galore -paired -phred33 -output_dir {outdir_name} -
stringency 1 -e 0.1 {reads1.fq} {reads2.fq}
```

7. For jumping libraries, use the command:

```
trim_galore -paired -nextera -phred33 -output_dir
{outdir_name} -stringency 1 -e 0.1 {reads1.fq}
{reads2.fq}
```

8. Run FastQC once again on trimmed reads. Verify that there are no overrepresented or adapter sequences according to the FastQC output.

9. Write the groups and libs files required by ALLPATHS-LG. Both are comma-separated files. They should be named {species}_groups.csv and {species}_libs.csv. The top line of each is the header and should appear as listed below. Asterisks are utilized in the file path to include both read directions. The library_name column must remain the same between the two files. It is underlined in this protocol.

10. Make the group csv file required by ALLPATHS-LG. This file contains three comma-separated columns, for example:

```
group_name,library_name,file_name
{species220},{frag},/{full_path_to}/{species_frag_}*.fastq.gz
{species3kb},{jump3kb},/{full_path_to}/{species_jump_}*.fastq.gz
```

11. Make the libs csv file required by ALLPATHS-LG. This file contains 12 comma-separated columns. Based on the type value (either fragment or jumping), fill frag_size and frag_stddev or insert_size and insert_stddev, respectively. Empty columns are allowed. An example appears below:

```
library_name,project_name,organism_name,type,paired,frag_size,
frag_stddev,insert_size,insert_stddev,read_orientation,genomic_-
start,genomic_end
{frag},{avianGenome},{genusSp},fragment,1,220,{22},,,inward,,
{jump3kb},{avianGenome},{genusSp},jumping,1,,,3000,{300},outward,,
```

12. Create the directory structure for ALLPATHS-LG. See below:

```
mkdir -p {REFERENCE}/{DATA}/{RUN}/
```

Here, REFERENCE is the same {species} used in the groups and libs file names, DATA is arbitrary (e.g., version), and RUN is the iteration (e.g., date).

13. Prepare the ALLPATHS-LG inputs. Run command:

```
PrepareAllPathsInputs.pl DATA_DIR={full_path_to/REFER-
ENCE/DATA/} PLOIDY=2 IN_GROUPS_CSV=${species}_groups.
csv IN_LIBS_CSV=${species}_libs.csv > ${species}_pre-
p_log.log
```

14. Assemble the genome. See below:

```
RunAllPathsLG PRE={full_path_to} REFERENCE_NAME={REFER-
ENCE} DATA_SUBDIR={DATA} RUN={RUN} HAPLOIDIFY=TRUE
THREADS={user_choice_based_on_resources} OVERWRITE=TRUE
1> ${species}_output(see Note 27)
```

15. Examine ALLPATHS-LG report (assembly.report) to view statistics.

4 Notes

1. Tissue can be fresh or frozen, but must be of high quality for the jumping library. DNA degradation can be examined on a gel as seen in Fig. 6.
2. The Apollo 324 NGS Library Preparation System was utilized alongside the Wafergen Biosystems PrepX ILM 32i DNA Library Kit for fragment library preparation in this protocol. This is not essential. The steps pertaining to use of the Apollo and this kit can be replaced with protocols from other library preparation kits if these libraries will be made entirely by hand, so long as the fragment library created this way is the appropriate size (i.e., the insert size is small enough, or the reads are long enough, that the paired-end sequencing reads will overlap).
3. When setting up the Apollo, only use Wafergen Apollo consumables unless otherwise instructed.
4. The compute required for ALLPATHS-LG is quite sizable, with our assemblies ranging from 257 to 389 GB of peak memory. If your lab does not have access to a supercomputer or server for this step, check with other departments at the university (e.g., physics and structural chemistry). Two additional options available in the United States are Amazon Web Services (available at https://aws.amazon.com/ec2/?nc2=h_m1) and the Extreme Science and Engineering Discovery Environment (XSEDE), funded by the National Science Foundation (available here: https://www.xsede.org/) [29]. To obtain ALLPATHS-LG, follow the link provided in the methods section to download the latest version and the ALLPATHS-LG manual. To install ALLPATHS-LG, consult the system administrator and the ALLPATHS manual. Regardless of compute resources utilized, speaking with a system or network administrator is always a good place to start when planning to install and/or run a memory-intensive program.
5. Prior to using any buffer, enzyme, etc. from a kit, ensure that it is completely thawed. Mix as directed and spin briefly (a few seconds) in a tabletop minicentrifuge to collect the volume in the bottom of the tube. Always keep enzymes on ice when in use and return them promptly to −20 °C when not in use (unless otherwise instructed).
6. Although recommended in the protocol for the E.Z.N.A. Tissue DNA Kit (as well as many other DNA extraction protocols), do not vortex your samples during DNA extraction as this could shear DNA, resulting in lower-quality jumping libraries. Flick-mix by holding the tube near the lid and flicking the bottom portion repeatedly.

7. As described in the E.Z.N.A. Tissue DNA Kit Protocol, different tissues will require varied incubation times for successful lysis. Mechanically cutting the tissue into small pieces prior to the lysis will speed the process up substantially. The incubation time and agitation method might require some optimization in your laboratory in order to get the highest-quality DNA from your available tissue. Decreasing lysis time and speed of agitation could reduce shearing of DNA during lysis.
8. To obtain the most accurate quantification of samples, aim to read each tube as close to 2 min after the addition of the sample/standard and the vortex mixing. Leaving approximately 30–60 s between mixing of each sample or standard will allow time for the reading and recording of measurements.
9. Do not allow the sample to run with visible bubbles, since these can interfere with Covaris shearing.
10. Throughout this protocol, it is recommended that Axygen Maxymum Recovery 1.7 mL microcentrifuge tubes should be used to minimize product loss. Care should be taken by the unfamiliar when opening and closing these tubes, since the same properties that allow for maximum recovery of the product also provide an environment from which the product will sometimes "jump" when agitated. It is recommended that time is spent practicing with small volumes of water in these tubes before beginning a library preparation.
11. Incubation in RSB prior to the elution spin with the Zymo kit has been increased to 5 min with no discernible differences (positive or negative) on the outcome of the protocol.
12. Prior to freezing (or continuing on with the protocol), the Nextera Mate Pair Reference Guide suggests running 1 μL of product mixed with 7 μL of water in an Agilent DNA 12000 Bioanalyzer run to assess the tagmentation (this protocol requires a different kit than the Agilent HS DNA Bioanalyzer run described in **steps 3–24** in Subheading 3.4). Our laboratory has never carried out this quality control step for the generation of many 3 kb jumping libraries (and one 6 kb jumping library), but it can be utilized to determine the distribution of sizes created by the tagmentation reaction in order to inform size selection ranges on the Pippin Prep. The protocol described herein appears to work well in our hands to generate 3 kb jumping libraries without the need to visualize the size distribution at this stage.
13. Add ethanol slowly to avoid disturbing beads on tube wall. If beads are disturbed, allow adequate time for beads to settle since your library is bound to the beads at this stage. Be mindful of the tendency for a sample (especially when combined with beads) to "jump" from the Axygen Maxymum

Recovery 1.7 mL microcentrifuge tubes when the tubes are snapped open or closed.

14. When removing and replenishing buffer to the Elution Modules of the Pippin Prep Cassette, gently insert pipette tip into module until the bottom is reached (if the tip is not long enough or the wrong shape, use different tips). The bottom of the module is sticky or tacky. Slowly dislodge the pipette tip from the bottom and pipette as close to the bottom as possible. Avoid introducing bubbles by following this procedure and not pushing past the first resistance on the pipette (do not blow out). This same strategy applies to removing sample once the run has completed. Additionally, the starting buffer volume of 40 μL does not fill the entire depth of the Elution Module (total space is ~65 μL). Do not overfill the Elution Module.
15. There are two columns on the Pippin Prep continuity test window. If the failure appears on the left (the separation lane), do not use the lane. If the failure appears on the right (the elution channel), it is okay to use that channel for reference since there will be nothing to collect. Replacing the 40 μL of electrophoresis buffer in the Elution Module and retesting the continuity might fix a failure on the elution lane (the right side of the window).
16. The Zymo purification step described here is a new addition to the Nextera Mate Pair Reference Guide. Our laboratory has not tested this step but based on the Nextera protocol, it should be included in here.
17. The Nextera Mate Pair Reference Guide provides an expected yield of 150–400 ng of DNA following the Pippin Prep protocol. In our laboratory, concentrations between ~1.25 and 5 ng/μL are common immediately following the Pippin Prep (prior to the new Zymo purification, while the sample is still in ~40–65 μL). This range of concentrations (including those below the expected 150 ng total listed in the Nextera Mate Pair Reference Guide) has worked well, and there does not appear to be any discernible difference in library quality between those at the low or high end of this range. Lower yields at this step might decrease the diversity of reads, but this has not been tested in our laboratory.
18. Since bubbles can interfere with Covaris shearing, ensure that an appreciable volume of sample has not escaped during placement of the tube into the tube holder. A small bubble is likely to form at this stage (which should not be a concern), but a large bubble could be problematic. Attempt to keep the bubble contained against the cap so that results are consistent across samples.

19. Some planning is useful at this step since base diversity in the adapters is necessary for proper demultiplexing of samples after sequencing is complete. The Nextera Mate Pair Reference Guide suggests the following options if you are multiplexing four or fewer libraries together for a sequencing run. Use AD006 (with AD012) or AD005 (with AD019) for 2-plexing. For 3-plexing, use AD002 (with AD007 and AD019), AD005 (with AD006 and AD015), or either of the 2-plex options with any third adapter. Use AD005 (with AD006, AD012, and AD019), AD002 (with AD004, AD007, and AD016), or any 3-plex option and any other adapter for 4-plexing. For any multiplex above four libraries, utilize any of the 4-plex options and then add in additional adapters as needed.
20. The number of cycles for the PCR step can be set between 10 and 15 depending on the amount of input DNA (following Pippin Prep) and concerns about library diversity. Fewer cycles will result in lower overall library yield, but more cycles can decrease library diversity by resulting in overrepresentation of highly amplified sequence. In our laboratory, 12 cycles are commonly used for most preparations with around 150–200 ng of DNA following Pippin Prep for 3 kb libraries. The Nextera Mate Pair Reference Guide suggests using only ten cycles if there are at least 200 ng of DNA following Pippin Prep and the jump is less than 8 kb, but 15 cycles if the input DNA was lower than 200 ng regardless of insert size.
21. The Agilent TapeStation can also be used for quantification and visualization of the Apollo library but is not compatible with the Nextera kit due to the Y-shaped adapters utilized in the jumping library.
22. A sonicator bath does an excellent job of cleaning the electrode. Following this, spray the electrode with 70% ethanol and use compressed nitrogen to dry. If a sonicator is not available, the electrode should be cleaned immediately after every run as follows: load 350 μL of nuclease-free water into electrode cleaning chips and rinse the electrode by placing the chip into the Bioanalyzer and closing the lid for 1 min. Repeat one to two more times with fresh water. Allow electrode to air dry.
23. Depending upon the success of all library preparation steps and the number of PCR cycles that were used for each library, 1–10 ng/μL is a reasonable starting point. Often the jumping library will require a 1:10 dilution to reach this concentration, but if some libraries are slightly lower than this range, they could still sequence well. The KAPA kit will give the best quantification, and these values should be discussed with a technician at the sequencing center to plan your multiplex.
24. Possibly the most important aspect of obtaining accurate and precise measures from the KAPA qPCR step is effective sample

mixing. Small pipetting errors will accumulate quickly, and if dilutions are not homogeneous at the time of pipetting, this will also add error. Pipette mixing 20–50 times at each mixing step appears to reduce noise in the final data.

25. Following KAPA library quantification, the technicians at our sequencing center recommend correcting the final nM concentration by a factor of 0.75 before multiplexing (e.g., library calculated to be 1 nM is reported to the sequencing center as 0.75 nM). In their experience, this factor produces optimal cluster density on their sequencers. Contact the technicians at your sequencing center to determine if they have recommendations for any correction following KAPA quantification.

26. Although we currently recommend Trim Galore! as written in the methods section for the adapter trimming, all genomes discussed in this chapter were trimmed using trimmomatic. It requires the generation of an adapter's file that provides trimmomatic with the specific adapter sequences utilized (e.g., Nextera and Apollo). The command for this is available here: github.com/tsackton/ratite-genomics/blob/master/assembly/trim_fastq.sh

27. `THREADS={user_choice_based_on_resources}` is the number of threads requested based on system infrastructure. Additionally, if assembling an inbred or low polymorphism genome, HAPLODIFY=FALSE might provide better results, but our genomes from wild and zoo specimens have been assembled with better success using HAPLODIFY=TRUE.

References

1. Stern CD (2005) The chick: a great model system becomes even greater. Dev Cell 8:9–17
2. Nagai H, Mak S-S, Weng W et al (2011) Embryonic development of the emu, *Dromaius novaehollandiae*. Dev Dyn 240:162–175
3. Padgett CS, Ivey WD (1960) The normal embryology of the coturnix quail. Anat Rec 137:1–11
4. Murray JR, Varian-Ramos CW, Welch ZS et al (2013) Embryological staging of the Zebra Finch, *Taeniopygia guttata*. J Morphol 274:1090–1110
5. Hillier LW, Miller W, Birney E et al (2004) Sequence and comparative analysis of the chicken genome provide unique perspectives on vertebrate evolution. Nature 432:695–716
6. Zhang G, Li C, Li Q et al (2014) Comparative genomics reveals insights into avian genome evolution and adaptation. Science 346:1311–1320
7. Jarvis ED, Mirarab S, Aberer AJ et al (2014) Whole genome analyses resolve the early branches in the tree of life of modern birds. Science 346:1320–1331
8. Zhang G (2015) Bird sequencing project takes off. Nature 522:34
9. Bonneaud C, Burnside J, Edwards SV (2008) High-speed developments in avian genomics. BioScience 58:587–595
10. Gnerre S, Maccallum I, Przybylski D et al (2011) High-quality draft assemblies of mammalian genomes from massively parallel sequence data. Proc Natl Acad Sci U S A 108:1513–1518
11. Bradnam KR, Fass JN, Alexandrov A et al (2013) Assemblathon 2: evaluating de novo methods of genome assembly in three vertebrate species. GigaScience 2:10
12. Langmead B, Salzberg SL (2012) Fast gapped-read alignment with Bowtie 2. Nat Methods 9:357–359
13. Goodwin S, Gurtowski J, Ethe-Sayers S et al (2015) Oxford nanopore sequencing and de

novo assembly of a eukaryotic genome. Genome Res 25(11):1750–1756
14. Rhoads A, Au KF (2015) PacBio sequencing and its applications. Genomics, Proteomics Bioinformatics 13:278–289
15. Gordon D, Huddleston J, Chaisson MJ et al (2016) Long-read sequence assembly of the gorilla genome. Science 352:aae0344
16. Mostovoy Y, Levy-Sakin M, Lam J et al (2016) A hybrid approach for de novo human genome sequence assembly and phasing. Nat Methods 13:12–17
17. Weisenfeld NI, Kumar V, Shah P et al (2017) Direct determination of diploid genome sequences. Genome Res 27(5):757–767
18. Siepel A, Bejerano G, Pedersen JS et al (2005) Evolutionarily conserved elements in vertebrate, insect, worm, and yeast genomes. Genome Res 15:1034–1050
19. Visel A, Prabhakar S, Akiyama JA et al (2008) Ultraconservation identifies a small subset of extremely constrained developmental enhancers. Nat Genet 40:158–160
20. Lowe CB, Kellis M, Siepel A et al (2011) Three periods of regulatory innovation during vertebrate evolution. Science (New York, NY) 333:1019–1024
21. Lowe CB, Clarke JA, Baker AJ et al (2015) Feather development genes and associated regulatory innovation predate the origin of dinosauria. Mol Biol Evol 32:23–28
22. Marcovitz A, Jia R, Bejerano G (2016) "Reverse genomics" predicts function of human conserved noncoding elements. Mol Biol Evol 33:1358–1369
23. Hiller M, Schaar BT, Bejerano G (2012) Hundreds of conserved non-coding genomic regions are independently lost in mammals. Nucleic Acids Res 40:11463–11476
24. Seki R, Li C, Fang Q, Hayashi S, Egawa S, Hu J, Xu L, Pan H, Kondo M, Sato T, Matsubara H, Kamiyama N, Kitajima K, Saito D, Liu Y, Thomas M, Gilbert P, Zhou Q, Xu X, Shiroishi T, Irie N, Tamura K, Zhang G (2017) Functional roles of Aves class-specific cis-regulatory elements on macroevolution of bird-specific features. Nat Commun 8:14229
25. Booker BM, Friedrich T, Mason MK et al (2016) Bat accelerated regions identify a bat forelimb specific enhancer in the HoxD Locus. PLoS Genet 12(3):e1005738
26. Eckalbar WL, Schlebusch SA, Mason MK et al (2016) Transcriptomic and epigenomic characterization of the developing bat wing. Nat Genet 48:528–536
27. Domyan ET, Kronenberg Z, Infante CR et al (2016) Molecular shifts in limb identity underlie development of feathered feet in two domestic avian species. elife 5:1–21
28. Adachi N, Robinson M, Goolsbee A et al (2016) Regulatory evolution of *Tbx5* and the origin of paired appendages. Proc Natl Acad Sci 113:10115–10120
29. Towns J, Cockerill T, Dahan M, Foster I, Gaither K, Grimshaw A, Hazlewood V, Lathrop S, Lifka D, Peterson GD, Roskies R, Ray Scott J, Wilkens-Diehr N (2014) XSEDE: accelerating scientific discovery. Comput Sci Eng 16(5):62–74

Chapter 3

A Step-by-Step Guide to Assemble a Reptilian Genome

Asier Ullate-Agote, Yingguang Frank Chan, and Athanasia C. Tzika

Abstract

Multiple technologies and software are now available facilitating the de novo sequencing and assembly of any vertebrate genome. Yet the quality of most available sequenced genomes is substantially poorer than that of the golden standard in the field: the human genome. Here, we present a step-by-step protocol for the successful sequencing and assembly of a high-quality snake genome that can be applied to any other reptilian or avian species. We combine the great sequencing depth and accuracy of short reads with the use of different insert size libraries for extended scaffolding followed by optical mapping. We show that this procedure improved the corn snake scaffold N50 from 3.7 kbp to 1.4 Mbp, currently making it one of the snake genomes with the longest scaffolds. Short guidelines are also given on the extraction of long DNA molecules from reptilian blood and the necessary modifications in DNA extraction protocols. This chapter is accompanied by a website (www.reptilomics.org/stepbystep.html), where we provide links to the suggested software, examples of input and output files, and running parameters.

Key words Genome sequencing, Genome assembly, Genomics, Corn snake, Optical mapping, Snake genome, DNA extraction, Transcriptome

1 Introduction

Since the publication of the human genome [1], major technological advances have greatly facilitated the de novo sequencing of any vertebrate genome (e.g., [2, 3]). These innovations were accompanied by an explosive growth of the bioinformatics field with the development of multiple algorithms and analytical tools for genome assembly, annotation, and quality assessment. Easier access to academic and commercial sequencing facilities resulted in a steady increase of the number of genomes deciphered each year, as illustrated by the growth of the "NCBI Genome database" (https://www.ncbi.nlm.nih.gov/genome) and the "Genomes OnLine Database" (https://gold.jgi.doe.gov). Furthermore, several extensive initiatives exist, from exploring genomic variation and uncovering the genetic basis of rare diseases by re-sequencing the

Guojun Sheng (ed.), *Avian and Reptilian Developmental Biology: Methods and Protocols*, Methods in Molecular Biology, vol. 1650, DOI 10.1007/978-1-4939-7216-6_3, © Springer Science+Business Media LLC 2017

human genome (1000 Genomes Project, 100,000 Genomes Project), to surveying the "Tree of Life," with projects such as the Genome 10K (assembly of 10,000 vertebrate genomes), i5k (assembly of 5000 arthropod genomes), or B10K (assembly of 10,000 bird genomes). Yet only a handful of genomes are of quality similar to that of the human one, with the large majority of new genomes being only drafts, comprising of scaffolds and not chromosomes. A number of studies pointed out that low-coverage genomes result in poor annotation and can introduce erroneous results in large-scale comparative analyses (e.g., [4, 5]). Nevertheless, genomes at various stages of build can be very useful, as was the case for the draft corn snake (*Pantherophis guttatus*) genome that assisted in uncovering the allele causing amelanism in that species [6, 7].

Reptilia comprise of more than 10,000 species, yet only the genomes of the anole lizard [8] and the western painted turtle [9] are assembled at the chromosome level. Nineteen additional genomes of this class are available (Table 1): three more lizards, eight snakes, four more turtles, and four Crocodilia, most of them solely assembled from Illumina reads and with a scaffold N50 ranging from 3.7 kbp to 96 Mbp. As shown in Table 1, even when similar sequencing depth and assembly strategies are used, the quality of the assembled genome, measured by its length and continuity, varies greatly. All available snake genomes are less than 1.9 Gbp long and the greatest scaffold N50 obtained with only Illumina reads is that of *Thamnophis sirtalis* at ~650,000 bp. On the other hand, turtle genomes are >1.9 Gbp and the scaffold N50 can reach 3.8 Mbp solely using Illumina reads. This difference in quality resulting from similar sequencing strategies on different taxa could be due to the great number of repetitive elements present in the genomes of Squamates (snakes and lizards), as illustrated by the observation that *Hox* clusters in this group are all 10–40% longer than in mammals, birds, or other reptiles [8, 10, 11].

Choosing an appropriate sequencing strategy is of paramount importance when de novo assembling a genome. For example, combining sequences from several individuals can significantly hinder the assembly process with the increased heterogeneity of the data. Besides sufficient coverage, the type of reads and the library preparation can substantially influence the final outcome. Major considerations also include using predominantly short reads with low error rate versus fewer long reads to span repetitive regions that have an error rate greater than 10%. Once the raw data collection is complete, a variety of software packages are available to perform the assembly, often yielding considerably different results [12, 13]. The hardware requirements for running such analyses need also to be taken into account, with access to computing clusters being essential. As demonstrated even for the human genome [14], currently no single technology can de novo recover a full genome. One

Table 1
Sequencing information for the available reptilian genomes (source: NCBI Genome database, September 2016)

Species	Length (Gbp)	Scaffolds	Scaffold N50 (kbp)	Scaffold L50	Coverage	Technology
Anolis carolinensis	1.8 (5%)	6646	4003	119	7.1	Sanger
Gekko japonicus	2.5 (3%)	191,500	708	963	95	Illumina
Ophisaurus gracilis	1.71	6715	1270	–	86	Illumina
Pogona vitticeps	1.8 (3.8%)	545,310	2291	219	86	Illumina
Crotalus horridus	1.5 (12%)	186,068	24	16,749	135	Illumina
Crotalus mitchellii	1.1 (0.2%)	473,380	5.3	58,778	40	Illumina
Ophiophagus hannah	1.6 (13%)	296,399	241	1750	28	Illumina
Pantherophis guttatus v1	1.4 (3%)	883,920	3.7	78,031	13	Illumina
Pantherophis guttatus v2	1.9 (13%)	114,644	1378	279	122	Illumina, BioNano
Protobothrops mucrosquamatus	1.7 (8%)	52,280	424	924	86	Illumina
Python bivittatus	1.4 (4%)	39,113	214	1939	20	Illumina, 454
Thamnophis sirtalis	1.4 (21%)	7930	648	639	72	Illumina
Vipera berus	1.5 (14%)	28,883	126	3408	121	Illumina
Apalone spinifera	1.9 (2.7%)	286,620	2307	225	33.4	Illumina
Chelonia mydas	2.2 (4%)	140,023	3864	158	110	Illumina
Chrysemys picta	2.4 (8%)	78,631	6606	102	15	Sanger, Illumina, 454
Malaclemys terrapin	2.4	21,683	437	1451	16	PacBio, Illumina
Pelodiscus sinensis	2.2 (4%)	19,904	3351	189	105	Illumina
Alligator mississippiensis	2.2 (2%)	7091	18,601	34	156	Illumina
Alligator sinensis	2.3 (3%)	9317	2188	325	109	Illumina
Crocodylus porosus	2 (5%)	69	84,438	7	74	Illumina
Gavialis gangeticus	2.6 (26%)	80	96,077	8	81	Illumina

For *Malaclemys terrapin*, statistics on the contigs rather than the scaffolds are provided. For *Pogona vitticeps* and *Ophisaurus gracilis*, genome statistics were extracted from the corresponding publications. In the "length" column, we also indicate in parenthesis the percentage of gaps in the assembly, when available.

approach could be to combine the great sequencing depth and accuracy of short reads, the extended scaffolds obtained with the use of long reads (>1 kbp) and libraries of different insert size, all anchored on the backbone provided by optical mapping. A final (chromosome-level) assembly would require resources that go beyond the capacities of small research groups, which are currently producing the greatest amount of non-classical model species genomic data.

Building up on our experience with the first version of the corn snake genome, we present here a step-by-step protocol for the sequencing and assembly of a snake genome that can be applied to any other reptilian or avian species. We show that this procedure improved the scaffold N50 of the corn snake from 3.7 kbp to 1.4 Mbp. Short guidelines are also given on the extraction of long DNA molecules from reptilian blood and the necessary modifications in existing DNA extraction protocols. This chapter is accompanied by a website (www.reptilomics.org/stepbystep.html), where we provide links to the suggested software, and examples of input and output files and running parameters.

2 Materials

2.1 Isolation of DNA from Reptilian Blood

1. Blood from a single individual; in the case of the *P. guttatus* genome, an adult male was the source of all biological material (*see* **Note 1**).
2. Blood preservation solution 1 (100 mM Tris, 100 mM EDTA, 2% SDS): Add about 800 mL water to a 1 L graduated cylinder. Weigh 12.11 g Tris–HCl, 37.22 g EDTA, and 20 g SDS and transfer to the cylinder. Make up to 1 L with water. Store at room temperature (*see* **Note 2**).
3. Blood preservation solution 2 (100 mM Tris, 100 mM EDTA): Add about 800 mL water to a 1 L graduated cylinder. Weigh 12.11 g Tris–HCl and 37.22 g EDTA and transfer to the cylinder. Make up to 1 L with water. Store at room temperature (*see* **Note 3**).

2.2 List of Input Data Used for Improved Genome Assembly (Table 2) (See Note 4)

1. An Illumina 100 base pair (bp) paired-end reads library with an average fragment size of 370 bp from the HiSeq2000 sequencer used for the draft genome [7]. A single lane was sequenced resulting in 24.56 gigabases (Gbp).
2. A PCR-free 250 bp paired-end reads library with an average fragment size of 415 bp from the HiSeq2500 Illumina sequencer in rapid run mode. Two lanes produced 210.3 Gbp.
3. Three mate-pairs libraries, with, respectively, an insert size of 3, 8, and 20 kilobases (kbp), prepared with the NxSeq Long Mate

Table 2
Statistics of the genome libraries used for the assembly of the corn snake genome version 2

Technology	Read length	Insert size	Lanes/ cells	Read pairs (millions)	Filtered read pairs (millions)	Coverage (1.8 Gbp)
HiSeq2000	2 × 100	370	1	122.8	105	11.7×
HiSeq2500	2 × 250	415	2	387.5	387.45	107.6×
MiSeq	2 × 250	3000	1	13.5	8.7	1.35×
MiSeq	2 × 250	6000	0.5	5.5	3.5	0.54×
MiSeq	2 × 250	20,000	0.5	7.7	4.6	0.72×
BioNano	>150 kb	–	3	–	–	235.3×

The "coverage" is based on an estimated genome size of 1.8 Gb.

Pair kit (Lucigen). The 3 kbp library was sequenced on a single MiSeq run and the 8 and 20 kbp were pooled in a second MiSeq run (Lucigen Corporation; http://www.lucigen.com). We obtained mate-paired reads of 250 bp facing inwards (Fig. 1).

4. High-molecular weight DNA molecules digested with the appropriate enzyme (for the corn snake, BsPQI), labeled and imaged on the Irys instrument (17 runs of 3 flow cells) (*see* **Note 5**). For the corn snake genome, we obtained 1.7 million molecules with an average length of 243.87 kbp, after filtering out molecules less than 150 kbp.

2.3 Software (See Notes 6–10)

1. FastQC v0.11.2 (run locally—optional graphical interface): http://www.bioinformatics.babraham.ac.uk/projects/download.html#fastqc
2. Skewer v0.1.123 (run on cluster or locally) [15]: https://github.com/relipmoc/skewer
3. DISCOVAR de novo v0.52219 (run on cluster): ftp://ftp.broadinstitute.org/pub/crd/DiscovarDeNovo/
4. BLAST+ v2.2.30 (run on cluster or locally) [16]: ftp://ftp.ncbi.nlm.nih.gov/blast/executables/blast+/LATEST/
5. LANE runner v2.0 (run locally) [17]: http://applications.lanevol.org/lanerunner/LANE_runner.html
6. Trimmomatic v0.32 (run on cluster or locally) [18]: http://www.usadellab.org/cms/?page=trimmomatic
7. bwa v0.7.10 (run on cluster or locally) [19]: https://github.com/lh3/bwa
8. samtools v1.1 (run on cluster or locally) [20]: https://sourceforge.net/projects/samtools/files/samtools/

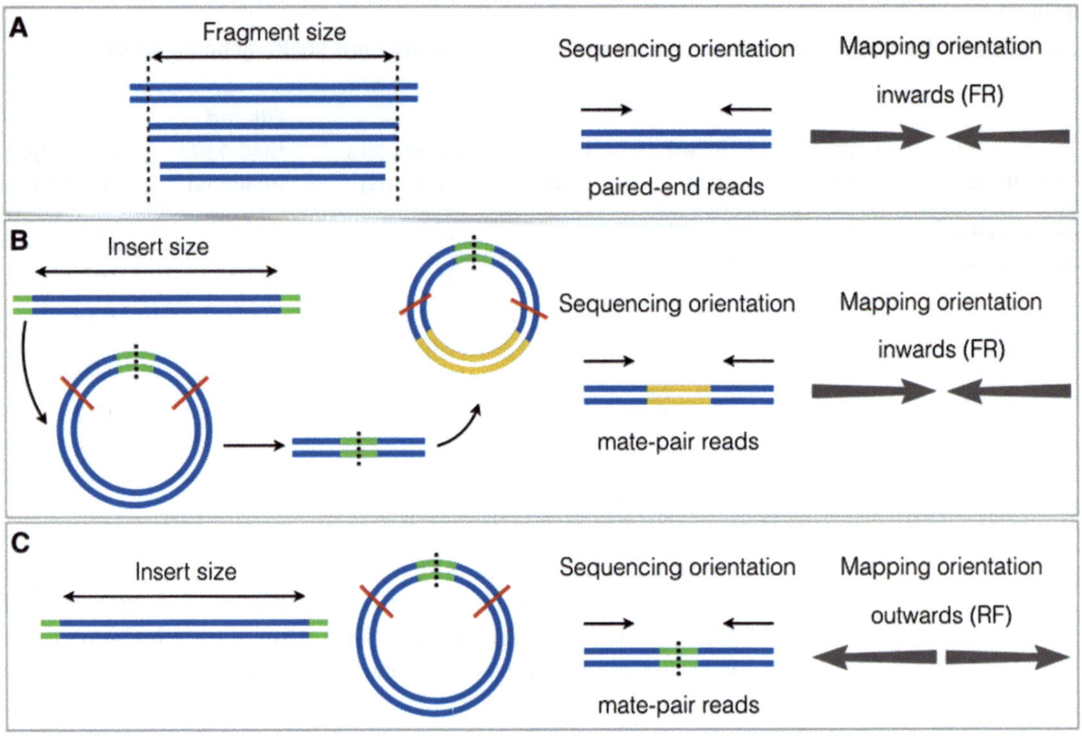

Fig. 1 Schematic representation of the sequencing and mapping orientation of the reads depending on the library preparation: (**a**) paired-end reads of directly sequenced fragmented DNA; (**b**) mate-pair reads facing inwards, such as those produced with the NxSeq Long Mate Pair kit (Lucigen); and (**c**) mate-pair reads facing outwards, such as those produced with the Nextera DNA Library Preparation Kit (Illumina). *Blue portions* of the DNA fragments correspond to the sequence of interest, *green/yellow portions* correspond to adaptor/coupler sequences, *dashed vertical lines* correspond to the ligation sites, and *red lines* on the circularized sequences indicate restriction sites

9. BESST v1.3.8 (run on cluster or locally) [21]: https://github.com/ksahlin/BESST
10. REAPR v1.0.16 (run on cluster or locally) [22]: http://www.sanger.ac.uk/science/tools/reapr
11. SSPACE standard v3.0, free for academics (run on cluster or locally) [23]: http://www.baseclear.com/genomics/bioinformatics/basetools/SSPACE
12. BLAT v36 (run on cluster or locally): http://genome-test.cse.ucsc.edu/~kent/src/
13. L_RNA_scaffolder (run on cluster or locally) [24]: http://www.fishbrowser.org/software/L_RNA_scaffolder/index.php?action=isin&do=download
14. IrysView v2.5.1, needs registration to download (requires Windows operating system): http://bionanogenomics.com/support/register-and-download-the-latest-copy-of-irysview/

15. IrysSolve v2.1.1 (run on cluster): http://bionanogenomics.com/support/software-updates/
16. CEGMA v2.5 (run on cluster) [25]: http://korflab.ucdavis.edu/datasets/cegma/#SCT3
17. BUSCO v1.1b1 (run locally) [26]: http://busco.ezlab.org/

3 Methods

3.1 Isolation of High-Molecular-Weight DNA from Reptilian Blood

1. Dilute 400 μL of reptilian blood in 5 mL of blood preservation solution 1 and mix well by inversion right away to prevent the blood from clogging. The mix can be stored for months at room temperature or at 4 °C.
2. Extract high-molecular-weight DNA using the "QIAGEN Genomic-tip 20/G" (reference 10223) or "500/G" (reference 10262) following the manufacturer's instructions with minor modifications (*see* **Note 11**). Approximately 10 μg of high-molecular-weight DNA can be extracted from 100 μL of diluted blood with the 20/G tips and ~200 μg from 1 mL of diluted blood with the 500/G tips. The quality and amount of DNA is sufficient for direct Illumina sequencing or mate-pair library construction with the NxSeq Long Mate Pair kit (Lucigen).

3.2 Isolation of Megabase DNA from Reptilian Blood for Optical Mapping

1. Dilute 250 μL of reptilian blood in 4.75 mL of blood preservation solution 2 and mix well (but gently) by inversion right away to prevent the blood from clogging. It is advised to proceed with **step 2** right away to avoid sample degradation.
2. To protect DNA from shearing, the blood is embedded in agarose plugs, using the "CHEF bacterial genomic DNA plug kit" (BioRad reference 1703592), and all subsequent steps are performed in this form. The aim is to have ~7 μg of DNA in 75 μL for plug preparation, so it is advised to first extract DNA from different blood dilutions with the "QIAGEN QIAamp DNA blood mini kit" (reference 51104) and check the yield (final elution in 200 μL), before preparing the plugs. We suggest starting with the following dilutions for the DNA extraction with QIAGEN kit: 50 μL blood mix from **step 1** with 150 μL PBS 1×, 100 μL blood mix with 100 μL PBS 1×, 150 μL blood mix with 50 μL PBS 1×, and 200 μL blood mix.
3. We isolated DNA following the "IrysPrep Experienced User Card Human Blood Protocol" (BioNano Genomics, revision 30026D) and the included technical note for the "Isolation of megabase DNA from nucleated blood." It is crucial to mix well the blood mix by inversion (and not pipetting) such that it is homogeneous just before preparing the agarose plugs, as the

red blood cells precipitate quickly in the preservation solution. For the corn snake genome, we prepared five plugs of each concentration as follows: 75 μL blood mix and 45 μL agarose 2% (plug "1:1" containing ~7 μg of DNA); 25 μL blood mix, 50 μL PBS 1×, and 45 μL agarose 2% (plug "1:3" containing ~2.3 μg of DNA); and 12.5 μL blood mix, 62.5 μL PBS 1×, and 45 μL agarose 2% (plug "1:6" containing ~1.2 μg of DNA). Consider preparing an excess of the blood and agarose mix, as the solution becomes quickly viscous and it is difficult to pipet in the cast. The proteinase K incubation was performed overnight in a hybridization oven with mild rocking agitation. Plug lysis, DNA digestion, labeling, and imaging were performed by the VIB Nucleomics Core (http://www.nucleomics.be).

3.3 De Novo Assembly of Contigs

1. Perform a quality control of all reads with the "FastQC" software (*see* **Note 12**). Use compressed or uncompressed FASTQ files as input. These files include the read sequences and a quality value per base. "FastQC" does not trim or filter libraries, but provides important information regarding:
   (a) Per base sequence quality to check if the quality is low in specific positions of the reads.
   (b) Per sequence mean quality scores.
   (c) Per base sequence content to check if there are overrepresented nucleotides in specific positions of the reads.
   (d) Per sequence GC content that should follow a distribution close to the theoretical one.
   (e) Sequence duplication levels.
   (f) Presence of overrepresented sequences that could correspond to contamination or adaptor sequences.
   (g) K-mer content, showing overrepresented k-mers in comparison to their expected count values.
2. (Optional) Detect and trim adaptor sequences in the 2 × 250 bp HiSeq2500 Illumina libraries using "skewer" and the following command:

```
skewer -x adaptor_R1 -y adaptor_R2 -k 1 -r 0.15 -l 0 -z -t
number_threads HiSeq2500_R1.fastq HiSeq2500_R2.fastq
```

3. Perform a contig assembly with the "DISCOVAR de novo" software. The assembler is optimized for PCR-free libraries of ~450 bp fragment size and 2 × 250 bp Illumina reads providing at least a 60× genome coverage. Due to the high memory, time, and space requirements to assemble vertebrate-sized genomes, in the order of Gbp, it is advised to install the software on a cluster/server with at least 48 cores, 0.5 TB of space, and "2 * genome size" bytes of RAM. We ran

"DISCOVAR de novo" on a machine with 48 cores and 512 GB RAM. It took 62 h to complete with two libraries from the two lanes in separate FASTQ files and we reached a memory peak of 422 GB (*see* **Note 13**). The command to run "DISCOVAR de novo" is as follows (*see* **Note 14**):

```
DiscovarExp NUM_THREADS=48 MEM_MONITOR=True READS="
HiSeq2500_R1_trimmed_lib{1,2}.fastq.gz,HiSeq2500_R2_-
trimmed_lib{1,2}.fastq.gz" OUT_DIR=DiscovarExp_out
```

"DISCOVAR de novo" provides as output the "a.lines.fasta" file, which includes the contig sequence with the highest coverage when multiple alternatives are present. Use the "sequence size selection" tool in the "LANE runner" software (*see* **Note 15**) to keep only contigs greater than or equal to 1 kbp in the final assembly (typing "999" as size), thus removing mostly non-informative sequences, as recommended by the software developers. An even more stringent threshold of 2 kbp can be considered [27]. Our assembly results at each stage of the build are presented in Table 3.

4. Search the contig assembly for contaminants against the latest UniVec database (ftp://ftp.ncbi.nlm.nih.gov/pub/UniVec/UniVec) from NCBI VecScreen and the species' mitochondrial DNA, if available (*see* **Note 16**). Run "blastn" from the "BLAST+" suite for each contaminant database with the same parameters as VecScreen, but a more stringent *e*-value, by using the following command:

```
blastn -task blastn -reward 1 -penalty -5 -gapopen 3 -
gapextend 3 -dust yes -soft_masking true -evalue 0.05 -
searchsp 1750000000000 -db contaminant_db -query con-
tigs_1Kb.fasta -outfmt 6 -out Results_contaminant_db.xls
```

Table 3
Genome assembly and scaffolding statistics at each step of the build

Genome assembly	Number of sequences	N50 (kbp)	L50 (sequences)	Total length (Gbp)
Contig assembly (all)	1,091,366	10.79	38,609	1.991
Contig filtering (≥1 kbp)	284,216	14.86	27,786	1.717
BESST 1st round	157,256	592.87	638	1.794
REAPR	160,783	351.89	1085	1.790
SSPACE 2nd round	116,918	773.48	503	1.870
L_RNA_scaffolder	114,753	844.59	463	1.88
BioNano	114,644	1378.44	279	1.94

The output (defined by the *-out* parameter) is a tabular file including the BLAST hits against the database. We consider as contaminants the hits with a sequence identity >95% and a match length similar to the hit or query sequences (or at least 1 kbp long, if the database consists of chromosomes or DNA from pathogens). These can be removed from the assembly using the "Remove sequences from a FASTA file" tool of "LANE runner." In the corn snake assembly, a single contig matched with the mitochondrial DNA and only one contaminant sequence was identified.

"DISCOVAR de novo" labels as "circular" contigs whose sequence ends share the same $K - 1$ bases ($K = 200$). These could correspond to artifacts, episomes, or repetitive genome regions. Circular contigs' identifiers can be listed in the "circular_contigs.txt" output using the command:

```
grep "circular" a.lines.fasta | cut -c 2- > circular_
contigs.txt
```

These sequences can then be extracted from the contig assembly using the "select sequences from a FASTA file" tool of "LANE Runner." The output FASTA file can be used as the query for a remote BLAST search against the non-redundant (nr or nt) database of NCBI:

```
blastn -task blastn -dust yes -soft_masking true -
evalue 0.0001 -db nt -max_target_seqs 1 -remote -query
Circular_contigs.fasta -outfmt 6 -out Results_circular.xls
```

The corn snake assembly included 542 "circular" contigs, 221 of which had a high-score BLAST hit against other reptilian genomes; thus, they were all retained in the final assembly. They would have been removed (using the same approach as for contaminants), if they had BLAST hits against sequences of distantly related species.

3.4 Scaffolding: First Round (Table 3)

1. Filter the mate-pair libraries using the Python scripts provided by Lucigen (http://lucigen.com/docs/sequencing/MatePairScripts.tar.gz) and the "ParseFastQ.py" python script from https://gist.github.com/xguse/1866279. All scripts should be in the same directory as the uncompressed FASTQ files. Make sure that the filenames for the FASTQ files are the same besides "_R1" for the forward (R1) library and "_R2" for the reverse (R2). Start by running the "IlluminaChimera-Clean5.py" script:

```
python IlluminaChimera-Clean5.py inputfile_R1.fastq
```

The output files with the prefix "mates_ICC5_" include the filtered sequences. Continue with the "IlluminaNxSeqJunction-Split9.py" script:

```
python IlluminaNxSeqJunction-Split9.py mates_ICC5_in-
putfile_R1.fastq
```

The output files with the prefix "R{1,2}_IJS9_mates_ICC5_" include the filtered and trimmed mate pairs.

2. Align all libraries to the filtered contig assembly (obtained at **step 4** of Subheading 3.3) separately to infer the most probable position of each read in the genomic sequence. It is important at this stage to include as many libraries as possible from the same individual. For the corn snake genome, we considered (a) a 100 bp paired-end Illumina library [7]; (b) the HiSeq2500 library used for contig assembly, after filtering out leading and trailing bases with quality lower than 3 with "Trimmomatic" (LEADING:3 and TRAILING:3 parameters (*see* **Note 12**)); and (c) the 3, 8, and 20 kbp mate-pair libraries. The contig assembly, used as reference sequence, needs to be indexed and then each library's FASTQ files (compressed or uncompressed) are aligned to it using the "bwa mem" software with default parameters:

```
bwa index contigs_1Kb_filtered.fasta
bwa mem -t number_threads contigs_1Kb_filtered.fasta
libray_R1.fastq library_R2.fastq > library_output.sam
2>out_library.err
```

The output is a SAM file ("library_output.sam") with information on the reads alignment (*see* **Note 17**). Using the "2>" command, the "out_library.err" file is created. In the latter file, find the lines that read: "*[M::mem_pestat] analyzing insert size distribution for orientation FR*. . ." Three lines below (in the line starting "*[M::mem_pestat] mean and std.dev:*") look for the estimated mean insert size and standard deviation of the pairs with forward-reverse (FR) orientation. As sequences are read in batches, check the mean insert size from a few of them to get a better estimate (*see* **Note 18**). The corn snake 8 kbp library turned out to have an estimated insert size of 6 kbp and was considered as such for the downstream analyses.

3. Convert the output SAM files from the "bwa mem" run into sorted binary BAM files using the *samtools view* command from "samtools":

```
samtools view -buS 3Kb_output.sam | samtools sort -@
number_threads -m memory_perthread - output_library.
sorted
```

and then index these using the *samtools index* command:

```
samtools index 3Kb.sorted.bam
```

The -@ parameter indicates the number of threads to use and -*m* the memory to allocate per thread (e.g. 4G).

4. Order, orient, and place contigs into scaffolds considering the reads paired-end/mate-pair alignment information from the

sorted and indexed BAM files using the "BESST" scaffolder with default parameters. In the command, libraries should be ordered from the shortest to longest fragment/insert size and the correct orientation must be indicated ("fr" for inwards and "rf" for outwards pairs). For the six corn snake libraries, the command reads as follows:

```
runBESST -c contigs_1Kb_filtered.fasta -f 100bp_v1.
sorted.bam HiSeq2500_lib1.sorted.bam HiSeq2500_lib2.
sorted.bam 3Kb.sorted.bam 6Kb.sorted.bam 20Kb.sorted.
bam -orientation fr fr fr fr fr fr -o BESST_scaffol-
ding_default
```

The scaffolds are stored in the output FASTA file from the last "BESST" pass (Table 3).

5. Identify errors and break mis-assemblies using "REAPR," a tool that identifies inconsistencies by mapping mate-pair reads to an assembly. First run:

```
reapr facheck scaffolds_assembly_firstRound.fasta
```

to check for errors in the sequence names. If it returns an error, modify the assembly FASTA file:

```
reapr facheck scaffold_assembly_firstRound.fasta new_-
assembly
```

As recommended in the "REAPR" manual, align only the mate-pair library with the largest insert size (20 kbp library for the corn snake) to the scaffold assembly using *reapr smaltmap*:

```
reapr smaltmap scaffold_assembly_firstRound.fasta
20Kb_R1.fastq 20Kb_R2.fastq firstRound_20Kb.bam
1>out_firstRound_20Kb.txt 2>err_firstRound_20Kb.txt
```

Finally, run the *reapr pipeline* command (Table 3):

```
reapr pipeline scaffold_assembly_firstRound.fasta
firstRound_20Kb.bam output_directory
```

"REAPR" outputs a FASTA file ("04.break.broken_assembly.fa") with the scaffolds broken at points where mis-assemblies are detected (*see* **Note 19**).

3.5 Scaffolding: Second Round

1. Perform a second round of scaffolding using only the mate-pair libraries (*see* **Note 20**). We recommend the "SSPACE" scaffolder for this step, as "BESST" yielded less optimal results in terms of gene completeness and genome continuity for the corn snake genome.

   Prepare a file with the following information, separated by spaces, for each mate-pair library: (a) the name of the library, (b) the aligner software to use (here "bwa"), (c) the FASTQ file names, (d) the estimated insert size, (e) the allowed error for the estimated insert size (we chose 0.6, i.e., "insert size" ± 0.6 × "insert size"), and (f) the pairs orientation (FR in this study). The command to run "SSPACE" is as follows:

```
perl SSPACE_Standard_v3.0.pl -l libraries_list.txt -s
Scaffolds_broken.fa -T number_threads -b SSPACE_output
1>&2 > SSPACE3.log
```

"SSPACE" runs sequentially, starting by scaffolding the mate-pair library with the smallest insert size, and then using the output to scaffold the next library. The final results (Table 3) correspond to the scaffolding of the library with the longest insert size ("SSPACE_output.final.scaffolds.fasta").

3.6 Scaffold Improvement Using Transcriptomic Data

1. If transcriptomic data is available for the species of interest, it can be used to improve the assembly. We recommend the "L_RNA_scaffolder" software that uses long reads, such as 454, Sanger or Ion Torrent, or assembled transcripts from shorter Illumina paired-end reads. For the corn snake, we used (a) the *P. guttatus* 454 libraries from the "Reptilian Transcriptomes v2.0" [17] and (b) the assembled contigs of the *P. guttatus* transcriptome, incorporating Illumina and 454 reads.

   First, all reads/transcripts in FASTA format need to be aligned to the genome assembly using "BLAT" with the *-noHead* option, a requirement for the downstream "L_RNA_scaffolder" run (*see* **Note 21**):

```
blat SSPACE_output.final.scaffolds.fasta RNA_read-
s_transcripts.fa output.psl -noHead
```

Perform the assembly (Table 3) with "L_RNA_scaffolder":

```
perl L_RNA_scaffolder.sh -d <path to L_RNA_scaffolder
application> -i output.psl -j SSPACE_output.final.
scaffolds.fasta
```

3.7 Optical Mapping with BioNano

1. Import the BNX files of the Irys runs to "IrysView" and merge them (*see* **Note 22**).
2. Filter the molecules for minimum average label signal-to-noise ratio (SNR), maximum molecule length (*see* **Note 23**), and maximum molecule average intensity using scripts of the VIB Nucleomics Core (https://github.com/Nucleomics-VIB/bionano-tools). Import the filtered BNX file to IrysView.
3. Run a first de novo assembly of the remaining molecules using the "optArguments_human.xml" configuration file. As it is a computationally intensive task, it is recommended to run it remotely on a Linux cluster/server with an "IrysSolve" installation. Specify the maximum memory and threads to allocate based on the server capabilities.

4. Run a second de novo assembly of the same molecules using the "optArguments_haplotype.xml" configuration file, for a diploid genome, and selected the output CMAP file from the previous run as reference for the "autonoise" parameter. This step produces the final genome maps. For the corn snake, we obtained 4996 maps with an N50 of 0.963 Mbp and an average coverage of 68.7x.
5. Perform the hybrid scaffolding using the genome maps and the genome assembly from Subheading 3.6 imported as a FASTA file (here called NGS assembly). Use either the "hybridScaffold_config.xml" or the "hybridScaffold_config_aggressive.xml" configuration file, adjusted for the memory and threads specifications. The aggressive parameters reduce the required length of pairwise alignments from 160 to 80 kbp and modify the merging *p*-value from $10e^{-13}$ to $10e^{-11}$. Select the option to resolve conflicts in both the BioNano and the NGS assembly. The hybrid scaffolding of the corn snake was performed using the "aggressive" configuration file.
6. In the "IrysView" workspace under the "HybridScaffold" directory, there is (a) a FASTA file containing the hybrid scaffolds and (b) a FASTA file with the non-scaffolded sequences of the NGS assembly. Most sequences <100 kbp are not used, as they do not exhibit enough labeling sites. The software removes leading or trailing Ns from the hybrid scaffolds introduced by flanking BioNano genome maps. For the final assembly, merge the two FASTA files and remove sequences less than 1 kbp (Table 3).

3.8 Quality Completeness Evaluation

1. "CEGMA" (*see* **Note 24**) searches in a genome for a group of 248 highly conserved eukaryotic genes from yeast to human and states if they are completely or partially retrieved (Table 4). Use the following command:

```
cegma -g final_assembly.fasta -o final_CEGMA -T number_threads -vrt -v
```

The *-vrt* option optimizes the "CEGMA" search parameters for vertebrate genomes.

2. "BUSCO" stands for Benchmarking Universal Single-Copy Orthologs, and it looks for the presence of a set of genes that are single copy in at least 90% of the species from a specific lineage in genomes, transcriptomes, or any gene set (*see* **Note 25**). We recommend using the "eukaryota", "metazoa", and "vertebrata" lineage profiles dataset for reptilian/avian genomes that include 429, 843, and 3023 genes, respectively (Table 4). An example command reads as follows (*see* **Note 26**):

```
python3 BUSCO_v1.1b1.py -c number_threads -in final_assembly.fasta -m genome -l eukaryota/ -o BUSCO_final_eukaryota 1> out.log 2> out.err
```

Table 4
Completeness statistics based on the CEGMA and BUSCO datasets

	CEGMA	BUSCO		
	Eukaryota	Eukaryota	Metazoa	Vertebrata
Complete	214 (86.3%)	278 (64.8%)	703 (83.4%)	1339 (44.3%)
Partial	31 (12.5%)	44 (10.3%)	87 (10.3%)	640 (21.2%)
Missing	3 (1.2%)	107 (24.9%)	53 (6.3%)	1044 (34.5%)
Total	248	429	843	3023

where *-l* corresponds to either "eukaryota", "metazoa", or "vertebrata" depending on the assessed lineage.

4 Notes

1. For genome assembly, genomic libraries from a single individual, if possible from an inbred line, reduces heterozygosity, hence, assembly problems. Selecting the homogametic gender will further facilitate the analyses, even though a chromosome will be missing from the final assembly. Reptiles have nucleated red blood cells, making blood a good source of high-quality DNA and its collection is minimally invasive.
2. Do not autoclave the blood conservation solution as SDS foams, but instead use milliQ or double-distilled water for its preparation. The final solution can also be filtered (pore size 0.45 μm) using an air pump. If the solution is stored at 4 °C, the SDS will precipitate, so warm it up before use.
3. The presence of SDS results in cell lysis, which can cause shearing of DNA, thus for the extraction of megabase-long DNA, SDS is omitted from the blood preservation solution.
4. Library construction protocols and sequencing technologies progress fast; thus, new and better combinations/types of sequencing data might become available. For the time being, we advice the sequencing of short fragment libraries along with multiple mate-pair libraries, each with a different average insert size of several kbp, complemented with optical mapping, to produce the longest possible scaffolds. Long reads >1 kbp, such as those produced by SMRT sequencing (PacBio), can assist even further the assembly but were not tested by the authors.
5. The selection of the digestion enzyme depends on the genome sequence and needs to generate 8–15 labels per 100 kbp.

BioNano (http://bionanogenomics.com) provides the tools to calculate these numbers. The number of runs and flow cells depends on the length of the DNA molecules, so great care needs to be taken for its preparation.

6. Most software considered for genome assembly and pre- or post-processing steps do not have a graphical interface and can only be run in a Unix environment. Therefore, some basic knowledge in scripting and Unix command-line applications is required.
7. "Run locally" means that the software can be run on a personal computer with Windows/MacOSX/Linux (mainly the latter) operating system installed. Software with high memory or storage requirements or multithreaded are better run on clusters/grids to reduce the running time or to have access to the necessary computational resources. Access to a computing cluster is required to successfully assemble a vertebrate-sized genome.
8. Software installation is out of the scope of this guide. Consult the installation documents to set up the software, taking into consideration the operating system requirements and any mandatory dependencies for a successful run.
9. Read the manual before running a software package, especially to understand the different options available. In this guide, we present the parameters that worked best for the corn snake assembly, but they might need to be adjusted for other genomes.
10. Software versions correspond to those we used for the corn snake genome assembly and are not necessarily the latest ones. When starting a project, use the latest stable version and do not use multiple versions for the same assembly, as it could lead to inconsistencies.
11. When extracting with 20/G genomic tips, digest with 50 μL of proteinase K, instead of 25 μL, for 60 min at 50 °C. Dilute the lysate with an equal volume of Buffer QBT prior to loading on the 20/G tip.
12. When the quality or contamination filters fail in "FastQC", "Trimmomatic" is a useful tool to filter out and trim low quality reads, keeping the paired-end information in the process. Check the software manual and [18] for running parameters, but a minimum quality of 20 is advised (1% chance for a base to be wrong).
13. Splitting the FASTQ files can reduce problems with memory usage. In addition, running any other process simultaneously on the server should be avoided.
14. The "DiscovarExp" executable is named "DiscovarDeNovo" in the latest version (v0.0.52488). The later includes improvements in memory usage.

15. Another option is to use the “SeqIO” module from “BioPython.”
16. Include any other database (e.g., pathogens, human, *Escherichia coli*, *Saccharomyces cerevisiae*), if a particular contamination is suspected.
17. Specifications regarding the SAM format can be found at https://samtools.github.io/hts-specs/SAMv1.pdf
18. Check the insert size distribution for the FR (forward-reverse) orientation only, if the read pairs are facing inwards. However, if the libraries include outward-facing pairs, then the insert size distribution for the RF (reverse-forward) is the correct one. RR and FF are erroneously aligned pairs. Note that reads are considered for the statistics only if both sequences in the pair align to the same reference sequence.
19. Another important output file from “REAPR” is “05.summary.report.txt,” which includes information on the nature of the errors found in the assembly. Check the “REAPR” manual for a more detailed explanation about the different types of errors.
20. The second round of scaffolding is optional, depending on the results obtained at the first round.
21. As “BLAT” is not multithreaded, large FASTA file can be split to perform the alignment faster and then combine the resulting PSL files. To split a FASTA file, “Biopython” is an option, as presented in http://biopython.org/wiki/Split_large_file. Then, use the ‘`cat *.psl > all.psl` command to concatenate all the output PSL files from the “BLAT” runs located in the same directory.
22. Detailed explanations on how to run IrysView are available at the BioNano Genomics forum (https://forums.bionanogenomics.com/forum/) under “DOCUMENTATION - Data Analysis”, along with the installation guide.
23. Molecules greater than the size of the chip, that is 2.5 Mbp, are merged molecules that are detected as a single one and thus need to be removed.
24. Note that “CEGMA” is no longer supported by its developers.
25. “BUSCOv1.1b1” only works with “Python3,” but following versions also work with “Python2.7.” The “BUSCO” script should be in the desired output directory.
26. Some symbols (“:” and “|”) in the FASTA file headers coming from previous assembly steps can disrupt the performance of “BUSCO” and “CEGMA.” Hence, they should be replaced to avoid problems:

```
sed 's/[:|]/_/g' final_assembly.fasta > final_assem-
bly_m.fasta
```

Acknowledgments

We would like to thank Adrien Debry for blood collection, William H. Beluch for the sample preparation of the HiSeq2500 run, and Carine Langrez for the isolation of megabase DNA. We thank the Lucigen team for their support and results. The VIB Nucleomics Core team for the preparation of the BioNano samples and initial downstream analyses. Most computations were performed at the Vital-IT Centre for high-performance computing (www.vital-it.ch) of the SIB Swiss Institute of Bioinformatics and the Baobab cluster of the University of Geneva. This work was supported by grants from the University of Geneva (Switzerland), the Swiss National Science Foundation (FNSNF, grants 31003A_140785 and SINERGIA CRSII3_132430), and the SystemsX.ch initiative (project EpiPhysX). The authors thank Michel Milinkovitch for comments on an earlier version of this manuscript. Yingguang Frank Chan was supported by the Max Planck Society. An iGE3 PhD award was granted to Asier Ullate-Agote.

References

1. Lander ES, Linton LM, Birren B, Nusbaum C, Zody MC, Baldwin J, Devon K, Dewar K, Doyle M, FitzHugh W, Funke R, Gage D, Harris K, Heaford A, Howland J, Kann L, Lehoczky J, LeVine R, McEwan P, McKernan K, Meldrim J, Mesirov JP, Miranda C, Morris W, Naylor J, Raymond C, Rosetti M, Santos R, Sheridan A, Sougnez C, Stange-Thomann Y, Stojanovic N, Subramanian A, Wyman D, Rogers J, Sulston J, Ainscough R, Beck S, Bentley D, Burton J, Clee C, Carter N, Coulson A, Deadman R, Deloukas P, Dunham A, Dunham I, Durbin R, French L, Grafham D, Gregory S, Hubbard T, Humphray S, Hunt A, Jones M, Lloyd C, McMurray A, Matthews L, Mercer S, Milne S, Mullikin JC, Mungall A, Plumb R, Ross M, Shownkeen R, Sims S, Waterston RH, Wilson RK, Hillier LW, McPherson JD, Marra MA, Mardis ER, Fulton LA, Chinwalla AT, Pepin KH, Gish WR, Chissoe SL, Wendl MC, Delehaunty KD, Miner TL, Delehaunty A, Kramer JB, Cook LL, Fulton RS, Johnson DL, Minx PJ, Clifton SW, Hawkins T, Branscomb E, Predki P, Richardson P, Wenning S, Slezak T, Doggett N, Cheng JF, Olsen A, Lucas S, Elkin C, Uberbacher E, Frazier M, Gibbs RA, Muzny DM, Scherer SE, Bouck JB, Sodergren EJ, Worley KC, Rives CM, Gorrell JH, Metzker ML, Naylor SL, Kucherlapati RS, Nelson DL, Weinstock GM, Sakaki Y, Fujiyama A, Hattori M, Yada T, Toyoda A, Itoh T, Kawagoe C, Watanabe H, Totoki Y, Taylor T, Weissenbach J, Heilig R, Saurin W, Artiguenave F, Brottier P, Bruls T, Pelletier E, Robert C, Wincker P, Smith DR, Doucette-Stamm L, Rubenfield M, Weinstock K, Lee HM, Dubois J, Rosenthal A, Platzer M, Nyakatura G, Taudien S, Rump A, Yang H, Yu J, Wang J, Huang G, Gu J, Hood L, Rowen L, Madan A, Qin S, Davis RW, Federspiel NA, Abola AP, Proctor MJ, Myers RM, Schmutz J, Dickson M, Grimwood J, Cox DR, Olson MV, Kaul R, Raymond C, Shimizu N, Kawasaki K, Minoshima S, Evans GA, Athanasiou M, Schultz R, Roe BA, Chen F, Pan H, Ramser J, Lehrach H, Reinhardt R, McCombie WR, de la Bastide M, Dedhia N, Blocker H, Hornischer K, Nordsiek G, Agarwala R, Aravind L, Bailey JA, Bateman A, Batzoglou S, Birney E, Bork P, Brown DG, Burge CB, Cerutti L, Chen HC, Church D, Clamp M, Copley RR, Doerks T, Eddy SR, Eichler EE, Furey TS, Galagan J, Gilbert JG, Harmon C, Hayashizaki Y, Haussler D, Hermjakob H, Hokamp K, Jang W, Johnson LS, Jones TA, Kasif S, Kaspryzk A, Kennedy S, Kent WJ, Kitts P, Koonin EV, Korf I, Kulp D, Lancet D, Lowe TM, McLysaght A, Mikkelsen T, Moran JV, Mulder N, Pollara VJ, Ponting CP, Schuler G, Schultz J, Slater G, Smit AF, Stupka E, Szustakowki J, Thierry-Mieg D, Thierry-Mieg J, Wagner L, Wallis J, Wheeler R, Williams A, Wolf YI, Wolfe KH, Yang SP, Yeh RF, Collins F, Guyer MS, Peterson J, Felsenfeld A, Wetterstrand KA, Patrinos A, Morgan MJ, de Jong P, Catanese JJ, Osoegawa K, Shizuya H, Choi S, Chen YJ,

Szustakowki J, International Human Genome Sequencing Consortium (2001) Initial sequencing and analysis of the human genome. Nature 409(6822):860–921. doi:10.1038/35057062

2. Bentley DR, Balasubramanian S, Swerdlow HP, Smith GP, Milton J, Brown CG, Hall KP, Evers DJ, Barnes CL, Bignell HR, Boutell JM, Bryant J, Carter RJ, Keira Cheetham R, Cox AJ, Ellis DJ, Flatbush MR, Gormley NA, Humphray SJ, Irving LJ, Karbelashvili MS, Kirk SM, Li H, Liu X, Maisinger KS, Murray LJ, Obradovic B, Ost T, Parkinson ML, Pratt MR, Rasolonjatovo IM, Reed MT, Rigatti R, Rodighiero C, Ross MT, Sabot A, Sankar SV, Scally A, Schroth GP, Smith ME, Smith VP, Spiridou A, Torrance PE, Tzonev SS, Vermaas EH, Walter K, Wu X, Zhang L, Alam MD, Anastasi C, Aniebo IC, Bailey DM, Bancarz IR, Banerjee S, Barbour SG, Baybayan PA, Benoit VA, Benson KF, Bevis C, Black PJ, Boodhun A, Brennan JS, Bridgham JA, Brown RC, Brown AA, Buermann DH, Bundu AA, Burrows JC, Carter NP, Castillo N, Chiara ECM, Chang S, Neil Cooley R, Crake NR, Dada OO, Diakoumakos KD, Dominguez-Fernandez B, Earnshaw DJ, Egbujor UC, Elmore DW, Etchin SS, Ewan MR, Fedurco M, Fraser LJ, Fuentes Fajardo KV, Scott Furey W, George D, Gietzen KJ, Goddard CP, Golda GS, Granieri PA, Green DE, Gustafson DL, Hansen NF, Harnish K, Haudenschild CD, Heyer NI, Hims MM, Ho JT, Horgan AM, Hoschler K, Hurwitz S, Ivanov DV, Johnson MQ, James T, Huw Jones TA, Kang GD, Kerelska TH, Kersey AD, Khrebtukova I, Kindwall AP, Kingsbury Z, Kokko-Gonzales PI, Kumar A, Laurent MA, Lawley CT, Lee SE, Lee X, Liao AK, Loch JA, Lok M, Luo S, Mammen RM, Martin JW, McCauley PG, McNitt P, Mehta P, Moon KW, Mullens JW, Newington T, Ning Z, Ling Ng B, Novo SM, O'Neill MJ, Osborne MA, Osnowski A, Ostadan O, Paraschos LL, Pickering L, Pike AC, Pike AC, Chris Pinkard D, Pliskin DP, Podhasky J, Quijano VJ, Raczy C, Rae VH, Rawlings SR, Chiva Rodriguez A, Roe PM, Rogers J, Rogert Bacigalupo MC, Romanov N, Romieu A, Roth RK, Rourke NJ, Ruediger ST, Rusman E, Sanches-Kuiper RM, Schenker MR, Seoane JM, Shaw RJ, Shiver MK, Short SW, Sizto NL, Sluis JP, Smith MA, Ernest Sohna Sohna J, Spence EJ, Stevens K, Sutton N, Szajkowski L, Tregidgo CL, Turcatti G, Vandevondele S, Verhovsky Y, Virk SM, Wakelin S, Walcott GC, Wang J, Worsley GJ, Yan J, Yau L, Zuerlein M, Rogers J, Mullikin JC, Hurles ME, McCooke NJ, West JS, Oaks FL, Lundberg PL, Klenerman D, Durbin R, Smith AJ (2008) Accurate whole human genome sequencing using reversible terminator chemistry. Nature 456 (7218):53–59. doi:10.1038/nature07517

3. Eid J, Fehr A, Gray J, Luong K, Lyle J, Otto G, Peluso P, Rank D, Baybayan P, Bettman B, Bibillo A, Bjornson K, Chaudhuri B, Christians F, Cicero R, Clark S, Dalal R, Dewinter A, Dixon J, Foquet M, Gaertner A, Hardenbol P, Heiner C, Hester K, Holden D, Kearns G, Kong X, Kuse R, Lacroix Y, Lin S, Lundquist P, Ma C, Marks P, Maxham M, Murphy D, Park I, Pham T, Phillips M, Roy J, Sebra R, Shen G, Sorenson J, Tomaney A, Travers K, Trulson M, Vieceli J, Wegener J, Wu D, Yang A, Zaccarin D, Zhao P, Zhong F, Korlach J, Turner S (2009) Real-time DNA sequencing from single polymerase molecules. Science 323(5910):133–138. doi:10.1126/science.1162986

4. Lewin HA, Larkin DM, Pontius J, O'Brien SJ (2009) Every genome sequence needs a good map. Genome Res 19(11):1925–1928. doi:10.1101/gr.094557.109

5. Milinkovitch MC, Helaers R, Depiereux E, Tzika AC, Gabaldon T (2010) 2× - Genomes—depth does matter. Genome Biol 11(2):R16. doi:10.1186/gb-2010-11-2-r16

6. Saenko SV, Lamichhaney S, Martinez Barrio A, Rafati N, Andersson L, Milinkovitch MC (2015) Amelanism in the corn snake is associated with the insertion of an LTR-retrotransposon in the OCA2 gene. Sci Rep 5:17118. doi:10.1038/srep17118

7. Ullate-Agote A, Milinkovitch MC, Tzika AC (2014) The genome sequence of the corn snake (Pantherophis guttatus), a valuable resource for EvoDevo studies in squamates. Int J Dev Biol 58(10-12):881–888. doi:10.1387/ijdb.150060at

8. Alfoldi J, Di Palma F, Grabherr M, Williams C, Kong L, Mauceli E, Russell P, Lowe CB, Glor RE, Jaffe JD, Ray DA, Boissinot S, Shedlock AM, Botka C, Castoe TA, Colbourne JK, Fujita MK, Moreno RG, ten Hallers BF, Haussler D, Heger A, Heiman D, Janes DE, Johnson J, de Jong PJ, Koriabine MY, Lara M, Novick PA, Organ CL, Peach SE, Poe S, Pollock DD, de Queiroz K, Sanger T, Searle S, Smith JD, Smith Z, Swofford R, Turner-Maier J, Wade J, Young S, Zadissa A, Edwards SV, Glenn TC, Schneider CJ, Losos JB, Lander ES, Breen M, Ponting CP, Lindblad-Toh K (2011) The genome of the green anole lizard and a comparative analysis with birds and mammals. Nature 477(7366):587–591. doi:10.1038/nature10390

9. Badenhorst D, Hillier LW, Literman R, Montiel EE, Radhakrishnan S, Shen Y, Minx P,

Janes DE, Warren WC, Edwards SV, Valenzuela N (2015) Physical mapping and refinement of the painted turtle genome (Chrysemys picta) inform amniote genome evolution and challenge turtle-bird chromosomal conservation. Genome Biol Evol 7(7):2038–2050. doi:10.1093/gbe/evv119

10. Di-Poi N, Montoya-Burgos JI, Miller H, Pourquie O, Milinkovitch MC, Duboule D (2010) Changes in Hox genes' structure and function during the evolution of the squamate body plan. Nature 464(7285):99–103. doi:10.1038/nature08789

11. Vonk FJ, Casewell NR, Henkel CV, Heimberg AM, Jansen HJ, McCleary RJ, Kerkkamp HM, Vos RA, Guerreiro I, Calvete JJ, Wuster W, Woods AE, Logan JM, Harrison RA, Castoe TA, de Koning AP, Pollock DD, Yandell M, Calderon D, Renjifo C, Currier RB, Salgado D, Pla D, Sanz L, Hyder AS, Ribeiro JM, Arntzen JW, van den Thillart GE, Boetzer M, Pirovano W, Dirks RP, Spaink HP, Duboule D, McGlinn E, Kini RM, Richardson MK (2013) The king cobra genome reveals dynamic gene evolution and adaptation in the snake venom system. Proc Natl Acad Sci U S A 110(51):20651–20656. doi:10.1073/pnas.1314702110

12. Bradnam KR, Fass JN, Alexandrov A, Baranay P, Bechner M, Birol I, Boisvert S, Chapman JA, Chapuis G, Chikhi R, Chitsaz H, Chou WC, Corbeil J, Del Fabbro C, Docking TR, Durbin R, Earl D, Emrich S, Fedotov P, Fonseca NA, Ganapathy G, Gibbs RA, Gnerre S, Godzaridis E, Goldstein S, Haimel M, Hall G, Haussler D, Hiatt JB, Ho IY, Howard J, Hunt M, Jackman SD, Jaffe DB, Jarvis ED, Jiang H, Kazakov S, Kersey PJ, Kitzman JO, Knight JR, Koren S, Lam TW, Lavenier D, Laviolette F, Li Y, Li Z, Liu B, Liu Y, Luo R, Maccallum I, Macmanes MD, Maillet N, Melnikov S, Naquin D, Ning Z, Otto TD, Paten B, Paulo OS, Phillippy AM, Pina-Martins F, Place M, Przybylski D, Qin X, Qu C, Ribeiro FJ, Richards S, Rokhsar DS, Ruby JG, Scalabrin S, Schatz MC, Schwartz DC, Sergushichev A, Sharpe T, Shaw TI, Shendure J, Shi Y, Simpson JT, Song H, Tsarev F, Vezzi F, Vicedomini R, Vieira BM, Wang J, Worley KC, Yin S, Yiu SM, Yuan J, Zhang G, Zhang H, Zhou S, Korf IF (2013) Assemblathon 2: evaluating de novo methods of genome assembly in three vertebrate species. Gigascience 2(1):10. doi:10.1186/2047-217X-2-10

13. Earl D, Bradnam K, St John J, Darling A, Lin D, Fass J, Yu HO, Buffalo V, Zerbino DR, Diekhans M, Nguyen N, Ariyaratne PN, Sung WK, Ning Z, Haimel M, Simpson JT, Fonseca NA, Birol I, Docking TR, Ho IY, Rokhsar DS, Chikhi R, Lavenier D, Chapuis G, Naquin D, Maillet N, Schatz MC, Kelley DR, Phillippy AM, Koren S, Yang SP, Wu W, Chou WC, Srivastava A, Shaw TI, Ruby JG, Skewes-Cox P, Betegon M, Dimon MT, Solovyev V, Seledtsov I, Kosarev P, Vorobyev D, Ramirez-Gonzalez R, Leggett R, MacLean D, Xia F, Luo R, Li Z, Xie Y, Liu B, Gnerre S, MacCallum I, Przybylski D, Ribeiro FJ, Yin S, Sharpe T, Hall G, Kersey PJ, Durbin R, Jackman SD, Chapman JA, Huang X, DeRisi JL, Caccamo M, Li Y, Jaffe DB, Green RE, Haussler D, Korf I, Paten B (2011) Assemblathon 1: a competitive assessment of de novo short read assembly methods. Genome Res 21(12):2224–2241. doi:10.1101/gr.126599.111

14. Pendleton M, Sebra R, Pang AW, Ummat A, Franzen O, Rausch T, Stutz AM, Stedman W, Anantharaman T, Hastie A, Dai H, Fritz MH, Cao H, Cohain A, Deikus G, Durrett RE, Blanchard SC, Altman R, Chin CS, Guo Y, Paxinos EE, Korbel JO, Darnell RB, McCombie WR, Kwok PY, Mason CE, Schadt EE, Bashir A (2015) Assembly and diploid architecture of an individual human genome via single-molecule technologies. Nat Methods 12(8):780–786. doi:10.1038/nmeth.3454

15. Jiang H, Lei R, Ding SW, Zhu S (2014) Skewer: a fast and accurate adapter trimmer for next-generation sequencing paired-end reads. BMC Bioinformatics 15:182. doi:10.1186/1471-2105-15-182

16. Camacho C, Coulouris G, Avagyan V, Ma N, Papadopoulos J, Bealer K, Madden TL (2009) BLAST+: architecture and applications. BMC Bioinformatics 10:421. doi:10.1186/1471-2105-10-421

17. Tzika AC, Ullate-Agote A, Grbic D, Milinkovitch MC (2015) Reptilian transcriptomes v2.0: an extensive resource for sauropsida genomics and transcriptomics. Genome Biol Evol 7(6):1827–1841. doi:10.1093/gbe/evv106

18. Bolger AM, Lohse M, Usadel B (2014) Trimmomatic: a flexible trimmer for Illumina sequence data. Bioinformatics 30(15):2114–2120. doi:10.1093/bioinformatics/btu170

19. Li H, Durbin R (2009) Fast and accurate short read alignment with Burrows-Wheeler transform. Bioinformatics 25(14):1754–1760. doi:10.1093/bioinformatics/btp324

20. Li H, Handsaker B, Wysoker A, Fennell T, Ruan J, Homer N, Marth G, Abecasis G, Durbin R, Genome Project Data Processing Subgroup (2009) The Sequence Alignment/Map format and SAMtools. Bioinformatics 25(16):2078–2079. doi:10.1093/bioinformatics/btp352

21. Sahlin K, Vezzi F, Nystedt B, Lundeberg J, Arvestad L (2014) BESST—efficient scaffolding of large fragmented assemblies. BMC Bioinformatics 15:281. doi:10.1186/1471-2105-15-281
22. Hunt M, Kikuchi T, Sanders M, Newbold C, Berriman M, Otto TD (2013) REAPR: a universal tool for genome assembly evaluation. Genome Biol 14(5):R47. doi:10.1186/gb-2013-14-5-r47
23. Boetzer M, Henkel CV, Jansen HJ, Butler D, Pirovano W (2011) Scaffolding pre-assembled contigs using SSPACE. Bioinformatics 27(4):578–579. doi:10.1093/bioinformatics/btq683
24. Xue W, Li JT, Zhu YP, Hou GY, Kong XF, Kuang YY, Sun XW (2013) L_RNA_scaffolder: scaffolding genomes with transcripts. BMC Genomics 14:604. doi:10.1186/1471-2164-14-604
25. Parra G, Bradnam K, Korf I (2007) CEGMA: a pipeline to accurately annotate core genes in eukaryotic genomes. Bioinformatics 23(9):1061–1067. doi:10.1093/bioinformatics/btm071
26. Simao FA, Waterhouse RM, Ioannidis P, Kriventseva EV, Zdobnov EM (2015) BUSCO: assessing genome assembly and annotation completeness with single-copy orthologs. Bioinformatics 31(19):3210–3212. doi:10.1093/bioinformatics/btv351
27. Love RR, Weisenfeld NI, Jaffe DB, Besansky NJ, Neafsey DE (2016) Evaluation of DISCOVAR de novo using a mosquito sample for cost-effective short-read genome assembly. BMC Genomics 17:187. doi:10.1186/s12864-016-2531-7

Chapter 4

Genomic and Transcriptomic Analyses of Avian Sex Chromosomes and Sex-Linked Genes

Jilin Zhang, Jing Li, and Qi Zhou

Abstract

Sex chromosomes and sex-linked genes usually show unusual features comparing to the rest of the genome and thus are of particular interests to evolutionary and developmental biologists. Here we describe recently developed bioinformatic methods for identifying sex-linked sequences, in a genome without priori linkage information. Some are developed during our course of studying avian genomes. These methods require sequence data, either assembled draft genome or raw sequences derived from the heterogametic sex (e.g., a female bird or a male mammal). Their application is not restricted to birds but can be used for any species with a sex chromosome pair that has diverged from each other for a substantial degree.

Key words Sex chromosome, Genome assembly, Sex-linked genes

1 Introduction

Sex chromosomes and sex-linked genes are at the center of evolution biology studies due to their disproportionately greater roles in sexual selection and speciation. They usually evolve faster in sequence and gene expression than the rest of the genome, which has been referred to as the "fast-X" effect [1, 2]. This pattern is driven by their biased transmission between the sexes, hemizygous state in the heterogametic sex (e.g., X chromosome in mammalian males), and distinctive regulation programs compared to autosomes. In the soma, many species have evolved dosage compensation (DC) to balance the expression difference between sex chromosomes and autosomes [3, 4]. In the germ line, by contrast, mammalian XY chromosomes undergo meiotic sex chromosome inactivation (MSCI) [5], presumably to prevent recombination between the nonhomologous X and Y chromosomes. Both DC and MSCI frequently involve noncoding RNAs and epigenetic modifications, making sex chromosomes also a hot spot for molecular and developmental biology studies.

Guojun Sheng (ed.), *Avian and Reptilian Developmental Biology: Methods and Protocols*, Methods in Molecular Biology, vol. 1650, DOI 10.1007/978-1-4939-7216-6_4, © Springer Science+Business Media LLC 2017

Most of our current knowledge on sex chromosomes is derived from the XY systems of human and *Drosophila*, which constitute only one of the many modes of sex determination. The other major sex chromosome type is a female heterogametic system as found in birds and butterflies, with male ZZ and female ZW karyotypes. Recent studies in chicken and other birds show that, despite many similar features to the XY systems (e.g., the "fast-Z" effect), avian ZW chromosomes have not evolved global dosage compensation [6]. Whether or not birds have MSCI is still under debate. The development and application of new sequencing technologies and genomic analysis tools further uncovered that birds' sex chromosomes have diverged to varying degrees, with some species even possessing sex chromosomes that resemble the ancestral autosomes in terms of sequence divergence level [7]. This is in stark contrast to the mammalian or *Drosophila* XY systems, which have completely differentiated from each other [4]. With the recent announcement of the "Bird 10,000 Genomes" (B10K) project, we are expecting waves of genomic data of the entire avian class to arrive within the next 5 years.

In this chapter, we introduce analytical methods that have been developed in our lab, as well as those published by other groups [8]. There are mainly two difficulties with identifying and assembling sex-linked sequences, regardless of the type of sex chromosomes. First, W or Y chromosomes, except for their recombining regions (called "pseudoautosomal regions" (PAR)), are usually highly degenerated in sequences and frequently interrupted by repetitive elements. Second, the sex chromosome on average only receives half of the sequencing coverage compared to autosomes. On the other hand, these factors constitute the main principles on which the pipelines introduced below are based. These pipelines aim to identify the sex-linked sequences without any priori linkage information, from de novo assembled draft genomes derived from female bird samples. More generally however, they can be applied to any other species with the heterogametic sex sequenced. The sex-linked sequences can then be subjected to a wide variety of downstream analyses like gene and repeat annotation, differential expression analyses between sex chromosomes vs. autosomes or between sexes, as well as characterization of sex-specific functional genes or regulatory elements. These analyses depend on the particular questions to be addressed, and the corresponding pipelines have been covered elsewhere [9, 10]. For example, sex-linked sequences can be further used to investigate recombination suppression events, which commonly result in a stratified sequence divergence pattern between sex chromosomes ("evolutionary strata") [7]. In addition, gene expression profiles can be used to assess the effect of dosage compensation throughout ontogenesis and phylogenesis [6, 11].

Depending on whether a reference genome is available or not, two types of methods have been developed to determine gene expression levels. If a reference genome is available, Tophat-Cufflinks (the "Tuxedo" suite) tools [12] or the HISAT-String-Tie-Ballgown (the "new Tuxedo" suite) tools [13] can be used for detecting sex-biased or sex-specific gene expression. Recently other tools discarding reads-alignment output (so-called pseudo-alignment) are emerging, e.g., *kallisto* [14]. For most species, however, a reference genome is unavailable. In these cases, gene expression level can be estimated by using a de novo assembler in order to recover transcript sequences, followed by an estimation of their abundance, as implemented in *Trinity* [15] and *SOAPdenovo-Trans* [16]. In this chapter, we focus on methods to identify sex-linked sequences, but not on these downstream analyses.

2 Materials

To identify sex-linked sequences, reads or assembled draft genome sequences from the heterogametic sex (e.g., a female bird or a male mammal) have to be available. The reason for this is that with the data of homogametic sex alone, sex chromosomes essentially show very similar sequence patterns with the autosomes, making the sex-linked sequences indiscernible without any other linkage information. The sex-linked sequences, specifically those from the sex chromosome regions exhibiting suppressed recombination, are expected to be different from each other in sequence. In the heterogametic sex, this results in a higher frequency of heterozygous sites of sex-linked regions than in the homogametic sex when the sex chromosomes are not too different from each other. Otherwise, one would expect a twofold reduction in read coverage relative to that of autosomes or PAR, as well as a sex-specific read coverage pattern along the sex-specific (W- or Y-linked) regions. These expectations constitute the principles of the methods detailed below. Figure 1 highlights these methods' major steps, their input data, as well as the software or customized scripts used in corresponding pipelines. A Linux or Mac OS computer cluster with at least 32 Gb physical memory is needed, predominantly for genome assembly jobs, depending on the data amount and genome size of the focal species.

2.1 Input Data

- Illumina sequencing reads: A minimum of 20-fold genome coverage of genomic DNA sequencing, i.e., at least tenfold coverage of each sex chromosome in the heterogametic sex, is recommended. If there is no draft genome available, the sequencing coverage should be at least 50-fold for the genome assembly purposes. The reads should be subjected to quality control processes using tools like *FastQC* [17].

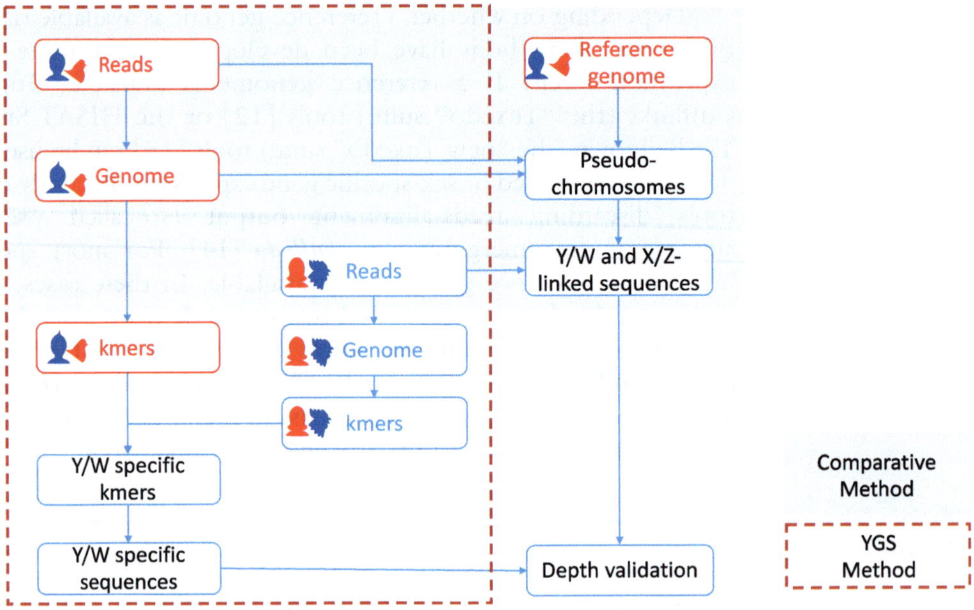

Fig. 1 Summary of methodologies. Main procedures of the two methods are illustrated depending on their starting data. Comparative method is highlighted in a *box* with background, while YGS method is indicated in the *dashed-frame box*. Data derived from different sex are color coded: *red* represents female and *blue* represents male. *Arrows* stand for the main analysis, which points to the major output

- If a draft genome from the heterogametic sex is available, we recommend to start with a genome with the scaffold N50 longer than 5 kb and read sequencing data of at least tenfold. N50 is a measurement of draft genome continuity, defined as the length of the shortest sequence that the sum of the fragments of equal length or longer is at least 50% of the total length of all fragments. Previous study in chicken showed that the longer the scaffold sequences, the lower the false positive rate associated with the identification of W-linked sequences [18].

2.2 Software

The details of installation and usage for the following software packages are included in their manuals and websites, which are provided below. We also provide the example command lines of running these programs in the next section.

- Reads data quality control: *FASTQ and cutadapt*:
- http://www.bioinformatics.babraham.ac.uk.
- https://cutadapt.readthedocs.io [19].
- Genome assembly: *SOAPdenovo* [20]:
- http://soap.genomics.org.cn.
- Repetitive sequences masking: *RepeatMasker* [21]:

- http://www.repeatmasker.org.
- Sequence alignment: *bwa* [22], *LASTZ* [23]:
- http://bio-bwa.sourceforge.net.
- http://www.bx.psu.edu/~rsharris/lastz/.
- Sequence alignment manipulation: *UCSC tools*, *SAMtools* [24], *and Picard* [25]:
- http://hgdownload.soe.ucsc.edu.
- http://www.htslib.org.
- https://broadinstitute.github.io/picard/.
- Coverage calculation: *BEDtools* [26]:
- http://bedtools.readthedocs.io/.
- Protein sequence alignment and reading frame prediction tool: *Wise2*:
- http://www.ebi.ac.uk/~birney/wise2/.
- Gene expression analysis: *Tophat*, *Cufflinks*, and *Kallisto*:
- http://cole-trapnell-lab.github.io/projects/.
- http://pachterlab.github.io/kallisto/.
- De novo transcriptome assembler: *Trinity* and *SOAPtrans*:
- https://trinityrnaseq.github.io/.
- http://soap.genomics.org.cn.
- K-mer counting: *Jellyfish* [27]:
- http://www.cbcb.umd.edu/software/jellyfish/.
- Customized code: *YGS.pl* [8]:
- Perl module: Carp::Clan and Bit::Vector.

3 Methods

3.1 Comparative Method

This method is suitable for identifying sex-linked sequences, when there are only reads from the heterogametic sex and a reference genome from the related species. In this case, reads data from the homogametic sex is not necessary, but they will be useful for validating the sex-linked sequences later. Take ostrich and chicken as example, with female ostrich reads data and chicken reference genome; this pipeline will enable the identification of Z- and W-linked sequence and genes in the ostrich genome. The method has two following assumptions: (1) there are very few interchromosomal rearrangements between the studied species and the species with the reference genome; (2) the sex-linked regions have been divergent enough to be discriminated by the read depth. In brief, one can start from assembling short sequencing reads into a draft genome into the form of scaffold sequences. Scaffold sequences are long genomic fragment sequences with gaps inside and without linkage group information. A genome comparison of them to the related species' chromosomal sequences will identify Z-linked and W-linked sequences, with the former showing a much higher level of sequence similarities than the latter with the reference genome, and both showing half of the read depths comparing to autosomes. When there are reads from the male bird available, we expect very few of them can be mapped to the identified W-linked sequences. A summary of this method is illustrated in Fig. 2. Ostrich data used for demonstration can be found at https://www.ncbi.nlm.nih.gov/bioproject/PRJNA212875. Complete chicken Z assembly is available at http://jura.wi.mit.edu/page/HIVODCRVRAFKKPXNUW.

Genome assembly (optional)

1. Collect genomic DNAs from a heterogametic individual (ZW sex-determining system in bird). Make appropriate sequencing libraries (we recommend a gradient of insert sizes like 200 bp, 500 bp, 800 bp, and 2 kb, etc.) for sequencing with Illumina protocol (*see* **Note 1**).
2. Assemble the sequencing reads into a draft genome, if not available (*see* **Notes 2** and **3**).

   Example commands:

   Quality control and filtering of reads

   ```
   fastqc -q -t 4 reads1.fq.gz read2.fq.gz read3.fq.gz read4.
   fq.gz read5.fq.gz read6.fq.gz
   cutadapt -m 90 -max-n=0 -q 30,20 -u 3 -U 1 -o read1.filt.fq.gz -
   p read2.filt.fq.gz read1.fq.gz read2.fq.gz
   cutadapt -m 90 -max-n=0 -q 30,20 -u 3 -U 1 -o read3.filt.fq.gz -
   p read4.filt.fq.gz read3.fq.gz read4.fq.gz
   cutadapt -m 90 -max-n=0 -q 30,20 -u 3 -U 1 -o read3.filt.fq.gz -
   p read4.filt.fq.gz read3.fq.gz read4.fq.gz
   ```

   Genome assembly

   ```
   SOAPdenovo-63mer all -s assembly.cfg -o bird_ostrich -K 27 -
   F -p 16 2> log_assem27.err
   ```

   Gap closing

   ```
   GapCloser -a bird_ostrich.scafSeq -b assembly.cfg -o bird_os-
   trich.scafSeq.fill -l 100 -t 16
   ```

3. Assess the quality of genome assembly regarding its N50, gap content, and also with external evidence, such as EST, representation of conserved proteins in the genome.

Construction of pseudo-chromosome sequence from closely related species' chromosomal sequence (*see* Notes 4 and 5)

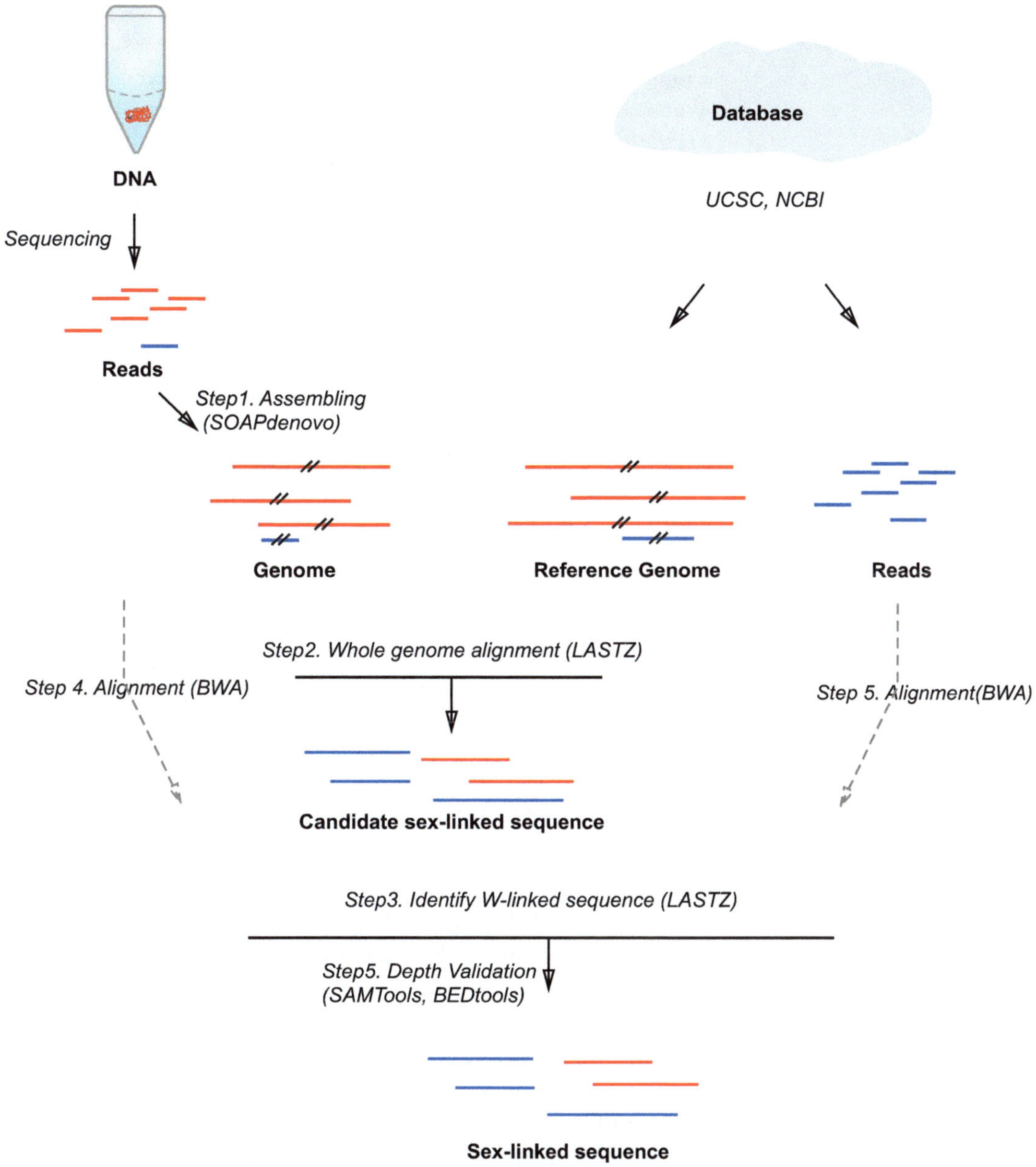

Fig. 2 Work flow of comparative method. Major steps and corresponding outputs are depicted, where each step pinpoints the key experiment or bioinformatic tool used to generate the intermediate output

4. Download the reference genome. Taking bird genomes as an example, we are using the sequence of chicken Z chromosome from [28].
5. Perform whole genome alignment with *LASTZ* using a specific parameter set for distant species' comparison (*see* **Note 6**). Example commands:

```
lastz chicken.fa bird_ostrich/Bird00.fa -step=19 -
hspthresh=2200 -inner=2000 -ydrop=3400 -
gappedthresh=10000 -format=axt > Bird00.axt
axtChain -minScore=5000 -linearGap=medium Bird00.axt
Bird00.2bit chicken.2bit Bird00.chain
chainMergeSort Bird*.chain > ./all.chain
chainPreNet ./all.chain chicken.sizes bird_ostrich.
scafSeq.fill.sizes ./all_sort.chain
chainNet ./all_sort.chain chicken.sizes bird_ostrich.
scafSeq.fill.sizes temp ./query.net
netSyntenic ./temp ./target.net
netToAxt target.net all_sort.chain chicken.sizes
bird_ostrich.scafSeq.fill.2bit all.axt
axtSort all.axt all_sort.axt
axtToMaf -tPrefix=chicken -qPrefix=bird_ostrich.scaf-
Seq.fill all_sort.axt chicken.sizes bird_ostrich.scaf-
Seq.fill.sizes all.maf
```

6. Construct pseudo-chromosomes by ordering scaffold sequences with at least 50% of the length aligned (*see* **Note 7**).

Identification of W-linked sequence

The following steps are based on the assumptions that (1) W-linked sequences cannot be aligned to autosomes and always form a worse alignment, if available, than the homologous Z-linked sequences with the reference genome; (2) there are still certain degrees (at least 70%) of similarity between Z- and W-linked sequences to allow for the identification of W-linked sequences.

7. Search for the W-linked candidate sequences by a second round of alignment with all the unaligned sequences (from **step 5**) against the Z chromosome sequence constructed from **step 6** (*see* **Note 8**).
8. Filtering the spurious alignment hits with the criteria that the sequence identity of the aligned region has to reach at least 70%, but lower than 95%, with the aligned sequences spanning at least 50% of the scaffold length (*see* **Note 9**).

Depth evaluation by reads alignment (*see* Note 10)

9. Map the original reads to the genome using *BWA*.
10. Compare the read-depth distribution of both Z-linked and W-linked sequences to autosomes (e.g., chromosome 1). Both Z-linked and W-linked sequence are expected to show a reduced depth level, which is approximately half of that of autosomes, except for PARs.

Validation with reads from homogametic individual(s) (*see* Notes 11 and 12)

11. Align the male reads (female reads in the case of XY sex system) to the newly identified sequences by the same parameter set. W-specific sequences are expected to have nearly no male reads mapped, whereas the Z-linked sequences and PARs are expected to show a read-depth level the same as autosomes. The identified sex-linked sequence will be subjected for downstream gene and transcriptomic analyses (*see* **Notes 13** and **14**).
    Example commands:
    Map female reads to pseudo chromosomes

```
bwa index -a bwtsw bird_ostrich.scafSeq.fill
bwa aln -o 1 -e 50 -m 100000 -t 4 -i 15 -q 10 -I -k 0 BirdPseudoChr read1.filt.fq.gz
bwa aln -o 1 -e 50 -m 100000 -t 4 -i 15 -q 10 -I -k 0 BirdPseudoChr read2.filt.fq.gz
bwa sampe read1.filt.sai read2.filt.sai read1.filt.fq.gz read2.filt.fq.gz | samtools view -bS - | samtools sort -m 4G BirdPseudoChr.sort
```

    Depth calculation

```
samtools depth -f BirdPseudoChrL1.sort.bam BirdPseudoChrL2.sort.bam |gzip -> BirdPseudoChr.depth.gz
```

12. Apply the Bayesian classifier for further validation [18].

3.2 Y Chromosome Genome Scan (YGS) Method

This method is developed by Carvalho and Clark [8] and has been applied to identify Y-linked sequences in human and *Drosophila*. Its modified version has been applied to identify W-linked sequences in chicken [18]. In brief, when the draft genome of heterogametic sex is available, sequencing reads of the homogametic sex can be first aggregated into an overlapping “k-mer” (sequence strings shorter than the read length, usually $k = 15$) dataset. This k-mer dataset can be used to index the entire raw read dataset and scan the heterogametic sex genome for their presence. Take bird genomes as an example, after masking the sequences or k-mers derived from repetitive regions, W-linked regions are expected to show almost no k-mers derived from the male reads (coded as “0” for absence of k-mer or “1” for presence in bit-arrays). A summary of this method is illustrated in Fig. 3. Here we use the data of Emu (*Dromaius novaehollandiae*) for demonstration, and this species' female genome and male and female reads are available at https://www.ncbi.nlm.nih.gov/sra/SRX252417 and https://www.ncbi.nlm.nih.gov/sra/SRX252416.

Reads filtering and k-mer extraction

1. Filtering rules: trim reads with Phred scores lower than 20 from both ends, and remove k-mers that present less than five times, which are likely to be derived from sequencing errors (*see* **Note 15**).

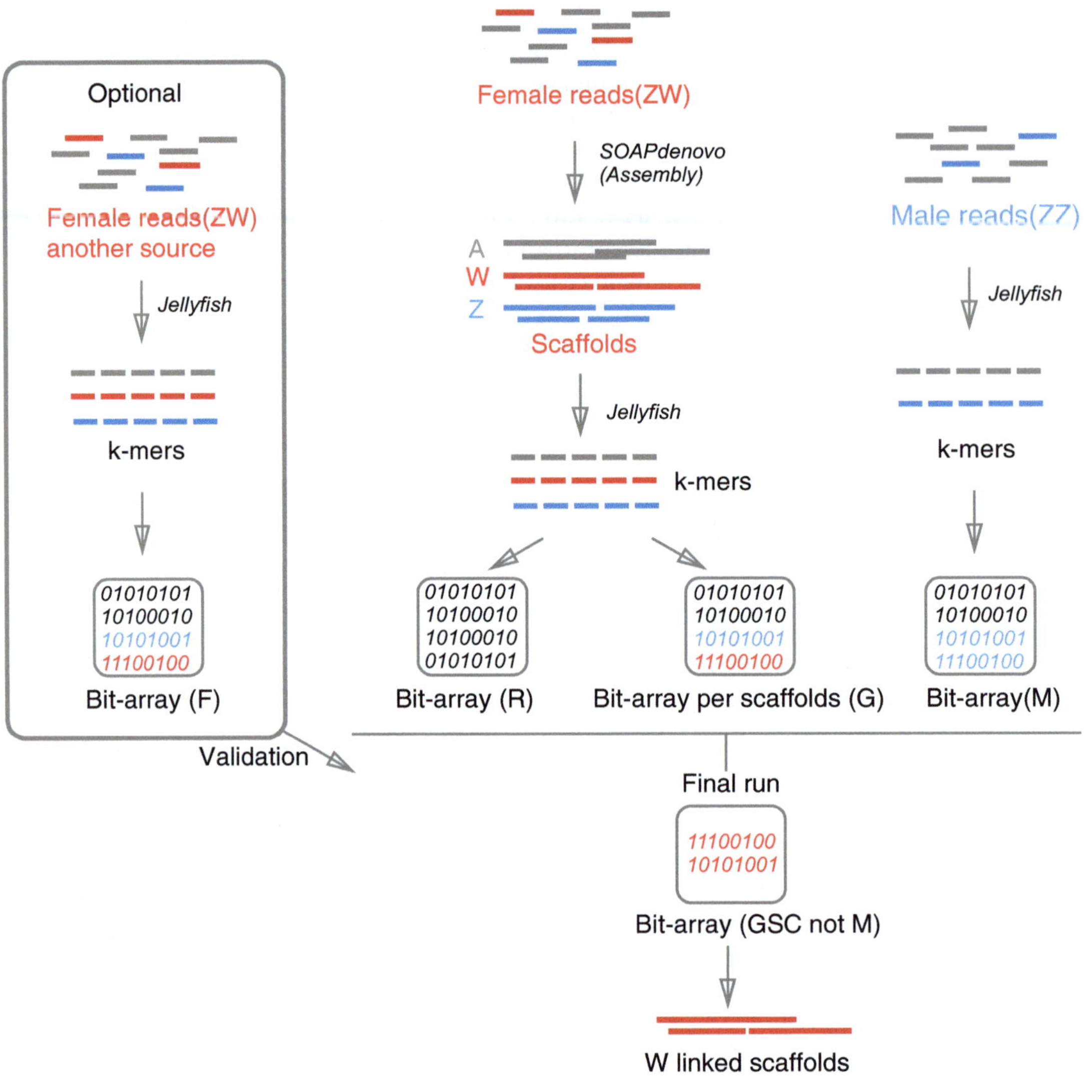

Fig. 3 Work flow of the YGS method [8]. Major steps and key intermediate outputs of YGS method are shown here. Marks for bit-arrays: F = female, R = repeat, G = genome, M = male, and GSC not M = genome single-copy kmers but not from male. Reads for validation are from the opposite sex compared to the reads used for the genome assembly

Example commands:
Reads filtering

```
cutadapt -q 20,20 -o Emu_M_gDNA_read1.fq -p Emu_M_gD-
NA_read2.fq Emu_M_gDNA_read1.filt.fq Emu_M_gDNA_read2.
filt.fq
```

Extract k-mers from all male reads with jellyfish

```
jellyfish count -t 32 -m 15 -s 3G -o Emu_M_gDNA_15mer.jf
-Q 5 -L 5 Emu_M_gDNA*.filt.fq
```

Dump k-mer table (.jf) into a fasta file for generating bit-array libraries.*

```
jellyfish dump -o Emu_M_gDNA_15mer.fa Emu_M_gDNA_15mer.jf
```

Generate k-mers from all female reads with jellyfish (optional, if the female reads are available)

```
jellyfish count -t 32 -m 15 -s 3G -o Emu_F_gDNA_15mer.jf
-Q 5 -L 5 Emu_F_gDNA*.fq
```

Dump the kmer table (.jf) into a fasta file (optional).*

```
jellyfish dump -o Emu_F_gDNA_15mer.fa Emu_F_gDNA_15mer.jf
```

Build bit-arrays to represent k-mers

2. Convert the k-mer tables in fasta format into bit-arrays (a data structure with bit strings that store the sequence data), in which each k-mer is coded by 1 for presence and 0 for absence. Note the k-mer size is chosen arbitrarily with a length shorter than the read length, here as 15 (*see* **Note 16**).
   Example commands:
   Build bit-array 'M' to store male Emu kmers

```
perl YGS.pl kmer_size=15 mode=trace trace=Emu_M_gDNA_
15mer.fa
```

Build bit-array 'F' to represent female Emu k-mers (optional) (*see* **Note 17**).

```
perl YGS.pl kmer_size=15 mode=trace trace=Emu_F_g
DNA_15mer.fa
```

Build bit-array 'R' with the repetitive k-mers derived from the heterogametic genome (*see* **Note 18**).

```
perl YGS.pl kmer_size=15 mode=contig contig=Emu.mix-
edsex.gapfilled.fa
```

Obtain W-linked scaffolds (*see* Notes 19 and 20)
The following steps are based on the assumptions that: (1) W-linked k-mers derived from non-repetitive regions only appear in the female genome; (2) W-linked regions have a certain degree of sequence divergence with their Z homologues.

3. In this step, an intermediate collection of bit-arrays "G" will be first generated for each scaffold from the genome. One can filter out repetitive k-mers to obtain a group of bit-array "GSC" representing unique k-mers for the processed scaffolds with the option "G not R." Following the above process, the comparison between the reference genome and male short reads can be obtained with the option "GSC not M." This procedure results in candidate female-specific k-mers, which are contained in W-linked sequences. The W-linked scaffolds

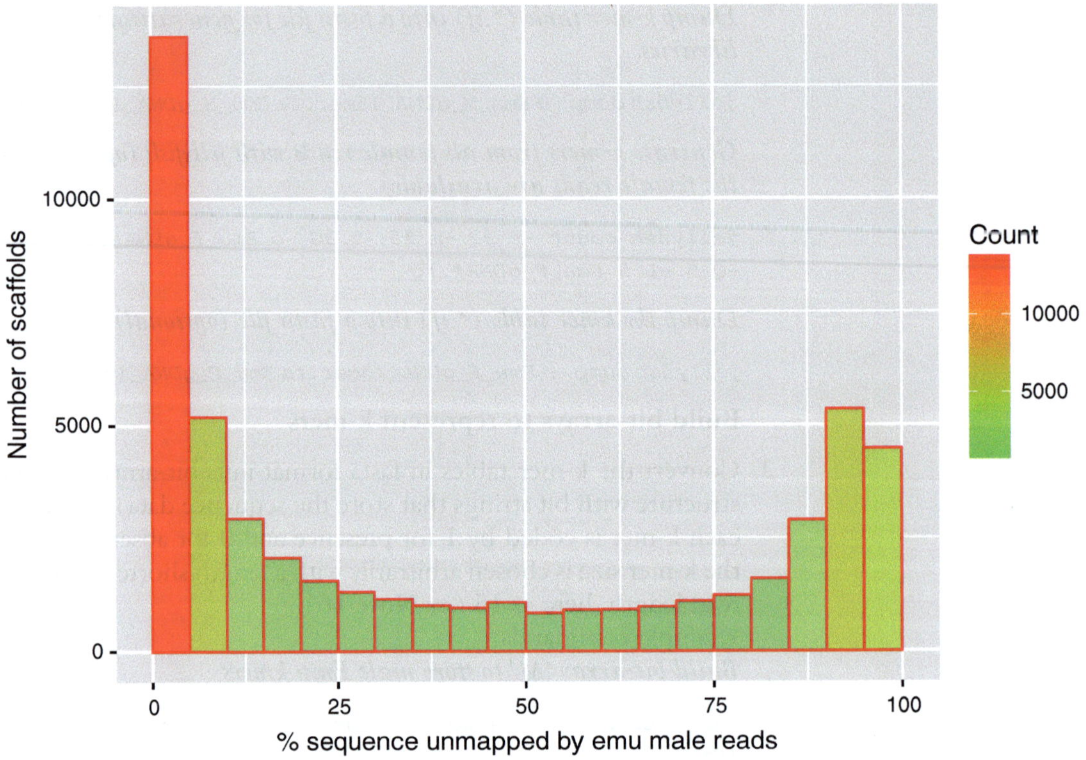

Fig. 4 Mapping rate of the male reads to the female scaffolds from YGS. Distribution of unmapped scaffold by male reads, indicated in percentage

can be distinguished from the proportion of unmatched male single-copy k-mers at each scaffold (*see* **Notes 21** and **22**). Example command:

```
perl YGS.pl kmer_size=15 mode=final_run contig=Emu.
mixedsex.gapfilled.fa trace=Emu_M_gDNA_15mer.trace.gz
gen_rep=Emu.gen_rep.gz male_trace=Emu_F_gDNA_15mer.
trace.gz
```

Visualize the results (*see* Note 23)

4. Generate histogram distribution plot with R (Fig. 4). Example codes:

```
library(ggplot2)
options(stringsAsFactors=FALSE)
data <- read.csv('final_file',header = TRUE)
data$P_VSC_UK<-as.numeric(data$P_VSC_UK)
ggplot(data=data, aes(data$P_VSC_UK)) +
geom_histogram(breaks=seq(0, 100, by =5),
col="red",
aes(fill=..count..),
alpha = .8) +
```

```
scale_fill_gradient("Count", low = "green", high =
"red")+
 labs(x="% sequence unmapped by emu male reads",
 y="Number of scaffolds")
```

4 Notes

Genome assembly

1. Different assemblers usually have different requirements for sequencing libraries. For example, *Allpath-LG* requires a sequencing library with overlapping reads (e.g., an insert size of 170 bp with sequencing length from both ends as 100 bp), while *SOAPdenovo* does not have such requirements. Thus, the choice of assembler depends on the input data type. For a full comparison and recipes of running different assemblers, one can refer to http://gage.cbcb.umd.edu/recipes/index.html.
2. It is suggested that any analyses should start from reads with as few errors as possible. Therefore, we recommend to filter the reads for adapter contamination, low-quality base pairs, as well as GC-content abnormalities. These can be evaluated by *FASTQC*, followed by a cleaning process by *cutadapt*.
3. PCR duplicates of the reads, i.e., reads with exactly the same sequences, should be subjected for removal as well with Picard tool.

Building pseudo-chromosomes

4. Repetitive sequences should be masked before the whole-genome alignment by *RepeatMasker*.
5. Installation of *LASTZ* (version 1.0.2, tested on Debian 7) requires removing the "-Werror" at line31 in the Makefile in src directory.
6. Parameters of *LASTZ* need to be carefully selected by considering evolutionary distance, and scoring matrix HoxD55 is used for distant species comparison.
7. Ordering the query sequences against the reference genome: the best syntenic alignment of each query in the *LASTZ* output (shown as the first level of "net" information) can be used for the subsequent construction of pseudo-chromosomes. Sort these best alignments in *.net file by their coordinates on the reference genome, and calculate the alignment ratio of each block for query sequences. Here a distribution of alignment identities of all the query sequences can be obtained to exclude those queries fallen into the lower 5% end. After this filtering step, the query scaffolds can be ordered into pseudo-chromosome.

Identification of W-linked sequence

8. In the filtered alignment file (*.maf file at **step 5**), the coverage of alignment ratio between the candidate W-linked fragments and Z-linked pseudo-chromosome sequences is required to be higher than 50%.
9. Any scaffolds that can be aligned to both autosomes and Z chromosomes are also excluded, because they can be derived from either unmasked repetitive elements or interchromosomal rearrangements.

Depth evaluation

10. A stringent parameter set (aln -e 50 -i 15 -q 10 -k 0) for *BWA* alignment is required in order to avoid ambiguous alignments in pair-end mode. Depth distribution is calculated with binned average effective depth (repetitive regions excluded). Positions with $\geq$1.5$\times$ whole genome depth are considered to be repetitive. Zero depth is omitted.

Validation

11. Mask the repetitive regions before any further analyses are essential. Depth distribution of W-linked sequences from female reads show peaks at the half level of autosomal depth, while the distribution from male reads is concentrated at zero.
12. One can infer PAR from the Z-linked sequence based on depth information.

Gene annotation

13. Some sequences cannot be properly assembled, probably due to the high repeat content on sex chromosomes, resulting in truncated long genes. For more information about gene and repeat annotation, please refer to [9].

Transcriptome analysis

14. For a full evaluation of different transcriptome assembly methods, please refer to [29, 30].

YGS

Filter short reads and produce k-mers

15. Filtering low-quality male reads is essential for obtaining reliable representative k-mers.
16. Too short k-mer size can lead to spurious matches and reduce the resolution of this method. The recommended k-mer size for insect genomes is 15 while 18 for vertebrate genomes.

Build bit-arrays representing k-mers

17. If reads of the heterogametic sex from other sources are available, it is recommended to validate the genome before the final run. For example, the W-linked sequences are expected to be mapped by the female reads but not male reads or male k-mers.
18. Usually scaffold sequences longer than 20 Mbp will take much more computation time when generating repetitive bit-arrays.

Obtain W-linked scaffolds

19. For a large genome ($\geq$1G), the whole process may take more than 48 h.
20. In the final run, one can accelerate the computation by using multiple threads, as each scaffold is processed separately.
21. The acronym meanings of final output file are as follows:

    GI: sequence name (as given in your fasta file).
    NUM: numerical index of the sequence.
    MAX_K: maximum number of k-mer.
    MAX_K = seq_length − kmer_size + 1.
    K: number of k-mers found in the sequence.
    UK: number of k-mers unmatched by the female reads.
    SC_K: number of single-copy k-mers found in the sequence.
    SC_UK: number of single-copy k-mers unmatched by the female reads.
    P_SC_UK: percentage of single-copy k-mers unmatched by the female reads.
    VSC_K: number of valid single-copy k-mers.
    VSC_UK: number of valid single-copy k-mers unmatched by the female reads.
    P_VSC_UK: percentage of valid single-copy k-mers unmatched by the female reads.
22. We can get the candidate W-linked scaffolds based on a threshold of P_VSC_UK value. But additional evidence are still needed to confirm such W-linkage, such PCR.

Visualize the results

23. Before plotting with *R*, the output file of final run should be truncated and transferred to csv format using basic perl or python.

References

1. Charlesworth B, Coyne JA, Barton NH (1987) The relative rates of evolution of sex chromosomes and autosomes. Am Nat 130:113–146
2. Meisel RP, Malone JH, Clark AG (2012) Faster-X evolution of gene expression in Drosophila. PLoS Genet 8:e1003013. doi:10.1371/journal.pgen.1003013
3. Kharchenko PV, Xi R, Park PJ (2011) Evidence for dosage compensation between the X chromosome and autosomes in mammals. Nat Genet 43:1167–1169. doi:10.1038/ng.991
4. Bellott DW, Hughes JF, Skaletsky H, Brown LG, Pyntikova T, Cho T-J, Koutseva N, Zaghlul S, Graves T, Rock S, Kremitzki C, Fulton RS,

Dugan S, Ding Y, Morton D, Khan Z, Lewis L, Buhay C, Wang Q, Watt J, Holder M, Lee S, Nazareth L, Rozen S, Muzny DM, Warren WC, Gibbs RA, Wilson RK, Page DC (2014) Mammalian Y chromosomes retain widely expressed dosage-sensitive regulators. Nature 508:494–499. doi:10.1038/nature13206

5. Huynh KD, Lee JT (2003) Inheritance of a pre-inactivated paternal X chromosome in early mouse embryos. Nature 426:857–862. doi:10.1038/nature02222
6. Julien P, Brawand D, Soumillon M, Necsulea A, Liechti A, Schutz F, Daish T, Grutzner F, Kaessmann H (2012) Mechanisms and evolutionary patterns of mammalian and avian dosage compensation. PLoS Biol 10(5):e1001328. doi:10.1371/journal.pbio.1001328
7. Zhou Q, Zhang J, Bachtrog D, An N, Huang Q, Jarvis ED, Gilbert MTP, Zhang G (2014) Complex evolutionary trajectories of sex chromosomes across bird taxa. Science 346:1246338. doi:10.1126/science.1246338
8. Carvalho A, Clark A (2013) Efficient identification of Y chromosome sequences in the human and Drosophila genomes. Genome Res 23:1894–1907. doi:10.1101/gr.156034.113
9. Yandell M, Ence D (2012) A beginner's guide to eukaryotic genome annotation. Nat Rev Genet 13:329–342. doi:10.1038/nrg3174
10. Parsch J, Ellegren H (2013) The evolutionary causes and consequences of sex-biased gene expression. Nat Rev Genet 14:83–87. doi:10.1038/nrg3376
11. Ellegren H, Hultin-Rosenberg L, Brunström B, Dencker L, Kultima K, Scholz B (2007) Faced with inequality: chicken do not have a general dosage compensation of sex-linked genes. BMC Biol 5:40. doi:10.1186/1741-7007-5-40
12. Trapnell C, Roberts A, Goff L, Pertea G, Kim D, Kelley DR, Pimentel H, Salzberg SL, Rinn JL, Pachter L (2012) Differential gene and transcript expression analysis of RNA-seq experiments with TopHat and Cufflinks. Nat Protoc 7:562–578. doi:10.1038/nprot.2012.016
13. Pertea M, Kim D, Pertea GM, Leek JT, Salzberg SL (2016) Transcript-level expression analysis of RNA-seq experiments with HISAT, StringTie and Ballgown. Nat Protoc 11:1650–1667. doi:10.1038/nprot.2016.095
14. Bray NL, Pimentel H, Melsted P, Pachter L (2016) Near-optimal probabilistic RNA-seq quantification. Nat Biotechnol 34:525–527. doi:10.1038/nbt.3519
15. Grabherr MG, Haas BJ, Yassour M, Levin JZ, Thompson DA, Amit I, Adiconis X, Fan L, Raychowdhury R, Zeng Q, Chen Z, Mauceli E, Hacohen N, Gnirke A, Rhind N, di Palma F, Birren BW, Nusbaum C, Lindblad-Toh K, Friedman N, Regev A (2011) Full-length transcriptome assembly from RNA-Seq data without a reference genome. Nat Biotechnol 29:644–652. doi:10.1038/nbt.1883
16. Xie Y, Wu G, Tang J, Luo R, Patterson J, Liu S, Huang W, He G, Gu S, Li S, Zhou X, Lam TW, Li Y, Xu X, Wong GKS, Wang J (2014) SOAPdenovo-Trans: de novo transcriptome assembly with short RNA-Seq reads. Bioinformatics 30:1660–1666. doi:10.1093/bioinformatics/btu077
17. Andrews S (2010) FastQC—a quality control tool for high throughput sequence data. http://www.bioinformatics.babraham.ac.uk/projects/fastqc/. Accessed 25 Aug 2013
18. Chen N, Bellott DW, Page DC, Clark AG (2012) Identification of avian W-linked contigs by short-read sequencing. BMC Genomics 13:183. doi:10.1186/1471-2164-13-183
19. Martin M (2011) Cutadapt removes adapter sequences from high-throughput sequencing reads. EMBnetjournal 17:10. doi:10.14806/ej.17.1.200
20. Li R, Yu C, Li Y, Lam TW, Yiu SM, Kristiansen K, Wang J (2009) SOAP2: an improved ultrafast tool for short read alignment. Bioinformatics 25:1966–1967. doi:10.1093/bioinformatics/btp336
21. Smit A, Hubley R, Green P (2016) RepeatMasker Open-4.0. http://www.repeatmasker.org/ (2013–2015)
22. Li H, Durbin R (2009) Fast and accurate short read alignment with Burrows-Wheeler transform. Bioinformatics 25:1754–1760. doi:10.1093/bioinformatics/btp324
23. She R, Chu JS, Uyar B, Wang J, Wang K, Chen N (2011) genBlastG: using BLAST searches to build homologous gene models. Bioinformatics 27(15):2141–2143. doi:10.1093/bioinformatics/btr342
24. Li H, Handsaker B, Wysoker A, Fennell T, Ruan J, Homer N, Marth G, Abecasis G, Durbin R (2009) The Sequence Alignment/Map format and SAMtools. Bioinformatics (Oxford, England) 25:2078–2079. doi:10.1093/bioinformatics/btp352
25. Wysoker A, Tibbetts K, Fennell T (2013) Picard-Tools 1.9.2. http://picard.sourceforge.net/. Accessed 28 May 2013
26. Quinlan AR, Hall IM (2010) BEDTools: a flexible suite of utilities for comparing genomic features. Bioinformatics 26:841–842. doi:10.1093/bioinformatics/btq033
27. Marçais G, Kingsford C (2011) A fast, lock-free approach for efficient parallel counting of

occurrences of k-mers. Bioinformatics 27: 764–770. doi:10.1093/bioinformatics/btr011

28. Bellott DW, Skaletsky H, Pyntikova T, Mardis ER, Graves T, Kremitzki C, Brown LG, Rozen S, Warren WC, Wilson RK, Page DC (2010) Convergent evolution of chicken Z and human X chromosomes by expansion and gene acquisition. Nature 466:612–616. doi:10.1038/nature09172
29. Huang X, Chen X-G, Armbruster PA (2016) Comparative performance of transcriptome assembly methods for non-model organisms. BMC Genomics 17:523. doi:10.1186/s12864-016-2923-8
30. Li B, Fillmore N, Bai Y, Collins M, Thomson JA, Stewart R, Dewey C (2014) Evaluation of de novo transcriptome assemblies from RNA-Seq data. Genome Biol. doi:10.1101/006338

Chapter 5

Systems Biology Analyses in Chicken: Workflow for Transcriptome and ChIP-Seq Analyses Using the Chicken Skin Paradigm

Yung-Chih Lai, Randall B. Widelitz, and Cheng-Ming Chuong

Abstract

With advances in molecular biology, various biological phenomena can now be explored at higher resolution using mRNA sequencing (RNA-Seq) and chromatin immunoprecipitation followed by high-throughput sequencing (ChIP-Seq), two powerful high-throughput next-generation sequencing (NGS) technologies. While methods are used widely in mouse, human, etc., less information is available in other animals, such as the chicken. Here we assemble a workflow of the RNA-Seq and ChIP-Seq analyses for the chicken studies using chicken skin appendage tissue as an example. We present guidelines for RNA-Seq quality control, alignment, quantification, normalization, and differentially expressed gene analysis. In the meantime, we outline a bioinformatics pipeline for ChIP-Seq quality control, alignment, peak calling, super-enhancer identification, and differential enrichment analysis.

Key words Next-generation sequencing, Feather, Scale, Regeneration, Embryo

1 Introduction

Advances in mRNA sequencing (RNA-Seq) and chromatin immunoprecipitation followed by high-throughput sequencing (ChIP-Seq) have shed new light in various areas of biological research. As next-generation sequencing (NGS) has become widely available, the major bottleneck is no longer based on data generation. Rather, bioinformatics analysis of NGS data has become more significant. However, the adaptation of this approach to other animals, such as the chicken, has not been developed like those in mouse or human. Here we wish to lay out the protocol for doing systems biology studies using the chicken model and hope to help scientists achieve a similarly high-resolution analysis.

Chicken skin appendages (e.g., feathers and scales) are an ideal system in which to study organ development and regeneration because of their diverse morphologies and their ability to undergo physiological regeneration. While analyzing chicken RNA-Seq and

Guojun Sheng (ed.), *Avian and Reptilian Developmental Biology: Methods and Protocols*, Methods in Molecular Biology, vol. 1650, DOI 10.1007/978-1-4939-7216-6_5,

ChIP-Seq data based upon skin appendage tissues, some special issues should be noted that differ from that seen in well-assembled and annotation mammalian genomes (*see* **Note 1**). For instance, bird genome assemblies are particularly problematic because of the presence of gene-rich small microchromosomes. Consequently, the chicken galGal4 reference genome is still fragmented with more than 10,000 fragmented scaffolds (*see* **Note 2**). For this reason, we have to include both the primary sequence (i.e., chr1–chr28, chrZ, chrW, and chrM) and unplaced scaffolds for the alignment (*see* **Note 3**). Currently, Ensembl has the most comprehensive annotation for the chicken genome encompassing 15,508 protein-coding genes [1] (*see* **Note 4**).

Besides whole-genome annotations, more and more independently curated chicken gene sets have been published. Some of them focus on highly expressed genes in skin appendage tissues, e.g., keratin (KRT) genes [2] and epidermal differentiation complex (EDC) genes [3] that have not been annotated by Ensembl, and can be manually added to a GTF annotation file (*see* **Note 5**). Because few chicken genes annotated by Ensembl have more than one alternative splicing form, even this most comprehensive annotation cannot be used for alternative splicing analysis (*see* **Note 6**). For this reason, we suggest using a union exon-based approach (i.e., gene level) but not a transcript-based approach for RNA-Seq quantification when studying chicken skin appendage tissues.

In our RNA-Seq analyses experience, about 30 million single-end 50 or 75 bp reads are sufficient to identify major differences in expression at the gene levels [4, 5]. In addition, only two biological replicates are acceptable to build a working hypothesis for downstream biological validation, if our (1) experimental design, (2) sample preparation, and (3) library preparation are sufficient to answer the biological questions we are exploring. For the budgets of most laboratories, increasing the number of biological replicates is more important than increasing the sequencing depth above 30 million reads [4]. It is important to note that, at the time we are submitting this manuscript, a new chicken genome (i.e., galGal5) has been released. This new assembly will be annotated by Ensembl and officially supported by the UCSC Genome Browser in the future that should greatly improve chicken RNA-Seq and ChIP-Seq analyses. This improved genome quality also will enable chicken studies to utilize other NGS applications, e.g., Chromatin Interaction Analysis by Paired-End Tag Sequencing (ChIA-PET) [6], that require high-quality genome assembly and annotation (*see* **Note 1**). In this chapter, we present a feasible framework to analyze chicken skin appendage RNA-Seq and ChIP-Seq data.

2 Materials

2.1 Genome Reference and Annotation

In this chapter, we will describe key steps used for both RNA-Seq and ChIP-Seq analyses of chicken skin appendage tissues (Fig. 1). Both the chicken galGal4 genome sequence and Ensembl release 81 genome annotation were downloaded from the UCSC Genome Browser [7]. All programs discussed below are run on a computer using the Macintosh OS X system, and facility with command-line programming is helpful.

2.2 Quality Analysis, Alignment, Quantification, Normalization, and Differential Expression Analysis

The trimming of raw reads and alignment are performed using Partek Flow (Partek Inc.), and we suggest aligning reads with STAR [8] for RNA-Seq and Bowtie2 [9] for ChIP-Seq. Then, htseq-count [10] is recommended to run the gene-level quantification (*see* **Note 6**) with the union parameter for rescuing reads only partially mapped on exons (*see* **Note 7**). After that, edgeR [11] is performed for trimmed mean of M-values (TMM) normalization [12] (*see* **Note 8**) and differential expression analysis. After getting a differentially expressed gene list, downstream analyses

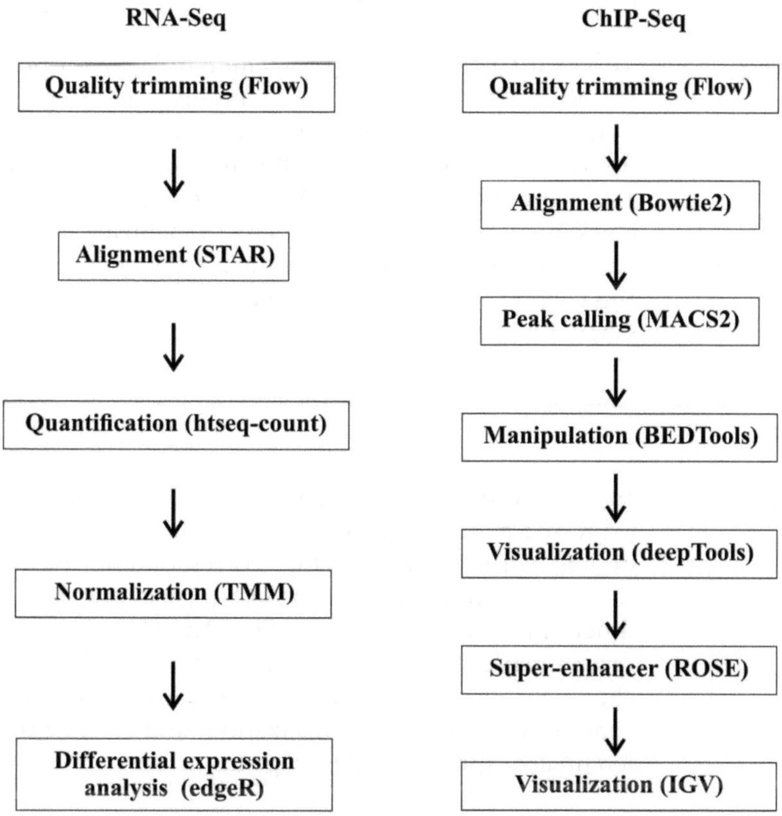

Fig. 1 A flowchart of the major steps involved in RNA-Seq (*left*) and ChIP-Seq (*right*) analyses. *Parentheses* indicate tools we used

Table 1
Chicken skin RNA-Seq and ChIP-Seq sample names and their corresponding attributes used in this chapter

Name	Character	Stage	Tissue	Type	Sample	Region
FE9D_R30	Feather	E9	Dermis	RNA-Seq	R30	Dorsum
FE9D_R34	Feather	E9	Dermis	RNA-Seq	R34	Dorsum
SE9D_R4	Scale	E9	Dermis	RNA-Seq	R4	Metatarsus
SE9D_R32	Scale	E9	Dermis	RNA-Seq	R32	Metatarsus
FE9S_H3K27ac_C1	Feather	E9	Skin	H3K27ac ChIP-Seq	C1	Dorsum
FE9S_Input_C15	Feather	E9	Skin	Input	C15	Dorsum
SE9S_H3K27ac_C26	Scale	E9	Skin	H3K27ac ChIP-Seq	C26	Metatarsus
SE9S_Input_C32	Scale	E9	Skin	Input	C32	Metatarsus

(e.g., pathway enrichment analysis) are beyond the scope of this chapter (but *see* **Note 9**).

2.3 Peak Calling, Manipulation, and Visualization

Tools for ChIP-Seq analyses include MACS2 for peak calling [13] (*see* **Note 10**), BEDTools for operation genomic intervals [14], and both deepTools [15] and Integrative Genomics Viewer (IGV) [16] for visualization. We use ROSE to identify super-enhancers [17] (*see* **Note 11**). However, ROSE does not directly support the chicken genome, so you will need to manually download an annotated file from the UCSC Genome Browser. The examples used in this chapter are unpublished chicken skin appendage RNA-Seq and ChIP-Seq datasets produced in our lab (Table 1).

3 Methods

3.1 RNA-Seq Analysis

In this section, we describe key steps to detect differentially expressed genes between metatarsal dermis with scales and dorsal dermis with feathers, using corresponding RNA-Seq datasets (Table 1). Here we use FASTQ format raw files to generate a differentially expressed gene list. An RNA-Seq analysis includes several major processes: (1) quality control of sequence reads, (2) alignment on the chicken genome, (3) quantification of gene expression levels, (4) normalization, and (5) identification of differentially expressed genes (Fig. 1).

3.1.1 Alignment

Trimming of sequence reads is important and can influence gene expression estimates [18]. Using Partek Flow (Partek Inc.), we trim

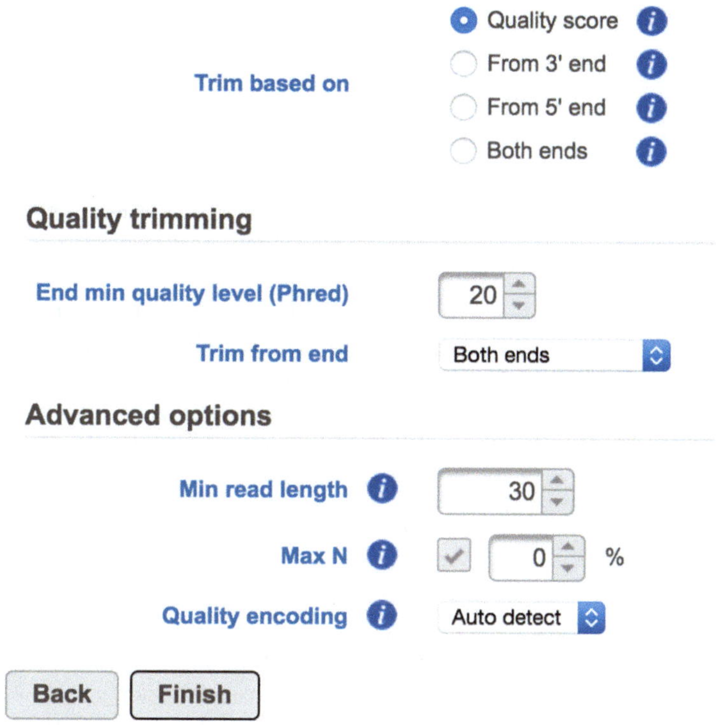

Fig. 2 Parameters in Partek Flow for trimming low-quality 50 bp single-end reads

low-quality bases of 50 bp single-end RNA-Seq reads based on the Phred quality score (>20) from both the 5′- and 3′-ends of reads. After trimming, reads are discarded if they are shorter than 30 bp or have one or more ambiguous bases (Fig. 2). After that, STAR [8] is suggested for the alignment with some adjusted parameters: (1) –outFilterMismatchNoverLmax 0.1, (2) –seedSearchStartLmax 30, (3) –outSAMstrandField None, and (4) –outSAMattributes None (*see* **Note 12**).

3.1.2 Quantification

We used the genePredToGtf script to extract the chicken galGal4 Ensembl Release 81 GTF file from the UCSC Genome Browser database. Here we named the saved GTF file as UCSCgalGal4ensGene81.gtf, using the command line below.

```
> genePredToGtf galGal4 ensGene UCSCgalGal4ensGene81.gtf
```

After that, the gene-level quantification of four chicken skin appendage RNA-Seq samples (Table 1) is run using htseq-count [10]. The parameter of -s no means our RNA-Seq data are not strand specific. The -m union is union mode, indicating that the union exon-based approach (i.e., gene level) is used for quantification.

```
> htseq-count -s no -m union SE9D_R4.bam galGal4UCSCens-
Gene81.gtf > SE9D_R4.txt
> htseq-count -s no -m union SE9D_R32.bam galGal4UCSCens-
Gene81.gtf > SE9D_R32.txt
> htseq-count -s no -m union FE9D_R30.bam galGal4UCSCens-
Gene81.gtf > FE9D_R30.txt
> htseq-count -s no -m union FE9D_R34.bam galGal4UCSCens-
Gene81.gtf > FE9D_R34.txt
```

3.1.3 Normalization

In edgeR [11], we input the four files generated by htseq-count (i.e., SE9D_R4.txt, SE9D_R32.txt, FE9D_R30.txt, FE9D_R34.txt). Totally, there are 17,108 genes. After filtering low-expressed genes (a count-per-million (CPM) above 1 in at least two samples), only 11,967 genes are retained. Then, TMM [12] is selected for sample normalization.

```
> library(edgeR)
> targets <- readTargets()
> targets
 files group description
1 SE9D_R4.txt Scale SE9D_R4
2 SE9D_R32.txt Scale SE9D_R32
3 FE9D_R30.txt Feather FE9D_R30
4 FE9D_R34.txt Feather FE9D_R34
> d <- readDGE(targets,skip=0)
> dim(d)
[1] 17108 4
> keep <- rowSums(cpm(d)>1)>=2
> dim(d)
[1] 11967 4
> d <- d[keep,,keep.lib.sizes=FALSE]
> d <- calcNormFactors(d,method="TMM")
```

3.1.4 Differential Expression Analysis

Using the edgeR commands below, 3349 genes are expressed at significant levels (defined as a false discovery rate (FDR) <0.05), including 1538 upregulated and 1811 downregulated genes in scale dermis compared to feather dermis. We use edgeR's plotSmear function to display differential expression (log fold change) versus expression strength (log average read count) (Fig. 3). Finally, the 3349 differentially expressed genes are saved to a text file named DEG_E9DScaleFeather_3349Gene.txt.

```
> d <- estimateCommonDisp(d)
> d <- estimateTagwiseDisp(d,trend="movingave")
> et <- exactTest(d,pair=c("Feather","Scale"))
> summary(de<-decideTestsDGE(et,p=0.05,adjust="fdr"))
  [,1]
-1 1811
```

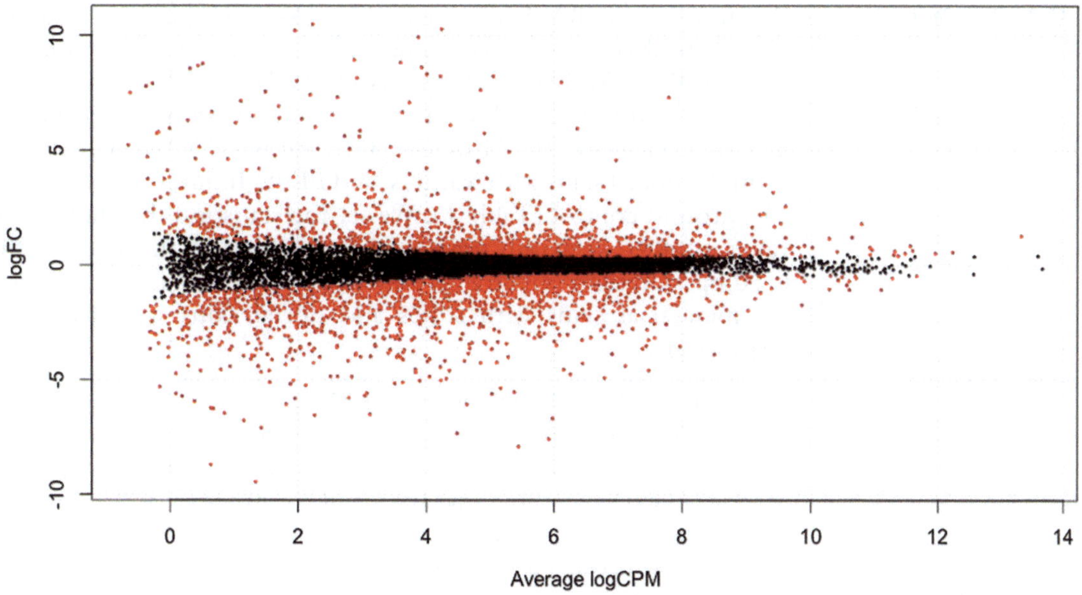

Fig. 3 Differentially expressed genes between metatarsal dermis with scales and dorsal dermis with feathers. The vertical axis indicates the log fold change (i.e., the log ratio of expression levels between scales and feathers), and the horizontal axis is the average gene expression level (i.e., the log counts per million, CPM). The differentially expressed genes are highlighted as *red dots*

```
0 8618
1 1538
> detags <- rownames(d)[as.logical(de)]
> plotSmear(et,de.tags=detags)
> deg <- topTags(et,n=3349)
> dim(deg)
[1] 3349 4
> write.table(deg,file="DEG_E9DScaleFeather_3349Gene.
txt")
```

3.2 ChIP-Seq Analysis

In this section, we describe a procedure to identify scale-specific enhancers and super-enhancers (Fig. 1) when comparing H3K27ac (an active enhancer marker) ChIP-Seq datasets between chicken E9 metatarsal skins with scales and chicken E9 dorsal skins with feathers (Table 1).

3.2.1 Alignment

The first step of ChIP-Seq analysis is to trim poor-quality reads and align them onto the chicken reference genome. Using Partek Flow (Partek Inc.), we trim low-quality bases of 50 bp single-end ChIP-Seq reads based on Phred quality score (>20) from both of the 5′- and 3′-ends of reads. After trimming, reads are discarded, if they are shorter than 30 bp or have one or more ambiguous bases (Fig. 2). Furthermore, we recommend Bowtie2 with a preset of very sensitive local to perform the alignment.

3.2.2 Peak Calling

When we have BAM alignment files, MACS2 [13] is used to identify peaks (i.e., candidate enhancers) using the below command lines. SE9S_H3K27ac_C26.bam is H3K27ac ChIP-Seq from the metatarsal skins with scales, and SE9S_Input_C32.bam is its background control. FE9S_H3K27ac_C1.bam is H3K27ac ChIP-Seq from the dorsal skins with feathers, and FE9S_Input_C15.bam is its background control (Table 1). We set the statistical threshold of q value = 0.001, keep only unique reads, tell MACS2 the read length is 50 bp, and then obtain 32,792 and 30,943 candidate enhancers for metatarsal skins with scales and dorsal skins with feathers, respectively.

```
> macs2 callpeak -t SE9S_H3K27ac_C26.bam -c SE9S_In-
put_C32.bam -format BAM -gsize 930000000 -keep-dup 1 -
outdir SE9S_H3K27ac_C26 -name SE9S_H3K27ac_C26 -tsize 50 -
nomodel -extsize 146 -qvalue 0.001 -slocal 1000 -llocal
10000 -cutoff-analysis
> macs2 callpeak -t FE9S_H3K27ac_C1.bam -c FE9S_Input_C15.
bam -format BAM -gsize 930000000 -keep-dup 1 -outdir
FE9S_H3K27ac_C1 -name FE9S_H3K27ac_C1 -tsize 50 -nomodel
-extsize 146 -qvalue 0.001 -slocal 1000 -llocal 10000 -
cutoff-analysis
```

After that, we use BEDTools [14] to identify scale-specific enhancers. Using the command below, obtain 11,595 scale-specific peaks.

```
> bedtools subtract -a SE9S_H3K27ac_C26.bed -b FE9S_-
H3K27ac_C1.bed -A > ScaleSpecificEnhancer.bed
```

3.2.3 Visualization

deepTools [15] can generate a heatmap for enriched profiles associated with scale-specific enhancers by the following commands. First, we have to build bigwig files for both scale and feather H3K27ac ChIP-Seq, using the bamCompare tool. The number of processors to use depends on the computing resources. Next, computeMatrix and plotHeatmap tools can estimate and visualize the enrichment profiles, respectively (Fig. 4). We set the length of 500 bp to fit all scale-specific enhancers and show an additional 500 bp upstream and downstream from the enhancers. In addition, enhancers are not sorted based on their enrichment levels.

```
> bamCompare -bamfile1 SE9S_H3K27ac_C26.bam -bamfile2 SE9-
S_Input_C32.bam -binSize 1 -extendReads 300 -scaleFactors-
Method readCount -ratio subtract -normalizeUsingRPKM -
minMappingQuality 1 -numberOfProcessors 6 -outFileFormat
bigwig -outFileName SE9S_H3K27ac_C26.bigwig
> bamCompare -bamfile1 FE9S_H3K27ac_C1.bam -bamfile2
FE9S_Input_C15.bam -binSize 1 -extendReads 300 -scaleFac-
torsMethod readCount -ratio subtract -normalizeUsingRPKM -
minMappingQuality 1 -numberOfProcessors 6 -outFileFormat
bigwig -outFileName FE9S_H3K27ac_C1.bigwig
```

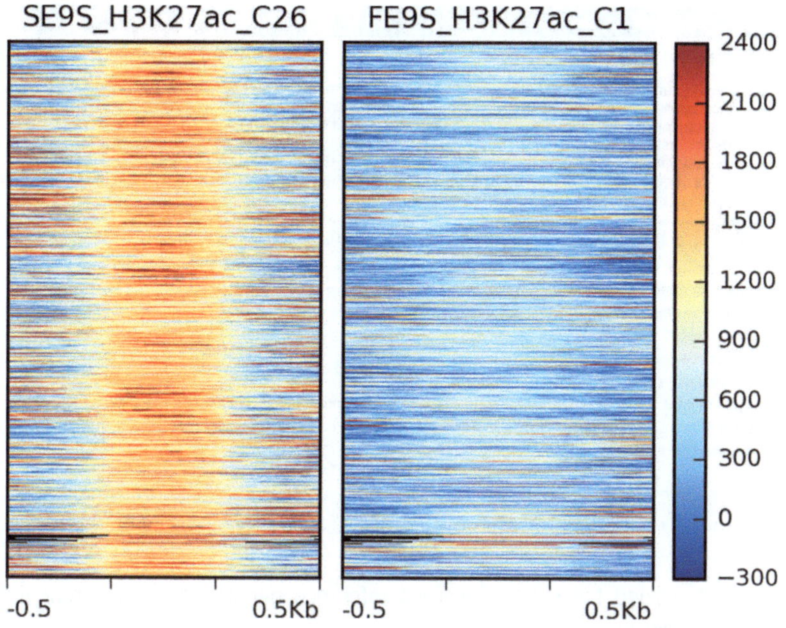

Fig. 4 Enrichment scores in scale-specific enhancers between metatarsal skins with scales (*left*) and dorsal skins with feathers (*right*) based on their H3K27ac ChIP-Seq enrichment profiles. The central regions contain 11,595 scale-specific enhancers fitted to 500 bp. The flanking regions show 500 bp upstream and downstream of the enhancers

```
> computeMatrix scale-regions -regionsFileName ScaleSpe-
cificEnhancer.bed -scoreFileName SE9S_H3K27ac_C26.bigwig
FE9S_H3K27ac_C1.bigwig -regionBodyLength 500 -upstream
500 -downstream 500 -binSize 10 -sortRegions no -number-
OfProcessors 6 -outFileName ScaleSpecificEnhancer.gz
> plotHeatmap -matrixFile ScaleSpecificEnhancer.gz -color-
Map 'RdYlBu_r' -heatmapHeight 7.5 -heatmapWidth 4 -zMin -
300 -zMax 2400 -whatToShow 'heatmap and colorbar' -sortRe-
gions no -averageTypeSummaryPlot mean -legendLocation best
-xAxisLabel " " -startLabel " " -endLabel " " -regionsLabel
" " -verbose -outFileName ScaleSpecificEnhancer.png
```

3.2.4 Super-Enhancers

Super-enhancers are large clusters of transcriptional enhancers that can be determined by ROSE [17]. However, ROSE does not directly support the chicken reference genome. Hence, we have to download the galGal4 database using Table Browser in the UCSC Genome Browser and save it as an unreal mouse mm8 annotation (personal communication with the software author). In the Table Browser (Fig. 5), we query the Ensembl gene track for the chicken galGal4 genome and download the entire table as an

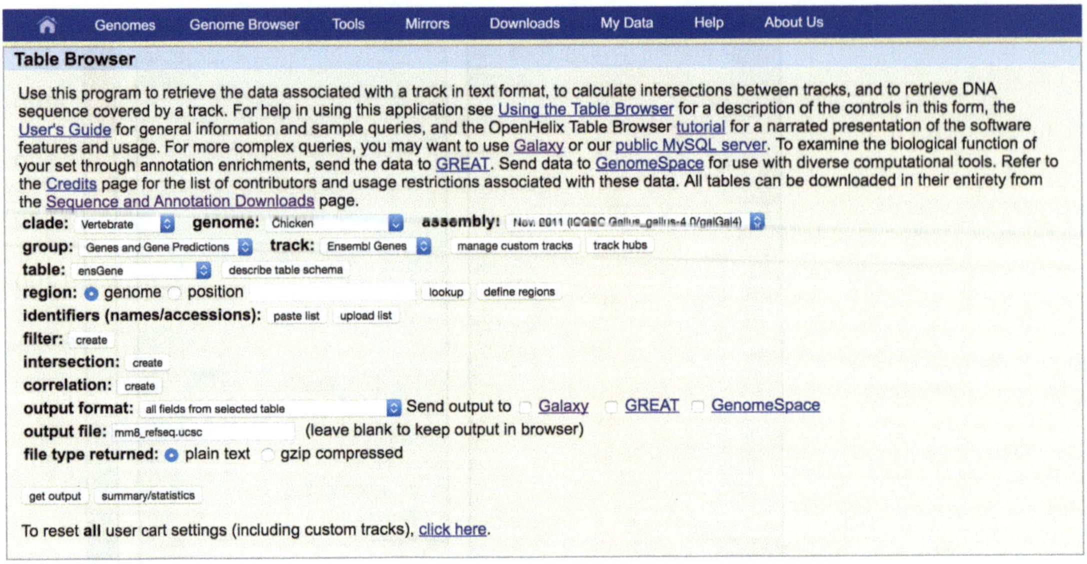

Fig. 5 Download the chicken galGal4 reference genome database from the Table Browser in the UCSC Genome Browser

all-fields output format named as mm8_refseq.ucsc. We use this unreal mouse mm8 genome build to run two commands below. We exclude candidate promoter regions upstream or downstream 2.5 kb from transcription start sites and set 12.5 kb as the maximum distance between two candidate enhancers that will be stitched together. Totally, ROSE detects 332 and 396 super-enhancers for metatarsal skins with scales (Fig. 6a) and dorsal skins with feathers (Fig. 6b), respectively.

```
> python ROSE_main.py -g mm8 -i ./FE9S_H3K27ac_C1.bed -r ./FE9S_H3K27ac_C1.bam -c ./FE9S_Input_C15.bam -s 12500 -t 2500 -o FE9S_H3K27ac_C1
> python ROSE_main.py -g mm8 -i ./SE9S_H3K27ac_C26.bed -r ./SE9S_H3K27ac_C26.bam -c ./SE9S_Input_C32.bam -s 12500 -t 2500 -o SE9S_H3K27ac_C26
```

BEDTools [14] is then used again as the below command line to identify scale-specific super-enhancers. There are 109 scale-specific super-enhancers based on our example datasets.

```
> bedtools subtract -a SE9S_H3K27ac_C26_SuperEnhancers.bed -b FE9S_H3K27ac_C1_SuperEnhancers.bed -A > ScaleSpecific-SuperEnhancer.bed
```

Finally, IGV is used to view one example of scale-specific super-enhancers overlapping with the TBX4 gene (Fig. 7).

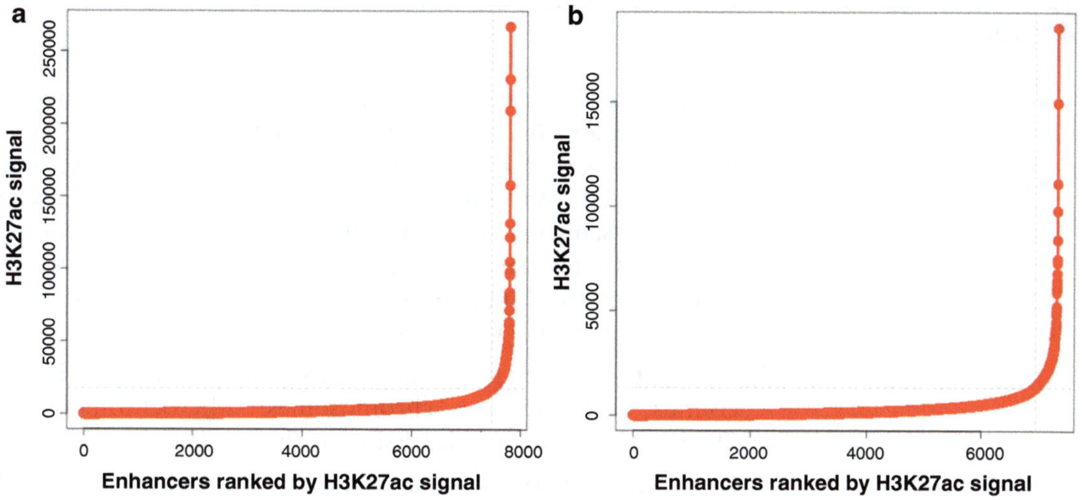

Fig. 6 ROSE identifies 332 super-enhancers for metatarsal skins with scales (**a**) and 396 super-enhancers for dorsal skins with feathers (**b**)

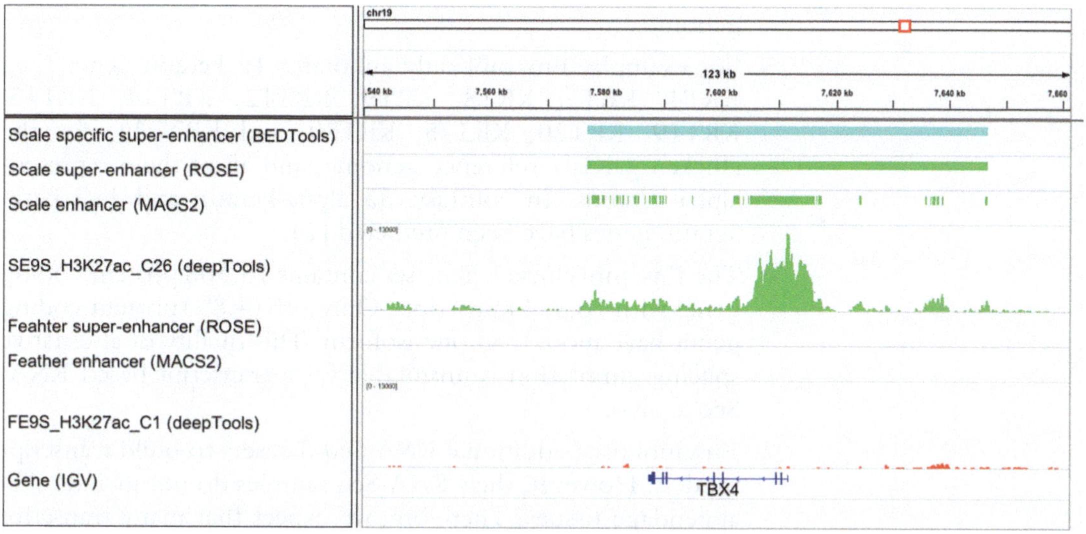

Fig. 7 Genome browser screenshot focused on a region of the TBX4 gene that is overlapping with a scale-specific super-enhancer. The *first track* shows the scale-specific super-enhancer generated by BEDTools. The *second* and *fifth tracks* are scale and feather super-enhancers identified by ROSE, respectively. The *third* (i.e., scale enhancer) and *sixth* (i.e., feather enhancer) *tracks* are candidate enhancers identified by MACS2. The *fourth* (i.e., SE9S_H3K27ac_C26) and *seventh* (i.e., FE9S_H3K27ac_C1) *tracks* show H3K27ac ChIP-Seq enrichment profiles for metatarsal skins with scales and dorsal skins with feathers calculated by deepTools, respectively. The *last track* is the chicken galGal4 RefGene provided by Integrative Genomics Viewer (IGV). The *parentheses* in the *left column* indicate the tool used to generate each track

4 Notes

1. Using the average depth of sequencing coverage as an example, the first draft chicken genome (WUGSC 1.0/galGal2) was built only with 6.6× coverage by Sanger sequencing. The second build (WUGSC 2.1/galGal3) includes an additional 198,000 Sanger reads. The third build chicken genome (ICGSC Gallus_gallus-4.0/galGal4) contains 12× coverage based upon 454 Titanium reads. These assembly quality levels are not suitable for ChIA-PET analysis [6].
2. Totally, the chicken galGal4 reference genome includes 15,932 contigs, scaffolds, or chromosomes. Microchromosomes 29–38 are missing in this version of chicken genome.
3. If only the primary sequence (i.e., chr1–chr28, chrZ, chrW, and chrM) is used, then many genes on unknown chromosomes will be omitted, e.g., DNMT1 on chrUn_JH375489.
4. For the chicken galGal4 genome annotation, Ensembl contains 17,954 transcripts, and RefGene (downloaded from the UCSC Genome Browser on 2016/07/21) only has 7376 transcripts. GENCODE, Vega, and AceView do not annotate the chicken genome.
5. For example, Ensembl only annotates 12 keratin genes (i.e., KRT4, KRT5, KRT8, KRT9, KRT12, KRT14, KRT15, KRT19, KRT20, KRT75, KRT80, and KRT222) for the chicken galGal4 reference genome, and all of them represent alpha-keratins. In contrast, 33 alpha-keratin and 149 beta-keratin genes have been predicted [2].
6. The Ensembl chicken gene set contains 15,508 protein-coding genes with 16,354 transcripts. Only 745 (4.8%) protein-coding genes have more than one isoform. This quality of alternative splicing annotation is unsuitable for a transcript-based RNA-Seq analysis.
7. Ensembl used additional RNA-Seq datasets to build transcript models. However, their RNA-Seq samples do not include skin appendage tissues. Therefore, we expect that many transcript models Ensembl annotated could be different from those existing in skin appendage tissues, because transcript isoforms are tissue specific. Consequently, using the union parameter in htseq-count is a better way to overcome the issue of incompatible tissue-specific transcript models.
8. Although Reads Per Kilobase per Million mapped reads (RPKM) normalization is still a popular method, it should be abandoned because it introduces a bias [19].
9. Many chicken-associated gene names adopted by Ensembl cannot be identified by various pathway analysis tools. Some of

them can be corrected by just removing the hyphen or converting the identifier to a HUGO gene symbol. For example, Ensembl chicken gene symbols of Alx-4, BMP-10, BMPR-II, c-src, cdermo-1, CRABP-I, Dkk-1, GDF-9, Hoxb-7, IGF-I, Lef-1, Pax-6, Pax-7, Pax-9, and vimentin could be identified by Ingenuity Pathway Analysis (IPA) [20], if they are converted into Alx4, BMP10, BMPR2, Src, Twist2, CRABP1, Dkk1, GDF9, Hoxb7, IGF1, Lef1, Pax6, Pax7, Pax9, and Vim, respectively.

10. We highly recommend using MACS2 but not MACS1.4 for peak calling of histone modification ChIP-Seq datasets. Based on our experiences on chicken skin tissues, the peaks called by MACS1.4 could be very broad compared to MACS2. Please see the detail explanation on Dr. Anshul Kundaje's lab website (https://sites.google.com/site/anshulkundaje/projects/idr).
11. The next version of ROSE, currently under development (i.e., ROSE2, https://github.com/BradnerLab/pipeline), can identify super-enhancers using biological replicate samples, but ROSE only can identify super-enhancers based on a single ChIP-Seq sample.
12. The default parameter of –outSAMattributes is standard that cannot be recognized by htseq-count. In fact, the extra SAM attributes are not required by htseq-count for RNA-Seq quantification.

Acknowledgments

This work is supported by a grant from Integrative Stem Cell Center, China Medical University Hospital, China Medical University, and by the National Institute of Arthritis and Musculoskeletal and Skin Diseases of the National Institutes of Health under Award Numbers AR 47364 and AR 60306 of NIH in USA. We thank the University of Southern California's Norris Medical Library Bioinformatics Service for assisting with sequencing data analysis. Some bioinformatics software and computing resources used in the analysis are funded by the USC Office of Research and the Norris Medical Library. The exemplary NGS data in this chapter were generated by Dr. Ping Wu and Dr. Yan Jie.

References

1. Smith J, Burt DW (2015) The Avian RNAseq Consortium: a community effort to annotate the chicken genome. Cytogenet Genome Res 145(2):78–179. doi:10.1101/012559
2. Ng CS, Wu P, Fan WL, Yan J, Chen CK, Lai YT, Wu SM, Mao CT, Chen JJ, Lu MYJ, Ho MR, Widelitz RB, Chen CF, Chuong CM, Li WH (2014) Genomic organization, transcriptomic analysis, and functional characterization of avian alpha- and beta-keratins in diverse feather forms. Genome Biol Evol 6 (9):2258–2273. doi:10.1093/gbe/evu181

3. Strasser B, Mlitz V, Hermann M, Rice RH, Eigenheer RA, Alibardi L, Tschachler E, Eckhart L (2014) Evolutionary origin and diversification of epidermal barrier proteins in amniotes. Mol Biol Evol 31(12):3194–3205. doi:10.1093/molbev/msu251
4. Williams AG, Thomas S, Wyman SK, Holloway AK (2014) RNA-seq data: challenges in and recommendations for experimental design and analysis. Curr Protoc Hum Genet 83:11 13 11–11 13 20. doi:10.1002/0471142905. hg1113s83
5. Wang Y, Ghaffari N, Johnson CD, Braga-Neto UM, Wang H, Chen R, Zhou H (2011) Evaluation of the coverage and depth of transcriptome by RNA-Seq in chickens. BMC Bioinformatics 12(Suppl 10):S5
6. Buisine N, Ruan XA, Bilesimo P, Grimaldi A, Alfama G, Ariyaratne P, Mulawadi F, Chen JQ, Sung WK, Liu ET, Demeneix BA, Ruan YJ, Sachs LM (2015) *Xenopus tropicalis* genome re-scaffolding and re-annotation reach the resolution required for in vivo ChIA-PET analysis. PLoS One 10(9):27. doi:10.1371/journal. pone.0137526
7. Speir ML, Zweig AS, Rosenbloom KR, Raney BJ, Paten B, Nejad P, Lee BT, Learned K, Karolchik D, Hinrichs AS, Heitner S, Harte RA, Haeussler M, Guruvadoo L, Fujita PA, Eisenhart C, Diekhans M, Clawson H, Casper J, Barber GP, Haussler D, Kuhn RM, Kent WJ (2016) The UCSC Genome Browser database: 2016 update. Nucleic Acids Res 44(D1): D717–D725. doi:10.1093/nar/gkv1275
8. Dobin A, Davis CA, Schlesinger F, Drenkow J, Zaleski C, Jha S, Batut P, Chaisson M, Gingeras TR (2013) STAR: ultrafast universal RNA-seq aligner. Bioinformatics 29(1):15–21. doi:10. 1093/bioinformatics/bts635
9. Langmead B, Salzberg SL (2012) Fast gapped-read alignment with Bowtie 2. Nat Methods 9 (4):357–U354. doi:10.1038/nmeth.1923
10. Anders S, Pyl PT, Huber W (2014) HTSeq—a Python framework to work with high-throughput sequencing data. Bioinformatics. doi:10.1093/bioinformatics/btu638
11. Robinson MD, McCarthy DJ, Smyth GK (2010) edgeR: a Bioconductor package for differential expression analysis of digital gene expression data. Bioinformatics 26(1):139–140. doi:10. 1093/bioinformatics/btp616
12. Robinson MD, Oshlack A (2010) A scaling normalization method for differential expression analysis of RNA-seq data. Genome Biol 11 (3):R25. doi:10.1186/gb-2010-11-3-r25
13. Zhang Y, Liu T, Meyer CA, Eeckhoute J, Johnson DS, Bernstein BE, Nussbaum C, Myers RM, Brown M, Li W, Liu XS (2008) Model-based analysis of ChIP-Seq (MACS). Genome Biol 9 (9):R137. doi:10.1186/gb-2008-9-9-r137
14. Quinlan AR, Hall IM (2010) BEDTools: a flexible suite of utilities for comparing genomic features. Bioinformatics 26(6):841–842. doi:10.1093/bioinformatics/btq033
15. Ramirez F, Dundar F, Diehl S, Gruning BA, Manke T (2014) deepTools: a flexible platform for exploring deep-sequencing data. Nucleic Acids Res 42(W1):W187–W191. doi:10. 1093/nar/gku365
16. Robinson JT, Thorvaldsdottir H, Winckler W, Guttman M, Lander ES, Getz G, Mesirov JP (2011) Integrative genomics viewer. Nat Biotechnol 29(1):24–26. doi:10.1038/nbt.1754
17. Whyte WA, Orlando DA, Hnisz D, Abraham BJ, Lin CY, Kagey MH, Rahl PB, Lee TI, Young RA (2013) Master transcription factors and mediator establish super-enhancers at key cell identity genes. Cell 153(2):307–319. doi:10.1016/j.cell.2013.03.035
18. Williams CR, Baccarella A, Parrish JZ, Kim CC (2016) Trimming of sequence reads alters RNA-Seq gene expression estimates. BMC Bioinformatics 17(1):103. doi:10.1186/ s12859-016-0956-2
19. Dillies MA, Rau A, Aubert J, Hennequet-Antier C, Jeanmougin M, Servant N, Keime C, Marot G, Castel D, Estelle J, Guernec G, Jagla B, Jouneau L, Laloe D, Le Gall C, Schaeffer B, Le Crom S, Guedj M, Jaffrezic F, French StatOmique C (2013) A comprehensive evaluation of normalization methods for Illumina high-throughput RNA sequencing data analysis. Brief Bioinform 14(6):671–683. doi:10. 1093/bib/bbs046
20. Kramer A, Green J, Pollard J, Tugendreich S (2014) Causal analysis approaches in Ingenuity Pathway Analysis. Bioinformatics 30(4): 523–530. doi:10.1093/bioinformatics/btt703

Chapter 6

Application of a CAGE Method to an Avian Development Study

Ruslan Deviatiiarov, Marina Lizio, and Oleg Gusev

Abstract

Cap analysis of gene expression (CAGE) is a convenient approach for genome-wide identification of promoter regions at single base-pair resolution level and accurate expression estimation of the corresponding transcripts. Depending on the initial biomaterial amount and sequencing technology, different computational pipelines for data processing are available, as well as variations of the CAGE protocol that improve sensitivity and accuracy. Therefore, this chapter elucidates the key steps of sample preparation, sequencing, and data analysis via an example of a promoter expression estimation study in chicken development. We also describe the applicability of this approach for studying other avian and reptilian species.

Key words Cap analysis gene expression (CAGE), Promoters, Chicken development, FANTOM

1 Introduction

During recent years, next-generation sequencing (NGS) technologies drastically accelerated the accumulation of genomic data. The Genomes OnLine Database (GOLD) lists more than 13.5 K sequenced eukaryotic genomes (https://gold.jgi.doe.gov), and the Genome 10K Project has already sequenced about 280 vertebrate genomes with a further goal of 10,000 [1, 2]. At the same time, different approaches for RNA signal detection have been developed, from methods like microarrays, applied on a relatively small set of genes, to whole-genome transcriptomics like RNA-Seq, CAGE, which is based on sequencing of short 5′ ends of transcripts, DGE (digital gene expression) for accurate tag-based expression estimation, and RACE (rapid amplification of cDNA ends), designed for identification of full-length RNA transcripts, allowing researchers to estimate gene expression on a genome-wide scale and to understand the principles of gene regulation [3–6].

Since utilization of high-throughput sequencing technologies became more affordable and accessible, many research groups

Guojun Sheng (ed.), *Avian and Reptilian Developmental Biology: Methods and Protocols*, Methods in Molecular Biology, vol. 1650, DOI 10.1007/978-1-4939-7216-6_6, © Springer Science+Business Media LLC 2017

started their own genome projects for specific species in a similar way to ENCODE [7], combining different omics data to annotate a genome and identify its functional elements. For such purposes, genome-wide bioinformatics solutions are required.

Similar to RNA-Seq, CAGE is applicable for comparison of RNA expression in different samples, including tissues and primary cell cultures under normal conditions, or under an external stimulation, or within a time course [5, 8]. Among the transcriptomic methods, CAGE is a powerful technique that allows identification of transcription start sites (TSS); for example, it was used for building a mammalian promoter-level atlas based on numerous human and mouse samples [5]. CAGE was also applied in other model organisms like fruit fly (*D. melanogaster*) [9], zebra fish (*D. rerio*) [10, 11], and chicken (*G. gallus*) (under review) to improve gene annotation. CAGE not only indicates TSS positions at 1 bp resolution but also allows to estimate the expression level of the promoter regions, and it is therefore useful for gene regulation studies as it allows identification of specific transcription factor (TF) binding sites [12]. The CAGE library preparation protocols keep improving [12–14], and even commercial kits for end users are available now (https://cage-seq.com/cage_kit/index.html).

A standard protocol for CAGE library preparation is applicable to samples with abundant RNA (1–5 μg) [12, 13]. The original CAGE library preparation protocol consists of several enzymatic reactions, PCR amplification, and a crucial step of cap-trapping. This step includes oxidation of RNA, further biotinylation of the oxidized ends, and capture on streptavidin-coated magnetic beads [15]. For nanoCAGE technology [14], which is a variation of CAGE for handling low RNA amount samples, cap trapper is replaced by template switching, based on specific riboguanosine-tailed adapters that ligate to capped RNA molecules only, and thereby allows library preparation from small amounts of the initial material (about 1000 cells). There is an automated version of the standard protocol, which is based on solid-phase reversible immobilization (SPRI) and AMPure purification [13]. The CAGE method was adapted for the most popular second-generation sequencers like Illumina, 454, SOLiD, and HeliScope. Usually a library of 4.0 pMoles/l is required for Illumina Hi-Seq 2000 (50 cycles protocol) [12] and ~80 pM for HeliScope [16].

A general overview of CAGE data analysis in the FANTOM project [17] could be useful for further studies; however, it is necessary to perform the computational steps with a clear understanding of the basic features of the CAGE approach, which is different from common RNA-Seq. This is why we decided to describe the methodology based on recent results obtained from an analysis of chicken embryo development.

2 Materials

2.1 Sample Collection

In our reference study, based generally on a chicken embryo development time course, samples were collected from fertilized hens' eggs (Shiroyama Farm, Kanagawa, Japan). Eggs were incubated at 38.5 °C, and then total RNA was isolated by using a standard TRIzol-based method [18] at several developmental stages. Samples from 1.5 h to 5 days of development (Hamburger and Hamilton stages HH1 ~ HH26) were collected from both embryonic and extraembryonic tissues. RNA samples corresponding to 7, 10, and 20 days (HH32, HH37, and HH45) were isolated only from the embryonic tissue. The extraembryonic sample consisted of amnion, chorion, allantois, and yolk sac, collected from 7-day (HH32) embryos. Fore- and hind limb bud samples were isolated from 3-day embryos (HH20) and separately dissected in cold PBS before RNA extraction. See Fig. 1 for the workflow overview.

2.2 Library Preparation Reagents and Equipment

Commercial reagents were used for total RNA extraction and cap-trapping: TRIzol reagent (Invitrogen), RNase-free glycogen (Invitrogen), SuperScript III Reverse Transcriptase (Invitrogen), Agencourt RNAClean XP (Beckman Coulter), Biotin (Long Arm) Hydrazide (Vector Lab), MPG Streptavidin (Takara Bio Inc.), RNase I (Promega), RNase H (Invitrogen), and OliGreen

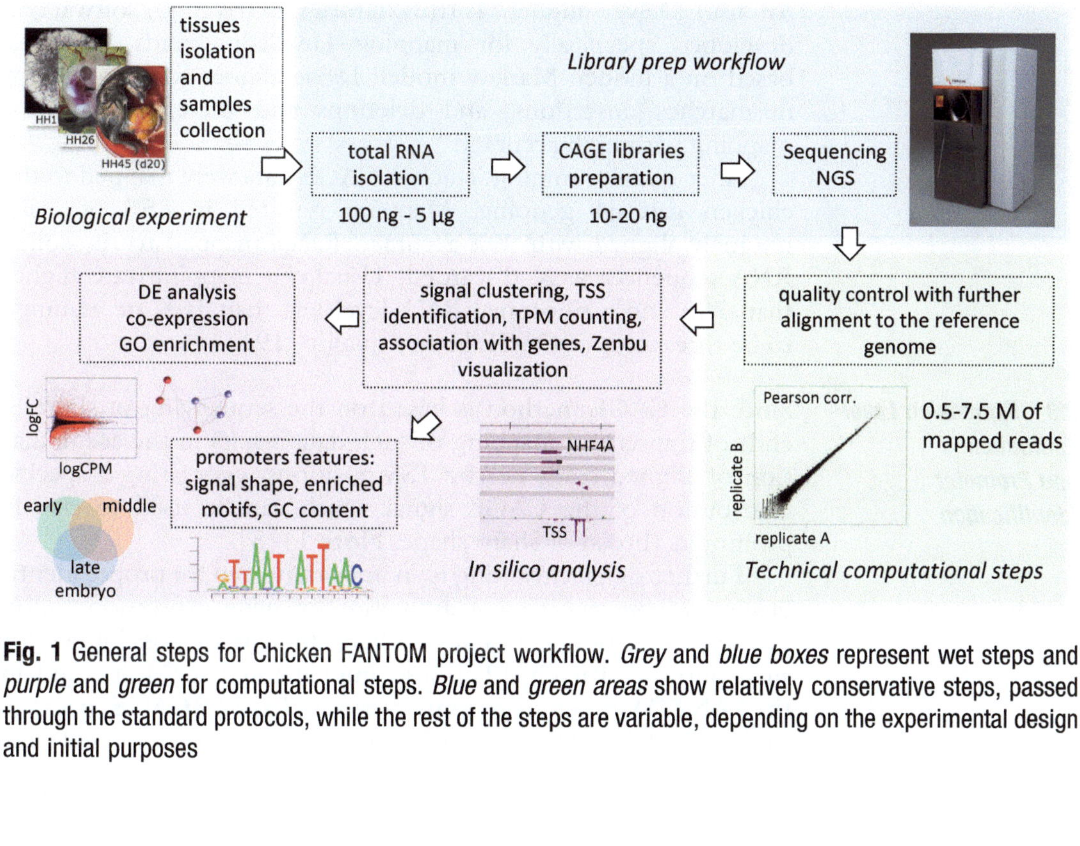

Fig. 1 General steps for Chicken FANTOM project workflow. *Grey* and *blue boxes* represent wet steps and *purple* and *green* for computational steps. *Blue* and *green areas* show relatively conservative steps, passed through the standard protocols, while the rest of the steps are variable, depending on the experimental design and initial purposes

ssDNA Quantitation kit (Invitrogen). The quality of the extracted RNA was tested on the Agilent 2100 Bioanalyzer.

3 Methods

3.1 CAGE Library Preparation and Sequencing

In case of the chicken CAGE project, each sample consisted of 5 μg of initial total RNA. Library preparation, including cap-trapping step and polyadenylation, followed a standard HeliScopeCAGE protocol [16]. The advantage of using such method is that PCR amplification step is omitted, excluding its side effects. In the broader FANTOM5 project, an automatized protocol for HeliScopeCAGE was also adopted [13]. Sequencing was performed on the HeliScope single molecule sequencer following the manufacturer instructions.

3.2 Mapping

Several standard pipelines for convenient CAGE data processing were developed at RIKEN DGT, all available via Moirai [19], a stand-alone software solution with a web-based interface that allows the design and execution of processing workflows. Moirai workflows mainly employ BWA aligner [20] for mapping the sequenced reads to the reference genome (mainly Illumina-sequenced reads). The pipelines also include quality control steps, and a visualization of the processes is possible via an html page [19].

The approach taken to process the FANTOM chicken data was to use Delve aligner (http://fantom.gsc.riken.jp/software/), developed specifically for mapping HeliScope reads, which is based on a hidden Markov model; Delve aligner can account for mismatches, insertions, and deletions and therefore performs mapping with high accuracy.

After initial trimming, chicken CAGE data were mapped to the chicken galGal4 genome. Mapping resulted in 85% of reads uniquely aligned with 99% accuracy. Reads mapped to ribosomal RNA sequences were discarded. Usually, a mapping rate higher than 70% and a ribosomal RNA level less than 10% are assumed to be thresholds of CAGE library quality [19].

3.3 Expression Level Estimation and Promoter Identification

Since the CAGE method is based on the sequencing of short 5′ ends of transcripts, mapping of such data results in the accumulation of aligned reads nearby TSS positions, generating a specific distribution of the CAGE signal which can be used to classify promoters (broad or sharp shape, **Note 1**) [5].

Further signal clustering is an important step for proper identification of promoters and estimation of gene expression (upon further associations to genes). There are several possibilities for such purpose: Paraclu [21], also implemented in Moirai and in the FANTOM5 Zenbu genome browser [19, 22]; CAGEr package for R [23]; RECLU pipeline [24]; and "decomposition-based peak identification" (DPI) that defines genomic coordinates of promoters, seen as clusters of CAGE read signal, based on "permissive" or "robust" TSS expression level thresholds on a set of samples [5]. For the chicken development data, the DPI method was applied in the same manner as for the FANTOM5 mammalian expression atlas [5]. A set of 24,131 robust CAGE clusters was defined for the chicken data set. For each cluster, the genomic position of the highest expressed TSS and the total number of reads are also reported. This information could be used, for example, for subsequent differential expression (DE) analysis or alternative promoter search, in a similar way to what has been done in the case of zebra fish [10, 11].

3.4 Association to Genes

Several analyses, such as DE analysis as mentioned before, may require gene-promoter associations. A common way to connect promoters to genes, which is also the strategy used to annotate the CAGE promoters in chicken, is based on intersection of the 5′ regions of genes, defined as an area ±500 bp from the gene start site, with the CAGE clusters [22]. Bedtools [25] is a convenient suite of programs for that purpose (*see* **Note 2**). In case a CAGE cluster overlaps multiple genes, the association to the closest gene is kept. The expected ratio of gene-associated promoters is about 50% or higher, but it varies across species as it depends on the level of annotation and characterization of their genomes. To increase the annotation rate, we used both RefSeq and Ensembl transcript annotations (*see* **Note 3**).

Based on the association rule described above, there can be a one-to-many relationship between genes and CAGE clusters, like in the case of gene isoforms. Since multiple CAGE clusters associated to a gene can be seen as an alternative promoter for the same gene (Fig. 2), the expression of that particular gene can be obtained by summing the read counts of its associated CAGE clusters. This allows gene-level DE and gene ontology (GO) enrichment estimation. Both analyses can be performed by edgeR package in R [26], for example.

3.5 Zenbu Genome Browser Navigation

Zenbu is a visualization tool developed specifically for FANTOM5 CAGE data browsing [22]. It is publicly available but has options for creating new, secured collaborative projects, like "FANTOM Chicken" where collaboration members can safely share genomic data. This browser allows upload of annotation files in BED, GFF, or OSC table (tabular file designed within FANTOM) formats but also genome alignment data such as BAM files for CAGE, RNA-Seq, and ChIP-seq experiments (*see* **Note 4**). Zenbu provides numerous predefined scripts for expression normalization, filtering, peak calling, and other basic processing and is therefore applicable for computational use as well as for visualization. The FANTOM Chicken collaboration contains a preconfigured genomic view, FANTOM Chicken

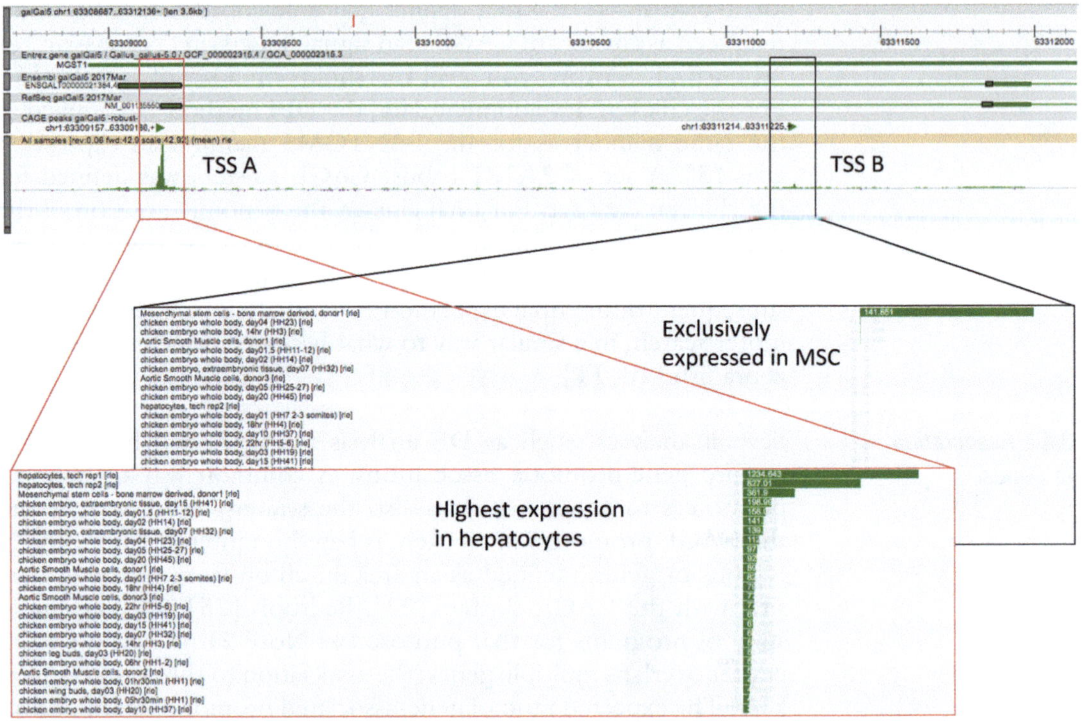

Fig. 2 Zenbu view on the promoter region of MGST1 (microsomal glutathione S-transferase 1). TSS A has hepatocyte-specific expression profile, while B is an alternative promoter with mesenchymal stem cell-related expression. CAGE signal is log scaled on the track. Promoter-related TPM counts are shown in the *boxes*

2016 (http://fantom.gsc.riken.jp/zenbu/gLyphs/#config=5CEx7HKVI5grfVGBjhMyzD;loc=galGal4::chr2:139,315,968..139322680+), displaying several tracks, including RefSeq and Ensembl transcripts, and CAGE promoter sets. Two tracks show birds and mammals related highly conserved elements (bird HCEs and AS-HCEs, respectively). These regions overlap 7.5% of the avian genome and are supposed to have functional constraints due to their conservation; such tracks could be useful for promoter-HCE cross-linking, since the majority of HCEs are located in intergenic regions and participate in the regulation of gene expression [27]. The uploaded mapped reads can be immediately used to visualize expression levels, either in a pooled (several experiments in one track) or single experiment fashion, with or without processing. The FANTOM Chicken 2016 view also displays four tracks representing CAGE data for primary cell cultures (aortic smooth muscle cells, hepatocytes, and mesenchymal stem cells), embryo development stages from HH1 to HH45 (1.5 h to 20 days), embryo development stages with additional extraembryonic and leg and wing bud samples, and pool of all chicken CAGE samples.

Another simple possible way for CAGE data visualization is by IGV browser [28]. It is not as convenient as Zenbu, but it has integrated tools like bedtools and therefore could be useful for non-computational researchers.

4 Potential Use for Avian/Reptilian Species

The Avian Phylogenomics Project already contains 48 assembled avian genomes available for downloading [29], and therefore they could be further annotated by using CAGE. Moreover, the Genome 10K Project includes a large number of avian and reptilian species (about 5000 and 3000, respectively), which are able for further studies [1, 2]. Genome projects like these can be interesting in the context of cross-species comparisons to reveal the molecular mechanisms responsible for determination of bird-specific traits such as vocal learning, flight adaptations, feathering, and other genomic features. Parthenogenesis of some reptile species, like lizards and snakes, makes these organisms a good model for further evolutionary studies regarding regeneration, sex determination processes, or effects of asexual reproduction [2]. CAGE methods could help to define new genes involved in reproduction or temperature-dependent regulatory networks, for example, or to explain some species-specific traits like adhesive toe pads, flying or gliding, tongue shape, or poisoning.

In summary, the Reptilian Transcriptomes database [30] and the Avian Phylogenomics Project could be further refined by integration with CAGE data, in order to improve gene annotations, to accurately define TSSs and enhancers [8, 31], and to improve our understanding of trait evolution.

The first CAGE chicken data produced within the FANTOM5 project revealed key regulatory genes at different developmental stages and their related pathways. Possible future analysis may involve studying the response to temperature changes during development.

5 Notes

1. Depending on the clustering method, different approaches for promoter classification are available. For interquantile-based methods, promoter classification is based on the length: <10 bp are sharp, while the rest are wide [5, 10]. Another simple way is based on the shape index (SI) calculation [9]. The advantage of this last one is that it is applicable to custom promoter regions, for example, ±100 bp from the TSS.
2. Generally, CAGE promoter regions are allocated within ±500 bp of a gene TSS, as noted before; however, these ranges are not strict and can be adapted to the transcripts' nature. For example, in the case of short transcripts, there is a risk to associate them with a downstream promoter that is actually related to another gene. In that case, in TSS, the window should be reduced. On the other side, in case of long genes,

some intragenic or intronic alternative promoters could be missed as they fall outside of the assigned window.

3. Some convenient tools for transcripts retrieval are BiomaRt by Ensembl (http://www.ensembl.org/biomart/), which is also available as a package for R, and the Table Browser tool in UCSC Genome Browser (https://genome.ucsc.edu/). Genscan, Augustus (also from UCSC Genome Browser), or other predicted gene sets can also be useful to try and improve promoter coverage, if available.

4. At this moment, Zenbu contains galGal3 and galGal4 assemblies for chicken. Genomes of several other model organisms are also available. For additional genome upload, it is necessary to contact RIKEN DGT [22].

Acknowledgments

This work was supported by the Russian Science Foundation, grant 14-44-00022.

References

1. Genome 10K Community of Scientists (2009) Genome 10K: a proposal to obtain whole-genome sequence for 10,000 vertebrate species. J Hered 100:659–674. doi:10.1093/jhered/esp086
2. Koepfli KP, Paten B et al (2015) The Genome 10K Project: a way forward. Annu Rev Anim Biosci 3:57–111. doi:10.1146/annurev-animal-090414-014900
3. Wu JQ, Rozowsky J, Zhang Z et al (2008) Systematic analysis of transcribed loci in ENCODE regions using RACE sequencing reveals extensive transcription in the human genome. Genome Biol. doi:10.1186/gb-2008-9-1-r3
4. Ozsolak F, Milos PM (2011) RNA sequencing: advances, challenges and opportunities. Nat Rev Genet 12:87–98. doi:10.1038/nrg2934
5. Forrest ARR, Kawaji H, Rehli M et al (2014) A promoter-level mammalian expression atlas. Nature 507:462–470. doi:10.1038/nature13182
6. Morrissy S, Zhao Y, Delaney A et al (2010) Digital gene expression by tag sequencing on the illumina genome analyzer. Curr Protoc Hum Genet 11(11):1–36. doi:10.1002/0471142905.hg1111s65
7. ENCODE Project Consortium (2004) The ENCODE (ENCyclopedia Of DNA Elements) Project. Science 306:636–640. doi:10.1126/science.1105136
8. Arner E, Daub CO, Vitting-Seerup K et al (2015) Transcribed enhancers lead waves of coordinated transcription in transitioning mammalian cells. Science 347:1010–1014. doi:10.1126/science.1259418
9. Hoskins RA, Landolin JM, Brown JB et al (2011) Genome-wide analysis of promoter architecture in *Drosophila melanogaster*. Genome Res 21:182–192. doi:10.1101/gr.112466.110
10. Nepal C, Hadzhiev Y, Previti C et al (2013) Dynamic regulation of the transcription initiation landscape at single nucleotide resolution during vertebrate embryogenesis. Genome Res 23:1938–1950. doi:10.1101/gr.153692.112
11. Haberle V, Li N, Hadzhiev Y et al (2014) Two independent transcription initiation codes overlap on vertebrate core promoters. Nature 507:381–385. doi:10.1038/nature12974
12. Takahashi H, Lassmann T, Murata M, Carninci P (2012) 5′ end-centered expression profiling using cap-analysis gene expression and next-generation sequencing. Nat Protoc 7:542–561. doi:10.1038/nprot.2012.005
13. Itoh M, Kojima M, Nagao-Sato S et al (2012) Automated workflow for preparation of cDNA for cap analysis of gene expression on a single

molecule sequencer. PLoS One. doi:10.1371/journal.pone.0030809
14. Salimullah M, Sakai M, Plessy C, Carninci P (2011) NanoCAGE: a high-resolution technique to discover and interrogate cell transcriptomes. Cold Spring Harb Protoc:pdb.prot5559. doi:10.1101/pdb.prot5559
15. Carninci P, Kvam C, Kitamura A et al (1996) High-efficiency full-length cDNA cloning by biotinylated CAP trapper. Genomics 37:327–336
16. Kanamori-Katayama M, Itoh M, Kawaji H et al (2011) Unamplified cap analysis of gene expression on a single-molecule sequencer. Genome Res 21:1150–1159. doi:10.1101/gr.115469.110
17. Lizio M, Harshbarger J, Shimoji H et al (2015) Gateways to the FANTOM5 promoter level mammalian expression atlas. Genome Biol 16:22. doi:10.1186/s13059-014-0560-6
18. Rio DC, Ares M Jr, Hannon GJ, Nilsen TW (2010) Purification of RNA using TRIzol (TRI reagent). Cold spring Harb Protoc:pdb.prot5439. doi:10.1101/pdb.prot5439
19. Hasegawa A, Daub C, Carninci P, Hayashizaki Y, Lassmann T (2014) MOIRAI: a compact workflow system for CAGE analysis. BMC Bioinformatics 15:144. doi:10.1186/1471-2105-15-144
20. Li H, Durbin R (2010) Fast and accurate long-read alignment with burrows-wheeler transform. Bioinformatics 26:589–595. doi:10.1093/bioinformatics/btp698
21. Frith MC, Valen E, Krogh A et al (2008) A code for transcription initiation in mammalian genomes. Genome Res 18:1–12. doi:10.1101/gr.6831208
22. Severin J, Lizio M, Harshbarger J et al (2014) Interactive visualization and analysis of large-scale sequencing datasets using ZENBU. Nat Biotechnol 32:217–219. doi:10.1038/nbt.2840
23. Haberle V, Forrest ARR, Hayashizaki Y, Carninci P, Lenhard B (2015) CAGEr: precise TSS data retrieval and high-resolution promoterome mining for integrative analyses. Nucleic Acids Res. doi:10.1093/nar/gkv054
24. Ohmiya H, Vitezic M, Frith MC et al (2014) RECLU: a pipeline to discover reproducible transcriptional start sites and their alternative regulation using capped analysis of gene expression (CAGE). BMC Genomics 15:269. doi:10.1186/1471-2164-15-269
25. Quinlan AR, Hall IM (2010) BEDTools: a flexible suite of utilities for comparing genomic features. Bioinformatics 26:841–842. doi:10.1093/bioinformatics/btq033
26. Robinson MD, McCarthy DJ, Smyth GK (2010) edgeR: a Bioconductor package for differential expression analysis of digital gene expression data. Bioinformatics 26:139–140. doi:10.1093/bioinformatics/btp616
27. Zhang G, Li C, Li Q et al (2014) Comparative genomics reveals insights into avian genome evolution and adaptation. Science 346:1311–1320. doi:10.1126/science.1251385
28. Thorvaldsdóttir H, Robinson JT, Mesirov JP (2013) Integrative Genomics Viewer (IGV): high-performance genomics data visualization and exploration. Brief Bioinform 14:178–192. doi:10.1093/bib/bbs017
29. Zhang G, Li B, Li C et al (2014) Comparative genomic data of the avian Phylogenomics project. Gigascience 3:26. doi:10.1186/2047-217X-3-26
30. Tzika AC, Ullate-Agote A, Grbic D, Milinkovitch MC (2015) Reptilian Transcriptomes v2.0: an extensive resource for Sauropsida genomics and Transcriptomics. Genome Biol Evol 7:1827–1841. doi:10.1093/gbe/evv106
31. Andersson R, Gebhard C, Miguel-Escalada I et al (2014) An atlas of active enhancers across human cell types and tissues. Nature 507:455–461. doi:10.1038/nature12787

Part II

Genetic Manipulation

Chapter 7

CRISPR/Cas9 in the Chicken Embryo

Valérie Morin, Nadège Véron, and Christophe Marcelle

Abstract

Genome editing is driving a revolution in the biomedical sciences that carries the promise for future treatments of genetic diseases. The CRISPR/Cas9 system of RNA-guided genome editing has been successfully applied to modify the genome of a wide spectrum of organisms. We recently showed that this technique can be combined with in vivo electroporation to inhibit the function of genes of interest in somatic cells of the developing chicken embryo. We present here a simplified version of the previously described technique that leads to effective gene loss-of-function.

Key words Chicken embryo, Electroporation, CRISPR, Cas9, gRNA

1 Introduction

Recent advances in the targeted modification of complex eukaryotic genomes using endonucleases have unlocked a new era of genome engineering. Early genome-editing efforts using zinc-finger nucleases (ZFNs [1, 2]) and transcription activator-like effector nucleases (TALEN [3, 4]) were complex, because each new target site required the design and construction of a site-specific nuclease. The recent development of the highly accessible CRISPR (clustered, regularly interspaced palindromic repeats)/Cas methodologies [5–8] considerably facilitated genome-editing since the specificity of the technique is provided by a "guide RNA" that directs the Cas endonuclease to its target DNA sequence. A modest limitation to the choice of the target sequence is that it must contain a "protospacer-adjacent motif" (PAM), and a short DNA sequence that is required for compatibility with the Cas protein being used (NGG in the case of the most widely used Cas9 from *Streptococcus pyogenes*). As a result of constant refinements of the CRISPR technology and the discovery of exciting novel tools and applications for this technique, we now possess an unprecedented ability to analyze biological processes using sophisticated designer genetic tools [9].

Guojun Sheng (ed.), *Avian and Reptilian Developmental Biology: Methods and Protocols*, Methods in Molecular Biology, vol. 1650, DOI 10.1007/978-1-4939-7216-6_7, © Springer Science+Business Media LLC 2017

CRISPR-mediated gene editing is widely used to generate loss-of-function in embryos and Primordial Germ Cells (PGCs) [10, 11], and this technology has also been used in mice to perform in vivo genome editing of somatic cells [12–15]. This approach creates genetic mutations in a subset of cells within a wild-type background, a technology that was used extensively in the *Drosophila* field to study complex biological processes [16]. The electroporation technique, extensively used in the chicken embryo [17–22], also results in the mosaic expression of constructs, which, combined to loss-of-function approaches, can generate somatic clones of mutant cells, thus providing the possibility to analyze the (cell-autonomous and non-autonomous) function of genes of interest during embryonic development. However, gene inactivation in the chicken has been limited to knockdown by RNA interference- and morpholino-based methodologies [22–28] that each have their own limitations, including variability in the level of knockdown, off target effects, and transient inhibition of transcripts.

We have previously shown that the in vivo electroporation technique can be used to achieve efficient CRISPR-mediated genome editing in a subset of cells of the chicken embryo [29]. We now present a significantly simpler version of the vectors we utilized and a detailed description of the various steps that should be followed to achieve efficient genome editing in the embryo.

2 Materials

2.1 Buffers

gRNA oligonucleotide hybridization buffer:

Tris–HCl pH 8.8	10 mM
NaCl	100 mM

2.2 Reagents and Kits

2.2.1 PCR Purification Columns

We use NucleoSpin Gel and PCR cleanup columns from Macherey-Nagel, but similar products can be purchased from Qiagen (e.g., QIAquick).

2.2.2 Genomic DNA Purification Columns

We use NucleoSpin Tissue purification columns to extract genomic DNA. Similar product (e.g., DNeasy kit) can be found at Qiagen.

2.2.3 Restriction and DNA Modification Enzymes, Taq Polymerase, Etc.

All enzymes (with their cognate buffer) are from New England Biolabs (NEB), but equivalent products can be found from a variety of companies.

2.2.4 DH5α Competent Bacteria

Chemically competent bacteria (NEB 5-alpha Competent *E. coli*) are purchased from NEB. Transformation is made according to manufacturer's instructions (details are given below).

2.2.5 Cell Transfection Reagent

We use Lipofectamine LTX transfection reagent from Invitrogen. The quantities of transfected plasmid for a successful transfection of DF1 fibroblasts are indicated below.

2.2.6 Chicken Fibroblast Cell Line UMNSAH/DF1

This is an immortalized fibroblast cell line that can be purchased from ATCC CRL 12203. They are grown in DMEM, supplemented with 10% Fetal Calf Serum (FCS), and 1× Penicillin Streptomycin (PS) at 37 °C in CO_2 incubator. Once established, DF1 cells doubling time is surprisingly short.

2.2.7 PCR Cloning Vector

We use the pGEM-T Easy PCR vector from Promega. Note that Taq polymerase, which leaves -A overhangs, should be used for the PCR amplification.

2.2.8 Opti-MEM

Opti-MEM reduced serum media can be purchased from a variety of sources (e.g., ThermoFisher Scientific).

3 Methods

3.1 CRISPR-Mediated Gene-Targeting Vectors

We designed an all-in-one CRISPR Cas9 vector, named 2gRNA-Cas9-GFPcx (*see* Fig. 1a), specifically adapted to in vivo electroporation of chicken embryos. Compared to the system we previously published [29], the major difference is that it is not inducible, but it is considerably more simple, since it contains in one plasmid the features that were previously distributed over four vectors. The main characteristics of 2gRNA-Cas9-GFPcx are the following: (1) the mammalian codon-optimized, nuclear-localized Cas9 (from Addgene # 480138 (30)) is driven by an ubiquitous CAGGS (CMV enhancer/chick beta actin promoter) promoter; (2) a self-cleaving P2A peptide directs to the co-expression of a membrane-localized eGFP (containing the prenylation CAAX sequence from H-Ras) together with the Cas9. The GFP fluorescence allows the detection and analyses of electroporated cells within the targeted tissue; (3) a cassette containing the human U6 and H1 RNA Pol III promoters allows the simultaneous expression of two gRNAs; (4) the construct is flanked by sequences for the Tol2 transposable elements, which permits the permanent integration of the intervening sequence in the presence of transposase, provided by a separate expression plasmid (CAGGS transposase, Fig. 1b).

3.2 Design of gRNAs

To maximize cleavage of the desired target sequence, we followed a strategy that we tested before [29] which consists in using two gRNAs per targeted gene. This was shown to be an efficient way to generate genomic deletions [30–32]. We typically target gene sequences located on two adjacent exons. This leads to the deletion of the intervening segment (of varying length depending on the gene of interest) by the introduction of two Double Stranded

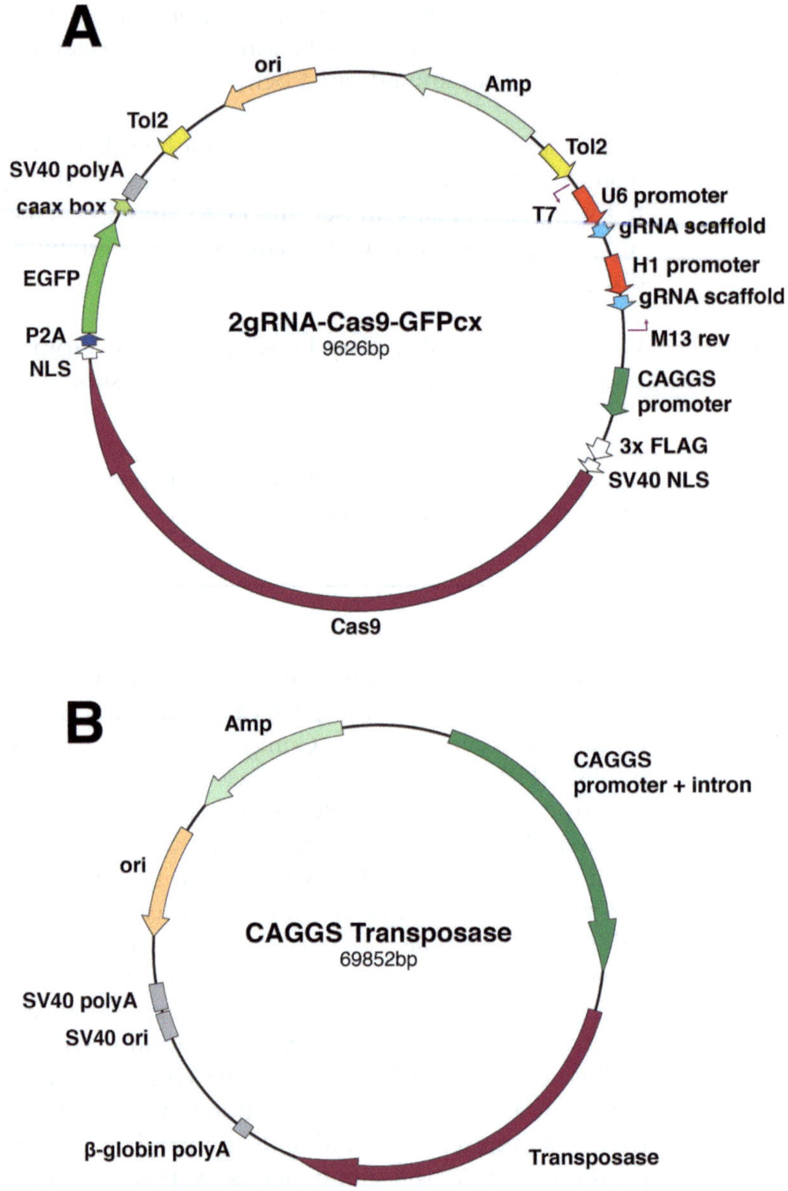

Fig. 1 Plasmids used in this study. (**a**) The main characteristics of 2gRNA-Cas9-GFPcx are that the Cas9 nuclease is placed under the CAGGS ubiquitous promoter, while two gRNA cloning sites are located under U6 and H1 promoters. Other features are indicated. (**b**) CAGGS transposase encodes the coding sequence for the *Tol2* transposase. It is co-electroporated with 2gRNA-Cas9-GFPcx

Breaks (DSB) followed by repair via non-homologous end joining (NHEJ). The position of the targeted sequences is chosen so that the resulting transcript is inactive. Thus, it is crucial to remove important regulatory domains that may be located close to the N-terminus of the protein. Of note, we have found that choosing two gRNAs that sit on opposite DNA strands increases the efficiency of the deletion substantially, as opposed to having both gRNAs on the same strand. Another important parameter that should be taken into account when choosing the gRNA target sequences is the method by which the gene editing will be evaluated. If antibodies against the gene product are not available, it is recommended that the deletion is sufficiently large to allow the design of an mRNA probe directed against the deleted region, which may be utilized for in situ hybridization (ISH). Since in our hands, probes of 300–400 bp result in an ISH signal of good quality, this should be the minimal size of the deletion at the mRNA level. In the case that antibodies are available, the size of the deletion is less critical, since immuno-fluorescence of electroporated tissues can be used to determine the efficiency of the procedure.

The design of guide RNAs (gRNAs) is made using the protocol of [33]. Target sequences (containing the 5′-NGG PAM sequence) are determined using the online website: http://crispr.mit.edu/. Note: this software shows the PAM sequence attached to the 20 bp long annealing sequence. The PAM sequence should not be included in the gRNA sequences. For the purification of oligonucleotides, *see* **Note 1**.

3.2.1 gRNA#1

The gRNA1 is inserted between two BbsI sites located immediately upstream of the first gRNA scaffold (*see* Fig. 2a). The digestion of the two sites creates an asymmetric digestion of the plasmid that allows directional cloning of the gRNA1.

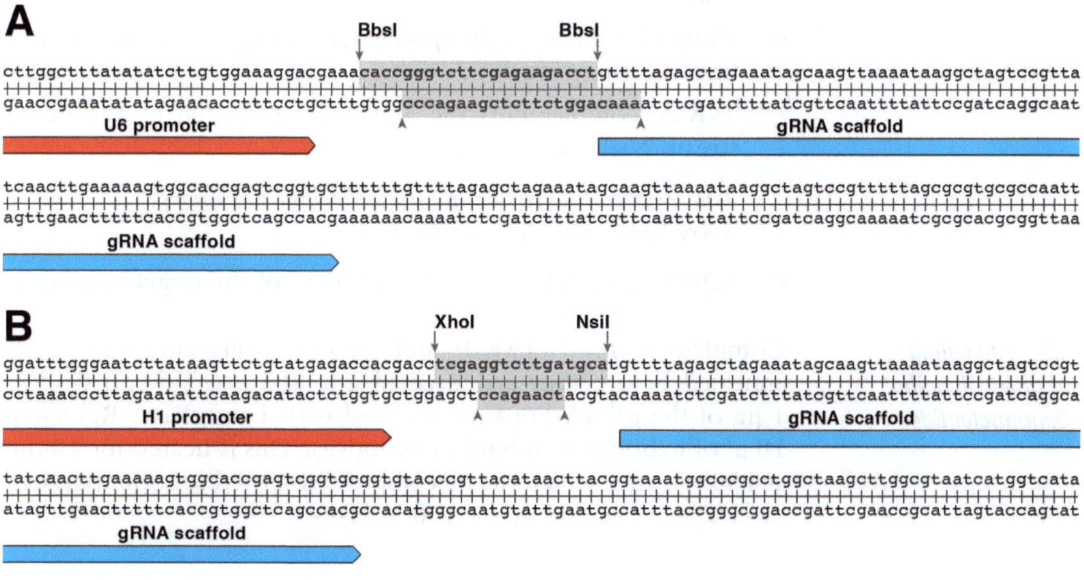

Fig. 2 Cloning sites for the guide RNAs 1 and 2. (**a**) It shows the two BbsI cloning sites upstream of the gRNA 1 scaffold (*blue box*). The *shaded area* indicates the plasmid fragment deleted by BbsI digestion. (**b**) It shows the XhoI and NsiI cloning sites upstream of the gRNA 2 scaffold (*blue box*). The *shaded area* indicates the plasmid fragment deleted by XhoI and NsiI digestions

The forward oligonucleotide is synthetized as follows:
5′ CACC-(G)-(20 bp gRNA1) 3′, where:

- CACC: complements the 5′ overhang created by the upstream BbsI site.
- (G): U6 RNA Pol III requires one G to initiate transcription [34]. If it is not present in the gRNA sequence, it is added at this position.
- 20 bp gRNA1: is the 20 bp-long target sequence determined by http://crispr.mit.edu/.

The reverse oligonucleotide is synthetized as follows:
5′ AAAC-(rc20 bp gRNA1)-(C) 3′, where:

- AAAC: complements the 5′ overhang created by the downstream BbsI site.
- rc20 bp gRNA1: reverse-complement of the target sequences.
- (C): added at this position if a G was added in the forward oligonucleotide.

3.2.2 gRNA#2

The gRNA2 is inserted between a XhoI and a NsiI sites located immediately upstream of the second gRNA scaffold (*see* Fig. 2b). The digestion of the sites allows directional cloning of the gRNA2.

The forward oligonucleotide is synthetized as follows:
5′ TCGA-(20 bp gRNA2)-TGCA 3′, where:

- TCGA: complements the 5′ overhang created by the upstream XhoI site.
- 20 bp gRNA2 is the 20 bp long target sequence determined by http://crispr.mit.edu/.
- TGCA: complements the 3′ overhang created by the downstream NsiI site.

The reverse oligonucleotide is synthetized as follows:
5′ (rc20 bp gRNA2) 3′, where:

- rc20 bp gRNA2: reverse-complement of the target sequences.

3.3 Annealing of gRNA-Specific Oligonucleotides

Complementary (forward and reverse) oligonucleotides are annealed before they are inserted into the plasmid. To do this, 1 μg of the gRNA-Forward is mixed with 1 μg gRNA-Reverse in 18 μl of hybridization buffer (*see* above). This is heated for 10 min at 100 °C, after which the mix is slowly returned to room temperature. This is then diluted 1:100 in H_2O.

3.4 Restriction Enzyme Digestion of the 2gRNA-Cas9-GFPcx Plasmid by BbsI (gRNA1) or by XhoI and NsiI (gRNA2)

The insertion of the annealed oligonucleotides into the Cas9 vector is done by classical restriction enzyme digestion and ligation. The cloning procedure to insert the gRNAs can be started by either the gRNA1 or the gRNA2. Restriction enzymes (10 μ/μl) are from NEB. 5 μg of 2gRNA-Cas9-GFPcx plasmid is mixed with 5 μl of (10×) restriction enzyme buffer (buffer 2.1 for BbsI or CutSmart for XhoI and NsiI). Water is added to bring the total volume to 47.5 μl. 2.5 μl of BbsI, XhoI, or NsiI are then added and this is incubated at 37 °C for 2 h. The digestion of the plasmid with XhoI and NsiI should be done sequentially (*see* **Note 2**).

3.5 Purification of Digested Plasmid

The purification of the digested plasmid is performed on purification columns (NucleoSpin Gel and PCR cleanup, Macherey-Nagel). The elution is made in 15 μl H_2O. The plasmid DNA concentration is verified with NanoDrop.

3.6 Ligation of the gRNA with the Cut Plasmid

(*See* **Note 3**) 10 ng of cut plasmid (from Subheading 3.5) are mixed with 1 μl of annealed gRNAs (from Subheading 3.3); 2 μl of T4 ligase buffer (NEB) are added and water is added to total 19 μl. 1 μl of T4 ligase (NEB) is then added and this is incubated at 16 °C overnight.

3.7 Transformation in DH5α Competent Bacteria

The ligation product is used to transform chemically competent bacteria. 5 μl of the ligation (from Subheading 3.6) are mixed with 50 μl of NEB 5-alpha competent bacteria and incubated 30 min on ice. They are placed for a 30″ heat shock at 42 °C, then back on ice. 250 μl of SOC bacterial growth medium are added and placed for 1 h at 37 °C, shaking. The bacteria are plated on LB-agar plates supplemented with ampicillin and incubated overnight at 37 °C.

3.8 Screening of Bacterial Colonies

The screening of bacteria colonies is done by PCR using gRNA [1, 2, or] primers together with external primers present in the plasmid (e.g., M13 reverse or T7, *see* Fig. 1). The positive colonies are grown and plasmid DNA is extracted and sequenced. The positive clones are then used for the second round of gRNA cloning (back to Subheading 3.4) and processed as above until both gRNA are inserted, and the final plasmid is verified by sequencing.

3.9 Testing CRISPR on a Chicken Fibroblast Cell Line

We routinely use transfection of a chicken fibroblast cell line (UMNSAH/DF1), followed by the sequencing of its genomic DNA to test that gene editing is taking place. This procedure is considerably faster and more reproducible than the technique we previously used [29], which consisted in sequencing the genomic DNA of electroporated chick embryo tissues.

Day 1

- Plate DF1 cells at about 40% confluency in 6-well plates in DMEM supplemented with 10% FCS + PS. They should be 80% confluent the next day for transfection.

Day 2

- Prepare Mix A: Mix 150 μl of Opti-MEM in an Eppendorf tube with 1 μg of CAGGS transposase and 4 μg of 2gRNA-Cas9-GFPcx, then add 2 μl PLUS reagent. Mix and leave 10′ at RT. During that time, change the cell medium to DMEM (no PS, no FCS).
- Prepare Mix B: Mix 150 μl of Opti-MEM with 6 μl Lipofectamine LTX. Transfer Mix A into Mix B, leave 10′ at RT.
- Add this to the cells. Place them back into the CO_2 incubator for at least 4 h. Then replace the medium by DMEM +10% FCS + PS.

Day 3

- Verify the efficiency of the transfection on an inverted fluorescence microscope. Trypsinize and transfer the cells to a 10 cm tissue culture plate.

Day 4

- Trypsinize cells, collect them in a 15 ml tube. Centrifugation of 10′ at 1000 rpm. Resuspend the cell pellet in 200 μl of buffer T1 (kit Macherey-Nagel NucleoSpin Tissue). Extract genomic DNA according to manufacturer's instructions. Verify DNA concentration with NanoDrop, dilute to 100 ng/μl.

3.10 PCR and Sequencing

We verify that the deletion of the DNA fragment is effective by performing a "nested" PCR on the genomic DNA extracted from the transfected DF1 cells. Two sets of primers (one inner, one outer) are designed that encompass the deleted fragment. We previously observed important variations in the size of the nucleotide insertion or deletions (InDels) after NHEJ, in particular at the 3′ end, where it was not rare that hundreds of base pairs passing the downstream gRNA were deleted [29]. We confirm this observation with the novel CRISPR system we designed: deletions on the 5′ end are in the range of a 50–300 bp, while insertions are limited to 10–30 bp. Surprisingly, we noticed that deletions 3′ of the downstream gRNA could reach 3000 bp, while no insertions were observed. Therefore, to verify that the CRISPR-mediated deletion took place in transfected DF1 cells, we chose sets of primers that were located about more than 300 bp 5′ and up to 3 kb 3′ of the expected deletion sites. The outer primers we designed are 25 bp long, while the inner primers are 20 bp long. The outer set of primers is used to perform the first PCR reaction on 100 ng of genomic DNA (from Subheading 3.9 above), with Taq DNA polymerase (NEB), according to manufacturer's instructions (60 °C annealing temperature, 40 cycles). 0.25 μl of the first PCR reaction is then used for a second round of PCR with the inner set of primers (60 °C annealing temperature, 40 cycles). 1/10th of the PCR reaction is run on a gel to verify that a fragment has been obtained (*see* **Note 4**). The PCR product is purified on columns (NucleoSpin Gel and PCR cleanup, Macherey-Nagel) and the DNA concentration is determined with NanoDrop. The DNA is cloned into the pGEM-T Easy (Promega) PCR cloning vector according to manufacturer's instructions and used to transform DH5α competent bacteria as above (Subheading 3.7). Blue/white screening of colonies allows easy identification of positive (white) colonies. Minipreps of plasmid DNA are performed on about ten colonies, which are then sequenced.

4 Notes

1. The purification of the oligonucleotides is made by desalting.
2. Since XhoI & NsiI restriction sites are only a few bases apart, it is recommended to do sequential digestions with the first enzyme, column extraction (*see* **step 3** of Subheading 3), followed by the second digestion.
3. Since oligonucleotides are not phosphorylated, the digested plasmid is not dephosphorylated prior to ligation.
4. Since "InDels" are variable and sometimes quite extensive with the two gRNA approaches we are using, the size of the amplified DNA band that will be obtained after PCR on genomic DNA cannot be predicted. This can be quite surprising. If the deletion is not too large, the non-deleted fragment will be amplified as well.

Acknowledgments

We thank Dr. Daniel Sieiro for critical reading of the manuscript. This work was supported by grants from the National Health and Medical Research Council (NHMRC, Australia) to C.M. and N.V. and by the Programme Avenir Lyon Saint-Etienne (PALSE) from the University of Lyon to C.M and V.M. The Australian Regenerative Medicine Institute is supported by grants from the State Government of Victoria and the Australian Government. The NeuroMyoGene Institute is supported by grants from the Association Française contre les Myopathies (AFM).

References

1. Porteus MH, Carroll D (2005) Gene targeting using zinc finger nucleases. Nat Biotechnol 23:967–973. doi:10.1038/nbt1125
2. Bibikova M, Golic M, Golic KG, Carroll D (2002) Targeted chromosomal cleavage and mutagenesis in Drosophila using zinc-finger nucleases. Genetics 161:1169–1175
3. Li T, Huang S, Jiang WZ et al (2011) TAL nucleases (TALNs): hybrid proteins composed of TAL effectors and FokI DNA-cleavage

domain. Nucleic Acids Res 39:359–372. doi:10.1093/nar/gkq704
4. Miller JC, Tan S, Qiao G et al (2011) A TALE nuclease architecture for efficient genome editing. Nat Biotechnol 29:143–148. doi:10.1038/nbt.1755
5. Jinek M, Chylinski K, Fonfara I et al (2012) A programmable dual-RNA–guided DNA endonuclease in adaptive bacterial immunity. Science 337:816–821. doi:10.1126/science.1225829
6. Pennisi E (2013) The CRISPR craze. Science 341:833–836. doi:10.1126/science.341.6148.833
7. Barrangou R (2014) Cas9 targeting and the CRISPR revolution. Science 344:707–708. doi:10.1126/science.1252964
8. Wu X, Scott DA, Kriz AJ et al (2014) Genome-wide binding of the CRISPR endonuclease Cas9 in mammalian cells. Nat Biotechnol 32:670–676. doi:10.1038/nbt.2889
9. Komor AC, Badran AH, Liu DR (2017) CRISPR-based technologies for the manipulation of eukaryotic genomes. Cell 168(1-2):20–36. doi:10.1016/j.cell.2016.10.044
10. Hsu PD, Lander ES, Zhang F (2014) Development and applications of CRISPR-Cas9 for genome engineering. Cell 157:1262–1278. doi:10.1016/j.cell.2014.05.010
11. Doudna JA, Charpentier E (2014) The new frontier of genome engineering with CRISPR-Cas9. Science 346:1258096
12. Sánchez-Rivera FJ, Papagiannakopoulos T, Romero R et al (2014) Rapid modelling of cooperating genetic events in cancer through somatic genome editing. Nature 516:428–431. doi:10.1038/nature13906
13. Xue W, Chen S, Yin H et al (2014) CRISPR-mediated direct mutation of cancer genes in the mouse liver. Nature 514:380–384. doi:10.1038/nature13589
14. Yin H, Xue W, Chen S et al (2014) Genome editing with Cas9 in adult mice corrects a disease mutation and phenotype. Nat Biotechnol 32(6):551–553. doi:10.1038/nbt.2884
15. Wang S, Sengel C, Emerson MM, Cepko CL (2014) A Gene regulatory network controls the binary fate decision of rod and bipolar cells in the vertebrate retina. Dev Cell 30:513–527
16. Blair SS (2003) Genetic mosaic techniques for studying Drosophila development. Development 130:5065–5072. doi:10.1242/dev.00774
17. Scaal M, Gros J, Lesbros C, Marcelle C (2004) In ovo electroporation of avian somites. Dev Dyn 229:643–650. doi:10.1002/dvdy.10433
18. Itasaki N, Bel-Vialar S, Krumlauf R (1999) Nature citation. Nat Cell Biol 1:E203–E207. doi:10.1038/70231
19. Nakamura H, Funahashi J (2013) Electroporation: past, present and future. Develop Growth Differ 55:15–19. doi:10.1111/dgd.12012
20. Yokota Y, Saito D, Tadokoro R, Takahashi Y (2011) Genomically integrated transgenes are stably and conditionally expressed in neural crest cell-specific lineages. Dev Biol 353:382–395
21. Serralbo O, Picard CA, Marcelle C (2013) Long term, inducible gene loss-of-function in the chicken embryo. Genesis 51(5):372–380. doi:10.1002/dvg.22388
22. Voiculescu O, Papanayotou C, Stern CD (2008) Spatially and temporally controlled electroporation of early chick embryos. Nat Protoc 3:419–426. doi:10.1038/nprot.2008.10
23. Das RM, Van Hateren NJ, Howell GR et al (2006) A robust system for RNA interference in the chicken using a modified microRNA operon. Dev Biol 294:554–563. doi:10.1016/j.ydbio.2006.02.020
24. Hou X, Omi M, Harada H et al (2011) Conditional knockdown of target gene expression by tetracycline regulated transcription of double strand RNA. Develop Growth Differ 53:69–75. doi:10.1111/j.1440-169X.2010.01229.x
25. Gros J, Serralbo O, Marcelle C (2009) WNT11 acts as a directional cue to organize the elongation of early muscle fibres. Nature 457:589–593. doi:10.1038/nature07564
26. Rios AC, Serralbo O, Salgado D, Marcelle C (2011) Neural crest regulates myogenesis through the transient activation of NOTCH. Nature 473:532–535. doi:10.1038/nature09970
27. Serralbo O, Marcelle C (2014) Migrating cells mediate long-range WNT signaling. Development 141:2057–2063. doi:10.1242/dev.107656
28. Norris A, Streit A (2014) Morpholinos: studying gene function in the chick. Dev Biol 66:454–465. doi:10.1016/j.ymeth.2013.10.009
29. Véron N, Qu Z, Kipen PAS et al (2015) CRISPR mediated somatic cell genome engineering in the chicken. Dev Biol 407(1):68–74. doi:10.1016/j.ydbio.2015.08.007
30. Canver MC, Bauer DE, Dass A et al (2014) Characterization of genomic deletion efficiency mediated by clustered regularly interspaced palindromic repeats (CRISPR)/Cas9 nuclease system in mammalian cells. J Biol Chem 289:21312–21324

31. Ran FA, Hsu PD, Lin C-Y et al (2013) Double nicking by RNA-guided CRISPR Cas9 for enhanced genome editing specificity. Cell 154:1380–1389. doi:10.1016/j.cell.2013.08.021
32. Zhou J, Wang J, Shen B et al (2014) Dual sgRNAs facilitate CRISPR/Cas9-mediated mouse genome targeting. FEBS J 281:1717–1725. doi:10.1111/febs.12735
33. Ran FA, Hsu PD, Wright J et al (2013) Genome engineering using the CRISPR-Cas9 system. Nat Protoc 8:2281–2308. doi:10.1038/nprot.2013.143
34. Ranganathan V, Wahlin K, Maruotti J, Zack DJ (2014) Expansion of the CRISPR–Cas9 genome targeting space through the use of H1 promoter-expressed guide RNAs. Nat Commun 5:4516. doi:10.1038/ncomms5516

Chapter 8

Fluorescent Quail: A Transgenic Model System for the Dynamic Study of Avian Development

David Huss and Rusty Lansford

Abstract

Real-time four-dimensional (4D, *xyzt*) imaging of cultured avian embryos is an ideal method for investigating the complex movements of cells and tissues during early morphogenesis. While methods that transiently label cells, such as electroporation, are highly useful for dynamic imaging, they can also be limiting due to the number and type of cells that can be effectively targeted. In contrast, the heritable, stable, and long-term expression of a fluorescent protein driven by the exogenous promoter of a transgene overcomes these challenges. We have used lentiviral vectors to produce several novel transgenic quail lines that express fluorescent proteins either ubiquitously or in a cell-specific manner. These lines have proven to be useful models for dynamic imaging and analysis. Here, we provide detailed protocols for generating transgenic quail with the emphasis on producing high titer lentivirus, effectively introducing it into the early embryo and efficiently screening for G1 founder birds.

Key words Transgenic, Avian, Lentivirus, Dynamic imaging, Japanese quail

1 Introduction

Recent advances in confocal laser scanning microscopy have made real-time four-dimensional (4D, *xyzt*) imaging of ex-ovo cultured avian embryos an ideal method for investigating the complex movements of cells and tissues during early morphogenesis. Effectively labeling the cells or tissue of interest with fluorescent reporter molecules or dyes is critically important to the success of this technique. While methods that transiently label cells, such as plasmid electroporation, are highly useful for dynamic imaging, they can also be limiting due to the number and type of cells that can be effectively targeted. In contrast, the stable, long-term expression of a fluorescent protein (FP) driven by an exogenous promoter in a transgenic avian model system potentially allows for every cell and tissue in a living embryo to be imaged and analyzed.

The ability of HIV-1-based lentivirus to stably integrate a transgene into the genome of both dividing and nondividing cells

Guojun Sheng (ed.), *Avian and Reptilian Developmental Biology: Methods and Protocols*, Methods in Molecular Biology, vol. 1650, DOI 10.1007/978-1-4939-7216-6_8,

is well known. Unlike other classes of retroviruses, the transgenes integrated by lentiviral vectors are not typically susceptible to genetic silencing [1]. These characteristics made lentivirus the most promising vehicle for producing the first transgenic avians. Initially, a number of transgenic chicken lines were generated using lentivirus that expressed eGFP behind various ubiquitous promoters such as CMV, PGK, CAG, and chick beta actin [2–5]. In order to address more specific embryological events, transgenic quail lines using cell-specific promoters, including mouse Tie1 for endothelial cells [6], human and rat Synapsin 1 for neurons [7, 8], and chicken RBP7 for adipose tissue [9], were generated using HIV-1-based lentivirus. By fusing the fluorescent protein with a localization signal, the FP can be targeted to a specific cell organelle. Labeling the cell membrane with eGFP made the detailed description of cellular movements and shape changes during primitive streak formation possible [10]. Recently, we generated a novel transgenic quail line, which used the human PGK1 promoter to drive expression of the fluorescent protein mCherry within the chromatin of every cell nucleus. This model system allowed for automated cell tracking analysis on thousands of cells during numerous morphological events such as gastrulation and axis elongation [11, 12].

The successful generation of a new transgenic quail line is dependent on two critical steps: production of very high titer replication-deficient lentiviral stocks followed by the efficient transduction of the primordial germ cells in vivo. Below, the protocol will be separated into roughly four distinct phases: (1) Production of infectious lentiviral particles. (2) Concentration and titer determination of the lentivirus. (3) Injection of the concentrated lentiviral solution into un-incubated (Stage X) embryos [13]. (4) Assessing the efficacy of the lentiviral injection in the production of mosaic founders. Mosaic G0 birds that hatch after this process are then bred to WT individuals and their G1 offspring are screened for the transgene. Like many protocols, this one has undergone small changes and revisions over the years in an attempt to incorporate our observations along with the latest findings in the scientific literature [14–16]. The following steps represent the current version of this protocol and we have included additional insights in Subheading 4 explaining the context and small details that may otherwise go un-reported.

2 Materials

2.1 Cell Culture

1. 293 T or 293FT HEK (human embryonic kidney) packaging cell line with low passage number.
2. DMEM, high glucose with glutamine and phenol red.
3. 1 M HEPES tissue culture media supplement.

4. 100× Pen-strep (5 U/l Penicillin, 5 μg/l Streptomycin) tissue culture media supplement.
5. 100× Non-Essential Amino Acids tissue culture media supplement.
6. 100× Sodium Pyruvate (100 mM) tissue culture media supplement.
7. Ethanol (70%, diluted in water).
8. Fetal bovine serum (FBS).
9. Gelatin (0.1%, w/v), prepared in H_2O and sterilized.
10. 0.05% Trypsin-EDTA solution.
11. Dulbecco's Phosphate Buffered Saline (PBS): 10 mM sodium phosphate, 2.7 mM potassium chloride, 137 mM sodium chloride (no calcium, no magnesium), sterile.
12. Bleach (diluted to 10% in water).
13. 6, 10, and 15 cm tissue culture-treated petri dishes.
14. 6-well tissue culture plates.
15. 9 in. glass Pasteur pipets (sterile).
16. Hemocytometer bright line counting chamber.
17. Tissue culture incubator, preset to 37 °C and 5% CO_2.

2.2 Transfection

1. Transfer plasmid: third-generation self-inactivating lentivirus expression construct, for example:

   pRRLSIN.cPPT. hUbC-H2B-Cerulean-2A-Dendra2.WPRE [1 μg/μl].
2. Packaging plasmid: pMDLg/pRRE, (gag/pol/rev response element) [1 μg/μl].
3. Packaging plasmid: pRSV-Rev., (rev) [1 μg/μl].
4. Envelope plasmid: pMD2G, (VSV-g glycoprotein) [1 μg/μl].
5. Lipofectamine 2000 Transfection Reagent (Invitrogen).
6. Opti-MEM I Reduced-Serum Medium.

2.3 Lentivirus Production and Concentration

1. 15 and 50 ml conical centrifuge tubes.
2. 0.5 ml microfuge tubes (sterile, RNase/DNase free).
3. Centrifuge (tabletop or similar).
4. 50 ml conical tube top filter units (0.45 um pore size).
5. Ultra Clear ultra-centrifuge tubes (38 ml, round bottom).
6. Dry ice and bucket.
7. Inverted epi-fluorescent microscope with CFP and GFP filter sets.
8. Ultra-centrifuge (Beckman Coulter Optima XE with SW32Ti swinging bucket rotor or similar).
9. Orbital or tilting shaker in 4 °C cold room.

2.4 Embryo Injection

1. Fertilized Japanese quail eggs (un-incubated, stored at 13 °C).
2. Paraffin wax (molten on hot plate).
3. Phenol Red: 0.5% stock solution for tissue culture.
4. Pannett-Compton Saline: Solution A: 121 g NaCl, 15.5 g KCl, 10.42 g $CaCl_2 \cdot 2H_2O$, 12.7 g $MgCl_2 \cdot 6H_2O$, H_2O to 1 l. Solution B: 2.365 g $Na_2HPO_4 \cdot 2H_2O$, 0.188 g $NaH_2PO_4 \cdot 2H_2O$, H_2O to 1 l. Autoclave for storage at room temperature (18–23 °C). Before use, mix (in order): 6 ml solution A, 135 ml sterile H_2O and 9 ml solution B in a sterile glass bottle.
5. 5 in. glass Pasteur pipets (sterile, with rubber bulb).
6. Quartz glass (1 mm OD, 0.7 mm ID, 10 cm length).
7. Laser glass needle puller (Sutter P-2000 or similar).
8. Micromanipulator with clamp for needle holder.
9. Pressure injector with needle holder for 1 mm OD glass.
10. Stereomicroscope with fiber optic light.
11. Hot plate.
12. High speed rotary tool with ½ in. diameter rubber sanding drum mandrel.
13. Sand paper drum, ½ in., #240 fine grit.
14. Egg holder: Foam pipe insulation cut into 2 cm thick sections and hollowed out.
15. Steri-strip (3 M Corp. 12 mm × 100 mm. Cut into 12 mm squares).
16. Kimwipes.
17. Fine forceps (No 5 or 55).
18. Blunt tip forceps with serrated ends.
19. Small iris scissors with curved blades.
20. Small stiff bristle brush.
21. Metal spatula.
22. Large plastic weighing dish (14 cm square).
23. Plastic bag.

2.5 Incubation, Hatching, and Assessment

1. Chukar egg flats (fiberboard).
2. Egg incubator with automatic tilting trays, preset to 37 °C and 60% relative humidity.
3. Egg candler.
4. Egg hatcher with stable trays, preset to 37 °C and 70% relative humidity.
5. Stereomicroscope with epi-fluorescent filter sets for CFP and GFP.

3 Method

3.1 Production of Infectious Lentiviral Particles

1. Working in a clean tissue culture hood, prepare the complete growth media for the 293FT HEK cells. Starting with a 500 ml bottle of DMEM, remove 50 ml of media into a new 50 ml conical tube, and store at 4 °C for later resuspension of the viral pellet. Add 50 ml FBS (final concentration 10%), 5 ml Pen/Strep stock solution, 5 ml 1 M HEPES stock solution, 5 ml Sodium Pyruvate stock solution, and 5 ml Non-Essential Amino-Acids stock solution. Store at 4 °C then warm to 37 °C before use.
2. Add 10 ml of warm complete growth media to a 10 cm tissue culture dish. Thaw a vial of 293FT or 293 T HEK cells from a liquid nitrogen storage dewar in a 37 °C water bath. Just as the final ice is melting in the vial, spray the exterior with 70% ethanol, transfer the contents into the pipet dish, and distribute the cells with a side-to-side and front-to-back linear motion. Quickly check for an even cell distribution on a bright-field inverted microscope and then incubate overnight in the tissue culture incubator at 37 °C with 5% CO_2 (*see* **Note 1**).
3. The following day, check that the cells are healthy and attached to the bottom of the plate. To remove residual DMSO that was used to freeze down the cells, aspirate off the growth media and gently replace with fresh 37 °C growth media. Continue to incubate the cells at 37 °C until they cover about 70–80% of the dish bottom (70–80% confluency).
4. Aspirate off the old growth media and gently replace with 2 ml of room temperature (RT) PBS to remove any residual FBS that may inhibit the trypsin reaction. Aspirate off the PBS and replace with 2 ml of Trypsin-EDTA solution. Incubate at 37 °C for 2 min. Agitate the dish and observe the cells on the bright-field inverted microscope to ensure that all the cells are rounded up and floating.
5. Transfer cell solution into a 15 ml conical tube. Wash the plate with 8 ml of growth media and add to the conical tube. Centrifuge at about 150 × *g* for 2 min at RT to pellet the cells. Aspirate off the media, add 8 ml of fresh growth media and triturate with a 10 ml pipet until there are no more cell clumps. Avoid producing excess bubbles (*see* **Note 2**).
6. Expand the cells by adding 26 ml of fresh media to each of two 15 cm tissue culture dishes. Add 4 ml of the suspended cells to each dish. Distribute evenly and return plates to the tissue culture incubator.
7. Grow the cells until they reach ~80% confluency. Do not let the cells become completely confluent, which can reduce the rate of cell division and adversely affect the health of the cells.

8. Once the cells are ready to passage again, coat eight 15 cm tissue culture dishes with gelatin. Add ~4 ml of a sterile solution of 0.1% gelatin in water to each dish. Tilt the dishes to distribute the gelatin across the entire surface area of the dish, then aspirate off the liquid. Leave the dishes in the tissue culture hood with the lids propped open until most of the gelatin solution has evaporated. Aspirate the remaining droplets from the dishes and let it dry completely. Add 29 ml of fresh growth media to each gelatin-coated dish (*see* **Note 3**).
9. Passage the cells as described above in **steps 4** and **5** with the following adjusted volumes for the increased surface areas of the 15 cm dishes. For each dish: aspirate the growth media, wash in 4 ml PBS, aspirate, incubate in 4 ml Trypisn-EDTA until the cells are floating and combine in a 50 ml conical tube. Collect the remaining cells in the dish by washing with 16 ml growth media. The final volume in the conical tube for the two dishes will be 40 ml. Centrifuge for 4 min to pellet the cells and remove the supernatant. Resuspend the cell pellet in 8.5 ml of fresh growth media.
10. Add 1 ml of the cell suspension to each 15 cm dish making the final volume in each 30 ml. This is a 1:4 split and translates into ~5×10^6 cells per dish. Distribute the cells evenly and return the plates to the 37 °C incubator. Use a portion of the remaining cells to seed a fresh 10 cm dish for continued cell passaging (*see* **Note 4**).
11. The following day, observe the cells under bright-field illumination. The cells should be at ~80% confluency, with a healthy morphology before proceeding with the transfection. Prepare a helper plasmid master mix by combining the following volumes of plasmid DNA in a new microfuge tube: 100 μl pMDLg/pRRE, 100 μl pRSV-Rev., and 50 μl pMD2G (2:2:1 ratio).
12. Set up two 50ml conical tubes. To each, add 24 ml of RT Opti-MEM Reduced Serum Media. To one, add 752 μl of Lipofectamine 2000 transfection reagent and mix by inversion (*see* **Note 5**).
13. To the second tube add the following amounts of plasmid DNA: 112 μl of transfer plasmid (pRRLSIN.cPPT. hUbC-H2B-Cerulean-2A-Dendra2.WPRE, Fig. 1a or similar) and 140 μl of the helper plasmid master mix. The final ratio of plasmids will be 4:2:2:1 (*see* **Note 6**).
14. Combine the DNA and Lipofectamine 2000 into a single 50 ml conical tube, final volume is 48 ml, mix by inversion, and let it stand at RT for at least 20 min (but no longer than 6 h).

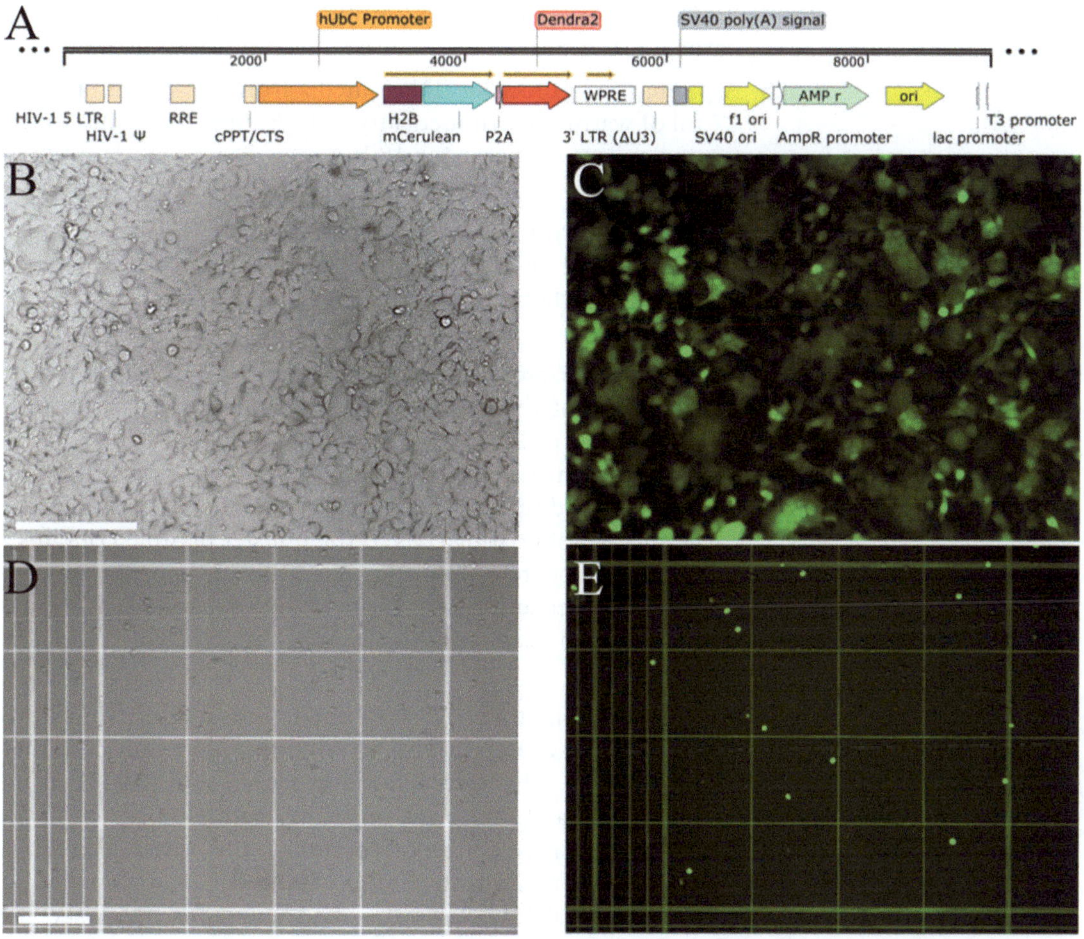

Fig. 1 Lentiviral production and titer determination. (**a**) Schematic drawing of the third-generation self-inactivating lentiviral transfer plasmid pRRLSIN.cPPT. hUbC-H2B-Cerulean-2A-Dendra2.WPRE used as an example in the protocol. Essential lentiviral elements include: long terminal repeat (LTR), encapsidation signal (psi), Rev.-response element (RRE), central poly-purine tract (cPPT), post-transcriptional regulatory element of wood chuck hepatitis virus (WPRE). Additional elements specific to the example include the human ubiquitin C promoter (hUbC), histone 2B localization signal (H2B), bicistronic linker (P2A). (**b**) Brightfield image of 293FT HEK packaging cells 24 h post-transfection with the transfer plasmid and three helper plasmids (pMDLg/pRRE, pRSV-Rev., and pMD2G). Cells will form multi-nucleated syncytia with rough cell membranes. Cell division will continue after the transfection with the cells approaching 100% confluencey after 48 h. (**c**) Epi-fluorescent image of the 293FT HEK cells shown in (**a**) using a standard GFP filter set to excite the Dendra2 fluorescent protein. In order to produce high titer lentivirus for embryo injections, the percentage of transfected cells must be very high. (**d**) Brightfield image of a hemocytometer counting grid with 293FT HEK cells harvested from the 1×10^4 well of the serially diluted concentrated lentiviral prep. (**e**) Epi-fluorescent image of D showing positive cells. Once the percentage of positive cells has been determined, this can then be used in a simple formula to calculate the total number of infectious particles produced. Scale bars: 200 μm

15. After at least 20 min, remove the 8×15 cm dishes from the tissue culture incubator, aspirate off the growth media, and add 16 ml of Opti-MEM to each dish Add 6 ml of the Lipofection/ DNA solution to each dish and mix gently with back to front

then side-to-side linear motion. Return the dishes to the incubator for 4–6 h.

16. Aspirate the lipofection/DNA media from the dishes and add 27 ml of warm growth media to each. Return the dishes to the tissue culture incubator for 48 h.
17. During the 2 days of viral production, seed a 6 cm tissue culture dish with a low density of fresh 293FT HEK cells. These cells will be infected with a small amount of viral media to indicate whether infectious virus particles have been produced.
18. The next day, observe the cells with bright-field illumination. The surfaces of the 293FT HEK packaging cells will appear rough, and the cells will have formed large multi-cellular aggregates called syncitia (Fig. 1b) due to the expression of VSV-g. The cells should remain firmly attached to the plate with only a small number of floating cells. If the transfer plasmid contains a fluorescent reporter, view the cells with an appropriate epi-fluorescent filter set. Ideally, more than 80% of the cells should be transfected and expressing the fluorescent reporters (Fig. 1c).
19. After 48 h, pipet the growth media containing infectious lentivirus, into a 0.45 μm 50 ml tube top filter unit and filter under vacuum. This step will remove any cellular debris from the viral media. Low-speed centrifugation can also be used. Combine the media from two dishes and filter through a single filter unit. Continue until all the media has been filtered into the conical tubes using as many filters as necessary (*see* **Note 7**).
20. Add 500 μl of the filtered viral media into the growth media of fresh 293FT cells that were passaged into a 6 cm dish previously. Mix to distribute the virus and then return the plate to the incubator for 48 h. Observe the cells with an inverted epi-fluorescent microscope to determine if infectious viral particles have been produced. If so, continue with the concentrating protocol described below. A small aliquot of viral media can be frozen at −80 °C if the titer needs to be determined.
21. Freeze the tubes of filtered viral supernatant by surrounding them with dry ice. Once frozen, transfer them into a −80 °C freezer for storage. The slight loss of titer caused by a single freeze thaw cycle is offset by the convenience of being able to start the concentration step when desired.
22. Soak the 15 cm dishes, filters, pipets, and any other plasticware that has come into contact with the lentiviral media in 10% bleach solution for at least 30 min to de-contaminate them. Dispose of them in accordance with the institutional safety procedures.

3.2 Concentration and Titer Determination of the Lentiviral Stock

1. One day before concentrating the virus, seed 293FT HEK cells into two 6-well tissue culture plates. Each well should contain 2×10^4 cells in 2 ml of growth media.
2. Thaw the 50 ml conical tubes of viral media which have been stored at −80 °C in a RT water bath. As this proceeds, refrigerate the rotor and buckets at 4 °C and allow the chamber of the ultra-centrifuge to equilibrate to 4 °C under vacuum.
3. In the tissue culture hood, sterilize and clean six Ultra Clear centrifuge tubes by adding 70% ethanol, letting it stand for 5 min, then aspirating them dry using a 9 in. glass Pasteur pipet.
4. Once the viral media has thawed, invert the tubes to mix. Then evenly distribute the media into the six ultra-centrifuge tubes taking care to add equal volumes to each tube to maintain the proper weight balance.
5. Load the centrifuge tubes into the prechilled buckets, close, and load the numbered buckets into the corresponding positions in the rotor. Run the ultra-centrifuge at 50,000 × *g* (17,100 rpm in a SW32Ti rotor) for 4 h at 4 °C (*see* **Note 8**).
6. Following the run, aspirate the media from the ultra-centrifuge tubes. A clear, yellowish pellet should be apparent in the center of the tube bottom. Tilt the tubes at an angle to allow the remaining media on the walls of the tube to pool away from the pellet. After several minutes, aspirate this small amount of media.
7. Add 20 μl of DMEM (no additives, saved from step 1) to each tube. Place each ultra-centrifuge tube inside a 50 ml conical bottom tube. This eliminates any viral material from escaping the tube during subsequent vortexing steps. Place the tubes on ice and let it stand for ~30 min. The pellet will turn whitish and the edges may start to loosen. Vortex the tubes gently until the pellet releases from the bottom of the tube and begins to break up into large clumps.
8. Place the tubes on an orbital or tilting shaker in a 4 °C cold room or refrigerator. Set the speed so that the DMEM and viral material in the bottom of the tube is slowly agitated. Leave overnight.
9. The next day, gently vortex the tubes to ensure that the viral pellet is dissolved in the DMEM with no clumps. Triturate with a pipettor if small clumps remain, being careful not to introduce air bubbles. Allow the tubes to sit upright several minutes on ice to pool the viral solution at the bottom.
10. Pipet the viral solution from all the ultra-centrifuge tubes into a single 0.5 ml microfuge tube on ice. To collect any residual viral material from the walls of the ultra-centrifuge tubes, add

10 μl of fresh DMEM to one of the tubes and vortex gently. Pipet this same solution sequentially into the remaining tubes, vortex and repeat, until all tubes have been washed. Combine the final aliquot into the viral solution already in the chilled 0.5 ml microfuge tube and vortex gently to mix. Store the tube on ice.

11. Set up six 0.5 ml microfuge tubes and add 9 μl of DMEM to each. Pipet 1 μl of the concentrated lentiviral solution into the first tube and mix to form a 1:10 dilution. Pipet 1 μl of this solution into the second tube containing DMEM and mix, thus making a 1:100 dilution. Repeat in sequence until the tubes contain serial lentiviral dilutions ranging from 1:10 to 1:1,000,000.

12. Pipet 2 μl of the 1:10 dilution into the 2 ml of media in the first well. This is the 1×10^4 dilution. Add 2 μl of the 1:100 dilution into the second well for the 1×10^5 dilution. Continue with the remaining dilutions until the plate contains serial dilutions between 1×10^4 and 1 × 10 [9]. Gently agitate the plate to evenly distribute the virus and return the plate to the tissue culture the incubator for 48 h.

13. Pipet 40 μl aliquots of the concentrated viral solution into separate 0.5 ml microfuge tubes and freeze at −80 °C. This volume is convenient for a single embryo injection session described below and helps to reduce repeated freezing and thawing which can lead to reduced titers.

14. Determine the number of cells present in the wells at the start of infection by aspirating the media from a single well in the second 6-well plate. Wash the cells with 1 ml of PBS, aspirate and add 1 ml of trypsin-EDTA solution, and return to the 37 °C incubator. Once all the cells have floated free of the bottom, collect the enzyme solution into a 15 ml conical tube. Wash the well several times with growth media and combine in the conical tube. Centrifuge at 150 × *g* for 2 min to pellet the cells, aspirate the supernatant and resuspend the cells in 1 ml of growth media.

15. Pipet 10 μl of the cell solution under each end of the coverslip on a hemocytometer counting chamber. Using bright-field illumination with an inverted microscope, count the number of cells in the five 1 mm^2 grids on both ends. The total number of cells in these ten grids multiplied by 1000 represents the total number of cells in the well at the start of the infection.

16. After 48 h of incubation, observe the serial dilutions in the 6-well plate under the inverted epi-fluorescent microscope using the appropriate filter sets. The number of fluorescent cells present should decrease at a linear rate until the final well, at 1×10^9 will likely have either few or no positive cells. Count

the number of positive cells in the second to last well and multiply by 1 × 10 [8]. This is a rough estimate of the infectious viral titer in transforming units per ml (TU/ml).

17. For a more precise titer determination, count the number of cells in the 1×10^4 dilution well following **step 14** above. Instead of 1 ml, resuspend the cell pellet in 5 ml of growth media. Continue with **step 15** above, counting the total number of cells in the $10 \times 1\ mm^2$ grids (Fig. 1d).
18. With the hemocytometer and cells still on the microscope, switch to epi-fluorescent illumination and count the number of fluorescent positive cells in the grids (Fig. 1e). Determine the percentage of positive cells by dividing the number of positives by the total number of cells. Next, multiply this percentage by the number of cells present at the start of infection as determined in **step 15**. This calculation takes into account the cell division that took place between the start of infection and 2 days later at counting. When multiplied by the dilution factor of 1 × 10 [4], the effective titer is determined. For example:

    Number of cells at the start of infection: 2.7×10^5.

    Total number of cells at the time of counting: 469.

    Number of positive cells at the time of counting: 232.

    Percentage of positive cells: 232/469 = 49.5%.

    $2.7 \times 10^5 \times 0.495 = 133{,}650$.

    $133{,}650 \times 1 \times 10^4 = 1.3 \times 10^9$TU/ml.

The described protocol typically yields titers between 1×10^8 and 1×10^9TU/ml. Only preps at the higher end of this range should be used for embryo injection (*see* **Note 9**).

3.3 Injection of Concentrated Lentiviral Solution into Quail Embryos

1. Remove about 75 wild-type (WT) fertilized Japanese quail eggs from the 13 °C storage refrigerator. Hold each egg in front of the egg candler and discard any eggs that are cracked. Using the fiberboard chukar egg flats, set the remaining eggs on their sides at RT. Use a Sharpie pen to mark the high point along the midline of the curved egg shell. Over about an hour, the embryo will float to a position just under the mark (*see* **Note 10**).
2. While the eggs come to RT, prepare the injection station. Warm a bottle containing solid paraffin wax on the hot plate until molten. Prepare 150 ml of fresh Pannett-Compton saline from the two stock solutions in a glass bottle. Insert a 5 in. glass Pasteur pipet with a rubber bulb into the saline. Open a plastic egg waste disposal bag inside a large beaker or ice bucket. Prepare a small beaker with a 10% bleach solution. Turn on

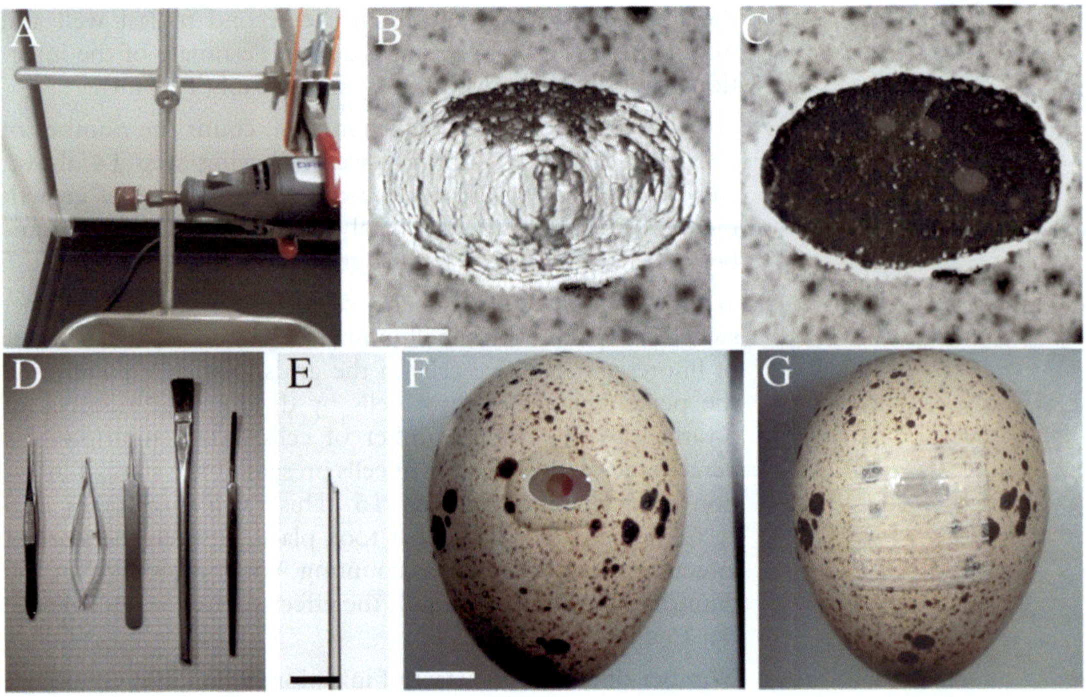

Fig. 2 Egg windowing and lentiviral injection. (**a**) The high-speed rotary tool, fitted with a sanding drum (fine grit), clamped to the scaffolding inside a fume hood to reduce vibration and to control the spread of the ground shell dust through the air. (**b**) Close-up image of the egg shell after grinding with the high-speed rotary tool. Often, fragments of the inner shell layers remain on the underlying soft shell membrane. Scale bar 1 mm. (**c**) View of the oval window after scraping the shell fragments away from the shell membrane. Ideally, the shell membrane has not been punctured during the removal of the shell. (**d**) The instruments needed to perform the egg windowing and lentiviral injection steps. From *left* to *right*: a small serrated forceps with rounded tips, a very fine iris scissors with curved blades, a fine forceps with sharp tips, a stiff bristle brush, and a flat metal spatula. (**e**) Broken end of the glass needle. Note the long taper of the shaft and the cutting edge of the broken end. The opening should be about 30 μm. Scale bar 100 μm. (**f**) A windowed egg. The window has been covered with a drop of Pannett-Compton saline to exclude air from entering the egg. A round hole has then been cut in the shell membrane, exposing a portion of the Stage X embryo beneath. The sub-germinal cavity has then been filled with concentrated lentivirus (spiked with phenol red for visibility). In this egg, the embryo is slightly offset to the *right* of the window. Note that the diameter of the hole in the shell membrane is just large enough to clearly view only a portion of the sub-germinal cavity and is much smaller than the diameter of the entire blastodisc. Scale bar 5 mm. (**g**) The same egg as in (**f**) with the window covered by an adhesive Steri-strip bandage. The window is positioned under the top half of the Steri-strip in order to avoid interfering with the cutting out process of the hatchling in 16 days. The glossy surface over the hole in the shell is the solidified molten paraffin wax used to eliminate buffer evaporation through the breathable Steri-strip

the pressure injector unit and open the air line. Lay out the forceps and scissors, shown in Fig. 2d, on the lab bench.

3. The high-speed rotary tool with ½ in. sanding drum attachment should be securely fastened to a stand, preferably inside a fume hood. The drum should hold fine sand paper of at least #240 grit. A pan beneath the rotary tool will help to collect the shell fragments (Fig. 2a).

4. Several quartz glass pipets should be pulled to a fine point needle using the laser puller. Under the stereomicroscope, use the fine forceps to break the sealed tip back to generate an opening of about 30 μm. A slight bevel in the broken end will help it cut cleanly into the embryo (Fig. 2e). Load the glass needle into the pipet holder and tighten the collar (*see* **Note 11**).
5. After about an hour at RT, clean the marked side of the egg using a Kimwipe soaked with 70% ethanol. Rotate the egg quickly along its short and long axes multiple times before replacing it in the egg flat with the marked area facing up. This aggressive rotation may help free some embryos that have not yet rotated into the proper position.
6. Thaw a 40 μl aliquot of concentrated lentivirus. Vortex well to mix. Centrifuge the tube in a tabletop microfuge at top speed for 5 s to pellet any large debris. Pipet the viral supernatant into a new 0.5 ml tube, add 2 μl of phenol red to give the viral solution a bright pink appearance and vortex again.
7. If the pressure injector has a fill function, pipet a 20 μl droplet of the viral solution onto a square of Parafilm and fill the pipet through the tip while watching through the stereoscope. If the tip clogs during filling, put a small amount of positive pressure on the pipet and draw the tip against a Kimwipe saturated with saline solution. Once full, clamp the needle holder into the micro-manipulator at about a 30° angle. Store the remaining virus on ice (*see* **Note 12**).
8. Turn on the high-speed rotary tool and allow it to come up to full speed. Hold the egg under the sanding drum with its long axis perpendicular to the shaft. Slowly bring the egg up until the shell touches the spinning sanding drum momentarily. Lower the egg and check the extent of the sanding. The outer, pigmented shell layers should be ground away in an oval pattern, exposing the white shell layers underneath. Grind the shell again in the same location until a small portion of the underlying soft shell membranes are visible as darker regions (Fig. 2b). The size of the oval window should not exceed about 5 mm wide × 4 mm high (*see* **Note 13**).
9. Place a large disposable plastic weigh dish under the stereomicroscope. Hold the egg inside the weigh boat and position it so that the partially ground window is easily observed. Using the small blunt-ended, serrated tip forceps, scrape the remaining hard white shell layers off the soft shell membrane below, rotating the egg as needed. Scrape any broken shell debris from the edges of the window to keep them from falling into the egg later. The weigh boat should collect the shell debris. When finished, the oval window should have clean edges with no remaining hard shell fragments (Fig. 2c).

10. Replace the weigh boat with a foam egg holder under the stereomicroscope. Position the egg on the holder with the window centered in the field of view. Place a drop of Pannett-Compton saline over the window. The drop should overlap slightly onto the surrounding shell and serves to keep air from being trapped inside the egg.
11. Working inside the drop of saline, use the fine curved bladed iris scissors to cut a round hole in the center of the soft shell membrane. The hole should be about 3 mm in diameter and centered within the oval window (Fig. 2f). Remove the loose circle of shell membrane with the fine forceps (*see* **Note 14**).
12. The embryo should be in view directly under the window. If it is not, rotate the egg back and forth then side to side vigorously to coax the embryo to the window. If only a portion of the embryo is visible under the window, position the egg at an angle so that the central area pellucida of the embryo is visible. An additional drop of saline may be required to keep the embryo floating high in the shell (*see* **Note 15**).
13. Set the pressure indicator to −0.1 psi. Using the controls on the micromanipulator, bring down the virus-filled quartz injection needle so that it just touches the vitelline membrane above the center of the area pellucida of the embryo. You may see a slight light refraction when the needle comes in contact with the vitelline membrane. Slowly drive the needle down through the vitelline membrane and epiblast and into the sub-germinal cavity of the blastoderm. The tip of the needle will disappear from view as it passes through the epiblast (*see* **Note 16**).
14. Retract the needle slightly. Increase the positive pressure to around 0.0 psi and continue to retract the needle. When the tip of the needle enters into the sub-germinal cavity, the pink viral solution will begin to flow out. Slowly increase the positive pressure to between 0.1 and 0.5 psi and watch the viral solution fill the entire sub-germinal cavity (1–2 μl) (*see* **Note 17**).
15. Once the cavity appears full (the diameter of the pink circle stops expanding), decrease the positive pressure back to −0.1 psi and simultaneously withdraw the needle from the embryo and out of the albumin.
16. Carefully wipe the area around the opening dry with a Kimwipe. Using the blunt-ended forceps, remove a precut Steri-Strip from its backing paper and position it on the eggshell so that the window is under the upper third of the bandage (Fig. 2g). Press down slightly with the tip of the forceps to adhere the Steri-Strip to the shell. Using your finger, rub across the Steri-Strip to ensure that if forms a tight seal from edge to edge.

17. With the small metal spatula, apply a thin layer of molten paraffin to the surface of the Steri-Strip which lies over the window (Fig. 2g). The paraffin prevents excessive evaporation through the window during the subsequent incubation period. Once the wax has initially cooled, set the egg upright into an egg flat for about 5 min at RT to allow the wax to completely harden.
18. Move the injected egg into the 37 °C egg incubator with the large end facing upright. Rotate the egg so that the window is not in contact with the egg holder and is not in line with the movement of the embryo inside the egg as the incubator trays tilt once an hour.
19. Repeat the viral injection procedure starting from **step 3.3.9** on the remaining eggs that have been previously windowed with the rotary tool sanding drum.

3.4 Assessing the Efficacy of the Lentiviral Injection in the Production of Mosaic Founders

1. Incubate the injected eggs at 37.5 °C and ~60% relative humidity for 7 days. Be sure that the incubator shelves tilt hourly to avoid detrimental effects on hatchability.
2. On day 7, candle the eggs to determine viable embryos. Embryos that are developing well will have a dark red or brown appearance from the development of the vascular system inside the eggshell. In addition, a viable egg will show the air sac as a distinct clear area that fills the large end of the egg. Embryos that have died may show blood rings or will appear bright or clear, having formed no blood. Discard any nonviable eggs into the biohazard waste.
3. On day 10 of incubation, harvest one viable embryo and dissect the gonadal tissue. Image the gonads using fluorescent microscopy. Mosaics with heavily labeled gonads indicate that the viral titer and injection procedure are effective (Fig. 3a–c). If the gonads show no fluorescent label, the procedure must be reviewed in detail. The most likely reason is low titer virus. Make sure the titer is at least 1×10^9 TU/ml.
4. On day 14 of incubation, set the eggs on their sides in a hatching incubator that is running with 37 °C with 70% humidity. The hatching trays should be lined with bench coat paper that has a soft, textured surface to prevent leg splaying in hatchlings. Hatching will take place over the course of days 16–17 (*see* **Note 18**).
5. Move the hatchlings to heated brooders and follow standard animal husbandry guidelines for Japanese quail. After 1 month, the mosaic birds and their WT breeding partners can be moved into normal cages. The birds will begin to lay fertilized eggs at about 2 months of age (*see* **Note 19**).

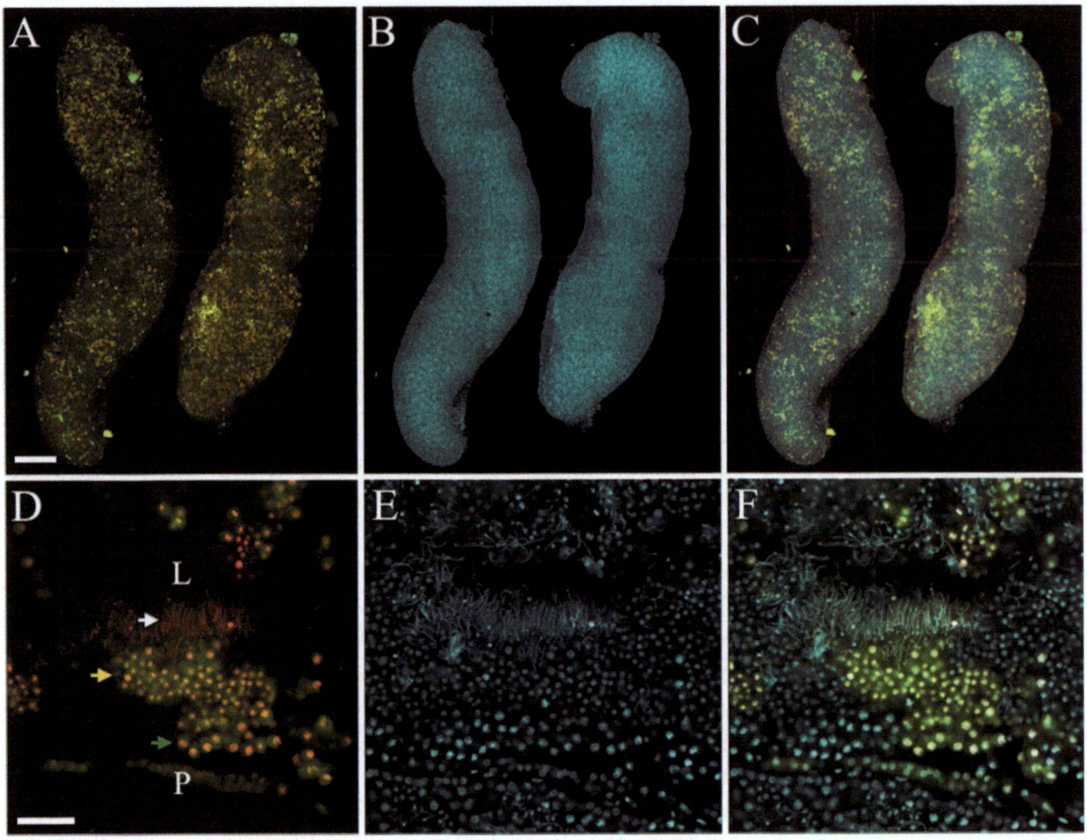

Fig. 3 Assessing the efficacy of the Stage X lentiviral infection by examination of the chimeric gonads for transgene expression. WT quail embryos were injected with 1–2 μl of concentrated lentivirus (hUbC:H2B-Cerulean-2A-Dendra2) at Stage X as described in the protocol, incubated until E10 (**a–c**) or grown to breeding age (2+ months, **d–f**). (**a**) Wholemount gonadal tissue dissected from a 10 day old hUbC:H2B-Cerulean-2A-Dendra2 chimera embryo showing a high percentage of transgenic cells. H2B-Cerulean in *red* and Dendra2 in *green*. (**b**) DAPI stain. (**c**) Merged image of (**a**) and (**b**). Scale bar is 200 μm. An adult chimera was euthanized and the testes harvested, fixed, cut into transverse slab sections, cleared, and mounted in a glass-bottom petri dish and imaged using confocal microscopy (**d**)–(**f**). (**d**) Longitudinal section of a seminiferous tubule in a chimeric adult quail testes. Lumen of the tubule (L), Periphery of the tubule (P). Nuclear ceruleanFP in *red*, Dendra2 in *green*. *White arrow* indicates a region of spermatozoa with the long, narrow heads characteristic of galliform birds such as quail and chickens. *Yellow arrow* indicates the region of primary and secondary spermatocytes. *Green arrow* indicates the region near the periphery of the tubule containing the larger cell nuclei of the spermatagonia. (**e**) DAPI staining. (**f**) Merge of (**d**)–(**e**). Scale bar: 50 μm

6. Screen the G1 hatchlings for transgenic birds. To detect the ubiquitous expression of a fluorescent protein, we quickly inspect the nictating membrane of the hatchling under an epi-fluorescent stereomicroscope. Some promoters such as PGK1 and mTie1 will also drive strong expression of the reporter molecule in the chorioallantoic membrane (CAM) which can be directly observed in the empty eggshell using an epi-fluorescent stereoscope. If the transgene cannot be

identified phenotypically, a standard genotypic analysis can be performed (*see* **Note 20**).

7. Once a transgenic G1 hatchling has been identified, grow the bird using standard husbandry practices. Breed the G1 transgenic with WT mates, expecting that 50% of these offspring should be hemizygous for the transgene. In later generations, hemizygotes can be mated together in order to produce homozygous birds (*see* **Note 21**).

4 Notes

1. The passage number of the 293 human embryonic kidney (HEK) cell line should be below 25. This is a somewhat arbitrary limit; however, 293 HEK cells should grow rapidly and will typically need passaging twice a week when using a 1:8 split. If the cells grow slowly or if they consistently de-laminate from the dishes after the transfection, starting with lower passage number cells is advisable.
2. Additional passages into 10 cm dishes before expanding up into 15 cm dishes may be needed after thawing a new vial from frozen stock. Always passage the cells when they reach 70–80% confluency. Healthy cells are an absolute prerequisite for generating the high lentiviral titers needed for avian transgenesis.
3. Gelatin is an in-expensive coating that promotes strong cell adhesion during the viral production phase. Other coatings such as poly-L-lysine can also be used. Eight 15 cm dishes will yield a volume of viral media that can be concentrated in a single ultra-centrifuge run when using a SW32Ti or SW28 swinging bucket rotor. Each rotor holds a total of 216 ml of viral supernatant (6 × 36 ml). Eight dishes, harvested one time, can therefore be filled with 27 ml of growth media. This volume of media is sufficient to ensure healthy cells and reasonable pH levels during the 48 h viral production phase.
4. This 1:4 split should be concluded 24 h before the transfection is scheduled. If conducted on dishes that are 70–80% confluent, we find that this ratio will produce cells at or near 80% confluency by the following day, which is the ideal density for an efficient transfection.
5. We prefer Lipofectamine 2000 (Invitrogen) as a transfection reagent. At a 3:1 ratio to DNA, it consistently gives transfection rates of 80% or better (based on fluorescence) which is vital to generate high viral titers. Maximum DNA uptake is the key given that the third-generation transfer vectors require a 4-way transfection for production of infectious viral particles.

However, for large preps, the cost can be prohibitive. Polyethylenimine (PEI, linear, MW = 25,000) also used at a 3:1 ratio to DNA is an inexpensive alternative that approaches the titers generated with Lipofection products.

6. All plasmid DNA should be high purity, prepared from fresh bacterial stocks and isolated using commercial spin column technology such as the Qiagen Maxi prep or similar. The plasmid DNA should be dissolved in TE buffer or water with 260/280 nm absorbance ratios at 1.9 or better and diluted to 1 μg/μl. Optimal ratios between transfer and helper plasmids vary widely in the literature but generally follow: transfer >gag/pol/rev>VSV-g. Likewise, the amount of DNA per cm^2 of surface area varies. We routinely use 0.2 μg DNA/cm^2. As the size of the transgene increases in the lentiviral transfer vector, the titer obtained will decrease. Therefore, it is beneficial to keep the size of the proviral integration sequence as small as possible, with 8 kb representing the upper size limit [17]. For detailed information on the third-generation, self-inactivating lentivirus vector, please *see* [18, 19]. While we do not routinely include a sodium butyrate induction step following transfection, it has been shown to increase the titer of some vectors [20].
7. Viral supernatants can be collected at 24, 48, and 72 h post-transfection; however, viral production begins to fall after about 48 h [21, 22]. In order to minimize the volume of viral media and maximize the titer, we have chosen to collect only once. By 48 h, some of the cells may be loosening around the edges of the plate, with some cells floating. This is normal. However, if large sheets of cells have de-laminated and are floating in the media, the final titer will not be optimal.
8. Ultracentrifugation is the standard method to concentrate lentiviral preps but other methods are effective including ultrafiltration, PEG precipitation, ion-exchange chromatography, and tangential flow filtration (TFF) [23–25]. If the lentiviral prep is scaled up with the use of multi-layer tissue culture flasks, ultracentrifugation becomes impractical. In that case, TFF can be used to concentrate down the large volumes (1 l or more) of serum-free viral media that are generated.
9. The literature suggests that a myriad of factors in the way the titer is determined will affect the final number generated [26–28]. Caution should be used when comparing titer numbers using different methods. This leads to the conclusion that not one particular titering protocol is necessarily best but that reproducibly using a single, consistent protocol to determine the viral titer is essential. This will allow for comparisons from prep to prep and help determine which preps should be used for embryo injection.

10. Fertile, freshly laid eggs can be obtained through local commercial vendors, or through your own breeding colony. The most important thing to consider when choosing the source of the eggs is how the eggs were handled after laying and during shipping. Eggs can be stored for 3–7 days at 13–15 °C with minimal consequences; longer storage or storage in suboptimal conditions tends to decrease hatchability. Eggs should always be stored in egg flats with the large end facing upward.
11. Borosilicate glass can also be used for the injection needles. These can be manufactured using automated pullers fitted with metal heating elements. If pinching the end of the pulled glass with a fine forceps fails to produce an acceptable cutting edge and tip, a pipet beveling instrument can be used to grind a very even bevel and sharp point at the opening. While more time consuming, a pipet beveller can help produce glass needles with consistent characteristics.
12. Alternatively, use a pipettor fitted with a disposable microloader tip. Fill the long thin tip with virus by slowly releasing the piston button of the pipettor. Insert the end of the long shaft of the microloader tip into the open end of the glass pipet and move it down into the narrow beveled end of the glass needle. Slowly start to fill the pipet from the bottom up, retracting the tip as the viral solution fills the lumen. Avoid any introduction of air bubbles into the column of viral solution.
13. Differences in the circumference of the egg will lead to various sized windows. Avoid over grinding, which will produce excessively large windows and decrease viability. Discard any eggs that show cracks radiating out from the window. If this happens repeatedly, tighten the rotary tool clamp to eliminate any excessive vibration. Do not allow the sanding drum to cut into the soft inner shell membranes [29]. This will result in a loss of thin albumin from the egg and the introduction of an air bubble. Such eggs should be discarded. Several eggs can be ground in a single run; however, prolonged exposure of the inner shell membranes to the air will lead to localized drying of the tissue and should be avoided. Grind only as many eggs as you can inject within a 15–20 min time frame.
14. To cut the hole in the soft shell membrane it is helpful to hold the iris scissors steady at a fixed shallow angle and rotate the egg with the opposite hand. The shallow angle will ensure that the embryo underneath is not damaged. In most eggs, the embryo sits down in the egg somewhat, surrounded by a thick albumin capsule that keeps it from being damaged in the windowing process. If the embryo or yolk is damaged during the windowing process, the egg should be discarded into the plastic egg waste bag.

15. Inspect the embryo closely and proceed with only the healthiest examples. Discard any eggs where the shape of the embryo is not round and clearly defined. Discard any eggs where the embryo appears as a small, concentrated, white circle as these are un-fertilized. Embryos that sit excessively deep within the shell are difficult to inject and should also be discarded. Finally, some embryos will simply not appear under the window despite aggressive rotations and must be discarded.
16. The upper cell layers of the epiblast will present some resistant to the tip of the needle as it is driven down. If the needle is not particularly sharp, the embryo and yolk may be pushed down slightly before the needle cuts into the tissue and embryo returns to its starting position. If this happens, the tip of the needle will be too deep, well into the yolk beneath the embryo and will have to be raised considerably before it enters the sub-germinal cavity. While workable, a dull needle tip makes injections more difficult and may lead to higher mortality. A glass pipet can be re-broken several times to get a better cutting tip (Fig. 2e).
17. Injection at 0.1–0.5 psi is recommended because higher pressures can lead to the disturbance of fragile embryonic cells. The exact pressure needed to slowly expel the virus will depend on several factors including the tip diameter, the viscosity of the viral solution, and the resistance presented by the embryonic tissue. During the injection, you should be able to see an area of diffuse pink color expanding in the area pellucida. If you do not see the pink virus solution, the needle may be too deep, below the embryo in the yolk. If the viral solution is very clearly evident, instead of diffuse, the needle is likely above the embryo and is expelling the viral solution into the albumin. In both cases, re-position the needle until the sub-germinal cavity is located.
18. Our rate of hatching in eggs that were injected with lentivirus at Stage X averages about 7%. This is similar to rates reported by other groups using this method [14, 30]. The use of serum-free media for lentivirus production and ultracentrifugation through a sucrose cushion has been shown to reduce the toxicity of the lentiviral prep in mammalian in-vivo injections [31]. However, these modifications failed to alter our hatch rate. There are two time periods during incubation when mortality appears to peak; in the first 2 days following injection and then again around days 12–14. The gonads from embryos that die late in development can be dissected and imaged for the presence of fluorescent markers. While not providing high-quality histology, these can help to determine if the lentiviral titer and injection process have successfully transduced germ cells.

19. For detailed information on Japanese quail husbandry in the laboratory environment, please *see* [32–34].
20. Conventional PCR genotyping can be performed on genomic DNA isolated from the CAM or small amount of blood collected from the wing vein or feather shaft. The sperm from mosaic males can be analyzed by PCR to gauge the amount of transgene in the ejaculate. This method will identify which males should then be bred with WT hens. Alternatively, the testes of a mosaic male can be examined for the fluorescent reporters of the transgene (Fig. 3d–f). Given the low percentage of chimeras hatched from eggs injected at Stage X, this method should be reserved for breeding age males that must be euthanized after sustaining injuries due to fighting or illness.
21. Our experience indicates that most transgenic lines generated using this method will carry one copy of the transgene. This can be confirmed by Southern blot analysis of genomic DNA. If multiple chimeric birds produce independent transgenic lines, positional effects on transgene expression can be examined. The line that gives the strongest fluorescent protein expression while not affecting normal development is the best choice for an imaging model system. Strong FP expression allows the use of lower laser power which reduces phototoxicity during long duration dynamic imaging.

Acknowledgments

The authors wish to thank Dr. Bertrand Benazeraf, Dr. Gerome Gros, and Greg Poynter for help in developing and optimizing this method.

References

1. Lois C, Hong EJ, Pease S, Brown EJ, Baltimore D (2002) Germline transmission and tissue-specific expression of transgenes delivered by lentiviral vectors. Science 295:868–872
2. McGrew MJ, Sherman A, Ellard FM, Lillico SG, Gilhooley HJ, Kingsman AJ, Mitrophanous KA, Sang H (2004) Efficient production of germline transgenic chickens using lentiviral vectors. EMBO Rep 5:728–733
3. Chapman SC, Lawson A, Macarthur WC, Wiese RJ, Loechel RH, Burgos-Trinidad M, Wakefield JK, Ramabhadran R, Mauch TJ, Schoenwolf GC (2005) Ubiquitous GFP expression in transgenic chickens using a lentiviral vector. Development 132:935–940
4. McGrew MJ, Sherman A, Lillico SG, Ellard FM, Radcliffe P, Gilhooley HJ, Mitrophanous KA, Cambray N, Wilson V, Sang H (2008) Localised axial progenitor cell populations in the avian tail bud are not commited to a posterior Hox identity. Development 135:2289–2299
5. Motono M, Yamada Y, Hattori Y, Nakagawa R, Nishijima KI, Iijima S (2010) Production of transgenic chickens from purified primordial germ cells infected with a lentiviral vector. J Biosci Bioeng 109:315–321
6. Sato Y, Poynter G, Huss D, Filla MB, Czirok A, Choi JM, Rongish BJ, Little CD, Fraser SE, Lansford R (2010) Dynamic analysis of vascular morphogenesis using transgenic quail embryos. PLoS One 5:e12674
7. Scott BB, Lois C (2005) Generation of tissue-specific transgenic birds with lentiviral vectors. Proc Natl Acad Sci 102:16443–16447

8. Seidl AH, Sanchez JT, Schecterson L, Tabor KM, Wang Y, Kashima DT, Poynter G, Huss D, Fraser SE, Lansford R (2013) Transgenic quail as a model for research in the avian nervous system: a comparative study of the auditory brainstem. J Comp Neurol 521:5–23
9. Ahn J, Shin S, Huh Y, Park JY, Hwang W, Lee K (2015) Identification of the avian RBP7 gene as a new adipose specific gene and PBP7 promoter driven GFP expression in adipose tissue of transgenic quail. PLoS One 10:e0124768
10. Rozbicki E, Chuai M, Karjalainen AI, Song F, Sang HM, Martin R, Knolker HJ, MacDonald MP, Weijer CJ (2015) Myosin-II-mediated cell shape changes and cell intercalation contribute to primitive streak formation. Nat Cell Biol 17:397–408
11. Huss D, Benazeraf B, Wallingford A, Filla M, Yang J, Fraser SE, Lansford R (2015) A transgenic quail model that enables dynamic imaging of amniote embryogenesis. Development 142:2850–2859
12. Benazeraf B, Beaupeux M, Tchernookov M, Wallingford A, Shirtz A, Shirtz A, Huss D, Pourquie O, Francois P, Lansford R. (2016) Multiscale quantification of tissue behavior during amniote embryo axis elongation. bioRxiv. doi: 10.1101/053124
13. Eyal-Giladi H, Kochav S (1975) From cleavage to primitive streak formation; a complementary normal table and a new look at the first stages of the development of the chick. I. General morphology. Dev Biol 49:321–337
14. Scott BB, Lois C (2006) Generation of transgenic birds with replication-deficient lentiviruses. Nat Protoc 1:1406–1411
15. Poynter G, Lansford R (2008) Generating transgenic quail using lentiviruses. In: Bronner-Fraser M (ed) Avian embryology, Methods in cell biology, vol 87, 2nd edn. Academic Press, San Diego, CA, pp 281–293
16. Poynter G, Huss D, Lansford R (2009) Japanese quail: an efficient animal model for the production of transgenic avians. Cold Spring Harb Protoc. doi:10.1101/pdb.emo112
17. Yacoub N, Romanowska M, Haritonova N, Foerster J (2007) Optimized production and concentration of lentiviral vectors containing large inserts. J Gene Med 9:579–584
18. Dull T, Zufferey R, Kelly M, Mandel RJ, Nguyen M, Trono D, Naldini L (1998) A third generation lentivirus vector with a conditional packaging system. J Virol 72:8463–8471
19. Miyoshi H, Blomer U, Takahashi M, Gage FH, Verma IM (1998) Development of a self inactivating lentivirus vector. J Virol 72:8150–8157
20. Ramezani A, Hawley RG (2002) Generation of HIV-1-based lentiviral vector particles. Curr Protoc Mol Biol:Unit 16.22.1–16.22.15
21. Reiser J (2000) Production and concentration of pseudotyped HIV-1-based gene transfer vectors. Gene Ther 7:910–913
22. Coleman JE, Huentelman MJ, Kasparov S, Metcalfe BL, Paton JFR, Katovich MJ, Semple-Rowland SL, Raizada MK (2003) Efficient large-scale production and concentration of HIV-1-based lentiviral vectors for use in vivo. Physiol Genomics 12:221–228
23. Kutner RH, Zhang XY, Reiser J (2009) Production, concentration and titration of pseudotyped HIV-1-based lentiviral vectors. Nat Protoc 4:495–505
24. Geraerts M, Michiels M, Baekelandt V, Debyser Z, Gijsbers R (2005) Upscaling of lentiviral vector production by tangential flow filtration. J Gene Med 7:1299–1310
25. Cooper AR, Patel S, Senadheera S, Plath K, Kohn DB (2011) Highly efficient large-scale vector concentration by tandem tangential flow filtration. J Virol Methods 177:1–9
26. Logan AC, Nightingale SJ, Haas DL, Cho GJ, Pepper KA, Kohn DB (2004) Factors influencing the titer and infectiviy of lentiviral vectors. Hum Gene Ther 15:976–988
27. Zhang B, Metharom P, Jullie H, Ellem KAO, Cleghorn G, West MJ, Wei MQ (2004) The significance of controlled conditions in lentiviral vector titration and in the use of multiplicity of infection (MOI) for predicting gene transer events. Genet Vaccines Ther 2:6–15
28. Geraerts M, Willems S, Baekelandt V, Debyser Z, Gijsbers R (2006) Comparison of lentiviral vector titration methods. BMC Biotechnol 6:34–43
29. Speksnijder G, Ivarie R (2000) A modified method of shell windowing for producing somatiic or germline chimeras in fertilized chicken eggs. Poult Sci 79:1430–1433
30. Agate RJ, Scott BB, Haripal B, Lois C, Nottebohm F (2009) Transgenic songbirds offer an opportunity to develop a genetic model for vocal learning. Proc Natl Acad Sci U S A 106:17963–17967
31. Baekelandt V, Eggermont K, Michiels M, Nuttin B, Debyser Z (2003) Optimized lentiviral vector production and purification procedure prevents immune response after transduction of mouse brain. Gene Ther 10:1933–1940
32. Huss D, Poynter G, Lansford R (2008) Japanese quail (*Coturnix japonica*) as a laboratory animal model. Lab Anim 37:513–519

33. Woodard AE, Abplanalp H, Wilson WO, Vohra P (1973) Japanese quail husbandry in the laboratory. Department of Avian Sciences, University of California-Davis, Davis, CA
34. Cheng KM, Bennett DC, Mills AD (2010) The Japanese quail. Chapter 42. In: Hubrecht R, Kirkwood J (eds) The UFAW handbook on the care and management of laboratory and other research animals, 8th edn. Universities Federation for Animal Welfare, Wiley-Blackwell, London

Chapter 9

Lentiviral-Mediated Transgenesis in Songbirds

Wan-chun Liu, Marian Hruska-Plochan, and Atsushi Miyanohara

Abstract

Transgenesis involves the insertion of an exogenous gene into an animal's genome, which allows the identification of the expressed phenotypes in brain function or behavior. Lentiviral-mediated transgenesis offers unique transduction potency making it possible to deliver and stably integrate transgenes into a wide variety of dividing and nondividing cells. The ability to establish long-term expression of such transgenes allows their use for transgenesis which is especially useful in organisms lacking quality pluripotent stem cell lines and which is otherwise difficult to produce via traditional pronuclear microinjection, such as songbirds. Here we describe a protocol to generate the transgenic songbird, the zebra finch, by producing and inserting lentiviral-mediated transgene into the blastoderm of freshly laid eggs. This protocol includes procedures for production of lentiviral vectors, injection of a virus into zebra finch embryos, and postinjection care. The implementation of the songbird transgenic approach provides a leap toward basic and translational neuroscience that uses an animal model for speech and language and their pathologies. Additionally, the highly quantifiable song behavior, combined with a well-characterized song circuitry, offers an exciting opportunity to develop therapeutic strategies for neurological disorders.

Key words Transgenesis, Songbird, Zebra finch, Lentiviral vector

1 Introduction

Songbirds provide an excellent animal model for the study of basic and translational neuroscience of vocal learning and production. Birdsong is a sequential and quantifiable motor-learning behavior. This behavioral phenotype allows us to quantitatively measure the subtle change, early onset, and developmental trajectory of behavior [1]. Additionally, vocal learning in songbirds has a well-defined, specialized cortical-basal ganglia "song circuit," making it possible to precisely define the circuit functions that underlie vocal and auditory learning [2–8]. Moreover, the many parallels in behavior, development, anatomy, and genes between vocal learning in songbirds and speech learning in humans [9, 10] make songbirds an ideal model for the study of neural basis of speech learning, vocal communication, and associated disorders [11–13].

Guojun Sheng (ed.), *Avian and Reptilian Developmental Biology: Methods and Protocols*, Methods in Molecular Biology, vol. 1650, DOI 10.1007/978-1-4939-7216-6_9, © Springer Science+Business Media LLC 2017

Currently available gene expression and genomic studies of songbirds [14, 15], combined with gene manipulation tools, can allow us to identify and understand the gene and circuit function. The recent development of songbird transgenesis [16–19] provides promising and powerful tools for many applications in basic and translational studies including speech learning and vocal communication, auditory processing, cortical-basal ganglia (CBG) circuitry for reward pathways and motor learning, evolutionary and developmental model, steroid hormone on the brain and behavior, avian cognition, disease model, and therapeutic treatment.

During the past three decades, lentiviral vectors based on HIV-1 have been developed and intensively used in both basic and clinical researches which eventually led to the development of the third-generation lentiviral packaging system which offers higher safety while allowing production of sufficient titers to be used in experiments and therapies. Development of *retrovirus/lentivirus* vectors pseudotyped with envelope glycoprotein of vesicular stomatitis virus (VSV-G) significantly expanded application of these vectors not only to varieties of mammalian cells but also to many nonmammalian cells such as songbird and fish cells. Our protocol is based on the third-generation lentiviral packaging system and produces VSV-G-pseudotyped lentiviral vectors using HEK293T producer cells.

In this protocol, we describe the procedures that include (1) the production of lentiviral vector, (2) the injection of lentiviral-mediated transgene, and (3) the postinjection husbandry to produce transgenic lines.

2 Materials

2.1 Plasmids

1. Vector plasmid (Fig. 1): As the third-generation lentiviral packaging system relies upon the chimeric 5′LTR containing a Tat-independent promoter [20], such HIV-1 vector backbone plasmid (transfer plasmid) pHIV7 was originally obtained from Dr. Yee (City of Hope Medical Center, Duarte, CA) [21]. In this vector, the HIV promoter in the U3 region of the 5′LTR was replaced with the CMV IE promoter, making it Tat protein independent, and in addition, removal of the HIV enhancer and promoter sequences in the U3 region of the 3′LTR resulted in self-inactivating (SIN) vector [22–24]. Furthermore, this plasmid contains cPPT and WPRE cis-enhancing elements [21].
2. Packaging plasmids: Packaging plasmids coding for HIV-1 genes *gag* and *pol* (pLP1) and *rev* (pLP2) were purchased from Invitrogen (currently Thermo Fisher Scientific). Expression of these genes is driven by constitutive CMV (pLP1) or RSV (pLP2) promoter.

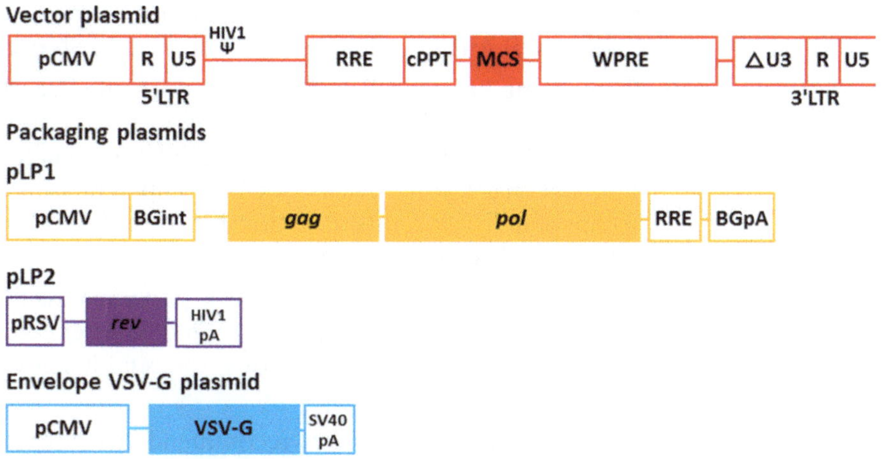

Fig. 1 Plasmids of the third-generation lentiviral packaging system. Multiple cloning site (MCS) in the vector plasmid represents the site of insertion of a promoter of choice along with the gene of interest (GOI) that will be integrated into the host genome. Packaging plasmid code for viral genes *gag* and *pol* (pLP1) and *rev* (pLP2). Envelope VSV-G plasmid codes for VSV-G

3. Envelope VSV-G plasmid (Fig. 1): In our pCMV-G plasmid, the VSV-G gene from vesicular stomatitis virus is used instead of the HIV-1 *env* gene to produce HIV-1 lentiviral vector pseudotyped with VSV-G envelope protein, thus allowing generation of high-titer, pantropic lentiviral vectors [25]. VSV-G gene expression is driven by constitutive CMV promoter and the plasmid was produced in-house [25].

2.2 Cloning and Plasmid Production

1. LB: For 500 ml of Luria Broth (LB) media, in glass bottle, mix each 5 g of tryptone, yeast extract, and NaCl, and add ddH_2O to bring up the volume to 500 ml (*see* **Note 1**). Cap the bottle with screw cap. Autoclave the bottle with LB under liquid cycle, preferably with closed bottles (if program for closed bottles is not available, do not screw the cap tightly). Store at 4 °C or room temperature (RT).
2. LB agar: In a glass bottle, add 7.5 g of agar to 500 ml of LB (*see* **Note 1**). Autoclave the bottle with LB agar under liquid cycle, preferably with closed bottles (if program for closed bottles is not available, do not screw the cap tightly). Remove the autoclaved LB agar from the autoclave, mix well to dissolve the agar completely, and cool it down to approx. 55 °C (*see* **Note 2**). After the agar cooled down, add appropriate antibiotic to the solution (*see* **Note 3**) and pour the LB into 10 cm polystyrene Petri dishes (approx. 20 ml per plate). When pouring plates or adding antibiotics, work near a flame, Bunsen burner, or in a laminar flow hood to keep your work space sterile. Let the LB agar cool down for 30–60 min and then invert the plates; let

them sit for few more hours at RT and then package into plastic bags or seal with parafilm and store at 4 °C (*see* **Note 4**).

3. Competent bacteria: For cloning, we usually use chemically competent bacteria DH5-alpha made in-house (using $CaCl_2$ and $MgCl_2$ solutions) or commercial NEB5-alpha from NEB. You can make your own stock of competent cells using Z-competent *E. coli* Transformation Kit from Zymo Research or $CaCl_2$ and $MgCl_2$ solutions.
4. Restriction enzymes: In general, it is preferable to use restriction enzymes from one source to keep the buffer compatibility which is especially advantageous for simultaneous double digests.
5. Plasmid preparations: We recommend to use kits from Macherey-Nagel or Zymo Research.
6. Site-directed mutagenesis kits: Q5 mutagenesis kit from NEB has been applied for many insertions (up to 100 bp), substitutions, and deletions.

2.3 Cell Culture and Plasmid Transfection

1. HEK293T cells: Producer cells are cultured in Dulbecco's modified eagle's medium (DMEM) supplemented with 10% fetal bovine serum (FBS) and 1× penicillin/streptomycin (P/S) in 15 cm culture dishes or T160 flasks.
2. Transfection: Polyethyleneimine (PEI) method is routinely used for its constant high efficiency, lower amounts of DNAs needed, and low cost.

2.4 Lentiviral Vector Production and Tittering

1. 0.45 μm filter.
2. Buffer for resuspending LVs: 10 mM Tris–HCl, pH 7.8, 1 mM $MgCl_2$, and 3% sucrose in ultrapure sterile ddH_2O.

2.5 Breeding Husbandry

1. Soft food (made fresh daily): including a mix of grinded hard-boil eggs with shells, overnight soaked seed, baby bird formula (Exact from Kaytee), and vitamins (Avia).
2. Dry finch seeds (Blue Seal).
3. Bird bands: L&M Bird Leg Bands, Inc.
4. Breeding cages: (size = 61 cm × 51.3 cm × 36.8 cm) Abba Products.
5. Dummy plastic egg.
6. Egg tray, handmade plastic tray to hold collected eggs.
7. Nest boxes and nest materials: Red bird Products.

2.6 Egg Injection

1. Glass micropipettes and plunger, Wiretrol I, Drummond Scientific.
2. Silicone elastomer (Kwik-Cast sealant), World Precision Instruments.
3. DPBS solution.
4. Phenol red (0.5% in DPBS), Sigma-Aldrich.
5. Incubator, Congo egg incubator and hatcher, HEKA.
6. Nunclon cell culture dishes (35 mm × 10 mm), Sigma-Aldrich.
7. Single-axis oil hydraulic micromanipulator, MO-10, Narishige.
8. Egg-holding mold (waterproof silicon from GE).
9. Micropipette puller P-97, Sutter Instrument.
10. BV-10 micropipette beveler, Sutter Instrument.
11. Stereomicroscope on a boom stand (SMZ800 from Nikon).
12. Microscope camera (DS-Fi1 from Nikon).
13. Light source for stereomicroscope (MKII fiber optic from Nikon).
14. Pipette (10 and 200 μl Pipetman from Gilson).
15. Surgical instruments: stab knife (5 mm blade), Fine Science Tools, and fine forceps (Dumont #55), Fine Science Tools.

3 Methods

3.1 Cloning

Prepare all solutions using ultrapure deionized distilled water and analytical grade reagents. All manipulations with final plasmid preparations, bacteria, producer cells, and lentiviral vectors should be carried out in sterile conditions of laminar flow hood (or next to the flame for bacteria) of BSL2-certified lab (*see* **Note 5**).

1. Ultrapure, nuclease-free ddH_2O.
2. pHIV7 was modified to contain MCS with the following restriction sites to be used for restriction cloning: 5′ BamHI-HpaI-MluI-NheI-NotI-PmeI 3′ and its inverted orientation (Fig. 1). If your cloning vector does not contain any of the restriction sites, proceed with **step 3**, and if yes, go directly to **step 5**.
3. High-fidelity DNA polymerase, such as Taq polymerase or Phusion from NEB, preferably as master mix including dNTPs, $MgCl_2$, and reaction buffers.
4. For PCR-based cloning, design forward and reverse primers so that they will consist of (1) leader sequence (3–6 bp) followed by (2) desired restriction site (usually around 6 bp) and (3) hybridization sequence (18–21 bp). Ensure that both start and stop codons lie within the hybridization sequence. Ensure that the restriction sites do not cut anywhere else within the hybridization sequence. Run PCR and purify the product using PCR

reaction using QIAquick PCR Purification Kit from Qiagen according to the manufacturer's instructions.

5. Proceed with restriction digestion with chosen restriction enzymes using the entire PCR product and 1 μg of recipient vector plasmid. Follow manufacturer's instruction for each restriction enzyme (it is important to use appropriate buffer, temperature and not to use too much of the enzyme to prevent star activity that may result because of increased volume of glycerol in reaction), bring up the volume with ultrapure ddH_2O, and allow the digest to last at least 1–2 h to overnight (*see* **Note 6**). To prevent re-circularization of the vector, treat digested recipient vector with a CIAP (calf intestinal alkaline phosphatase, NEB) after digestion, before gel purification step.
6. Mix appropriate amount of standard agarose for electrophoresis to make 1% solution in 1× TBE buffer (to prepare 1 l of 5× TBE buffer, in a glass bottle, add 54 g of Tris base, 27.5 g of boric acid, and 20 ml of 0.5 M EDTA (pH 8.0), add ddH_2O up to 1 l, and stir; store at RT) in Erlenmeyer flask and heat up in short microwave cycles until all agarose is dissolved. Cast the gel, put in gel comb(s), and wait until solidified, then pipette DNA ladder (such as 1 kb DNA ladder Thermo Fisher Scientific), plasmid digests mixed with loading dye (usually comes together with restriction enzymes), and run the electrophoresis at 5–8 V/cm (distance between electrodes), constant current, until the digests are well resolved. Immerse gel into 1× SYBR® Gold nucleic acid gel stain, incubate approx. 10–40 min, remove from staining solution, and visualize using blue-light transillumінator, 300 nm ultraviolet or 254 nm epi- or transillumination, or laser scanner. Identify the digest bands, cut out from the gel removing all the DNA while trying to reduce the amount of gel surrounding the band, and purify using QIAquick Gel Extraction Kit (*see* **Note 7**).
7. Elute the digests in ddH_2O and measure using NanoDrop. If plasmid concentration is too small, use vacuum concentrator to concentrate the DNA for subsequent DNA ligation.
8. Proceed with DNA ligation. It is generally recommended to use a 3:1 (insert/vector) ratio (if insert is smaller than the vector) while ending up having up to 100 ng of total DNA in reaction. Each DNA has only two ends meaning that in order to keep the 3:1 ratio (three inserts are available for one vector), it is important to calculate the molarity of each DNA used in ligation using online ligation calculator (such as NEBiocalculator).

   Measure and calculate molarity of both insert and vector DNAs; pipette the DNAs, 1 μl of T4 DNA ligase and 2 μl of 10× ligase buffer; bring the volume up to 20 μl with ddH_2O; and let the reaction run for 5–10 min at RT (*see* **Note 8**).

9. Bacteria transformation. Work next to flame to keep sterile conditions of your workspace. From −80 °C freezer, remove tubes of One Shot NEB 5-alpha chemically competent cells for normal cloning and leave them to thaw on ice for approx. 20 min. Pipette 1–2 μl of the ligation reaction (up to 100 ng of DNA) directly into the tube (*see* **Note 9**), mix by gently flicking the tube, and put it on ice for 30 min. Heat shock each transformation tube by submerging the bottom half of the tube into a 42 °C water bath for 30 s. Remove the tube(s) and put them back on ice for 2–5 min; then add 250–500 μl of SOC or RT LB media. In case that other than ampicillin-resistant gene is present in your plasmid, add 250–500 μl of RT SOC media directly into the One Shot tube (*see* **Note 10**), place the tube horizontally into bacteria incubator, and incubate for 45–60 min at 37 °C, 180 RPM.
10. Plate bacteria onto pre-warmed antibiotic selection LB agar plates. Plate 20 and 100 μl of the transformation using sterile single-use polypropylene cell spreaders (such as from VWR), put them back into the incubator, and incubate at 37 °C O/N (*see* **Note 11**). Leave the rest at 4 °C O/N so that in case no colonies are present on the plates on the following day, one could still recover some transformants from the leftover aliquot.
11. Pick colonies using sterile pipette tips. Place the colony into 5 ml of LB containing selection antibiotic by discharging it into the LB and then pipetting up and down few times before trashing the tip. Place the minipreps into bacterial incubator at 37 °C O/N at 180 RPM (*see* **Note 12**).

3.2 Plasmid Preparation and Sequencing

1. It is recommended to prepare glycerol stocks of each miniprep to shorten the plasmid generation. Remove 500 μl of the grown miniprep culture and mix with 500 μl of autoclaved 50% glycerol (to prepare 50% glycerol solution, mix glycerol 1:1 with ultrapure ddH_2O, put into glass bottle, and autoclave using a program for closed liquids) in sterile 1.5 ml screw tube and place into −80 °C.
2. Spin down the minipreps at 4000 RPM (RCF 2900 × *g*) for 10 min at 4 °C and discard the supernatant. Follow the instructions from the kits (mini, midi, maxi) for plasmid preparation which are highly recommended as the kits are very cost-effective, and the produced plasmids are usually clean and directly suitable for downstream applications.
3. Most of the cloning plasmids have primer sequences flanking the MCS so that standard primers such as T7, M13, etc., can be used for sequencing of the insert. If no such sequences are present or the insert is too big (over 2000 bp), design primers

using one of the web tools (such as OligoPerfect™ Designer from Thermo Fisher Scientific). General recommendations are to keep the GC ratio between 45% and 60%, melting temperature 55–65 °C, and length 18–30 nucleotides. Submit samples for sequencing.

4. Align sequencing reads to in silico build vector and select the correct clone (*see* **Note 13**).
5. Remove the glycerol stock of the selected clone from −80 °C, thaw on ice, and pipette 20–100 μl into autoclaved Erlenmeyer flask with 200–1000 ml of LB containing selection antibiotic. Cover the flask with sterile aluminum foil and place into bacterial incubator and incubate at 37 °C, 180 RPM O/N.
6. Spin down the bacteria and discard supernatant. Use midi-/maxiprep plasmid preparation kit based on anion exchange to achieve transfection grade plasmid DNA. Perform the last step in the flow hood (after the wash with 70% EtOH, transfer the tube into flow hood, remove the EtOH supernatant, dry the pellet, and resuspend in ultrapure ddH_2O). Take a small aliquot, measure DNA concentration on NanoDrop, then aliquot the plasmid into sterile tubes inside the flow hood, and store at −20 °C.

3.3 Culture and Transfection of HEK293T Cells

1. Get the producer HEK293T cells from a cell bank or company (ATCC or Thermo Fisher Scientific) (*see* **Note 14**). Thaw the vial in a water bath at 37 °C and immediately transfer into sterile laminar flow hood and put the thawed cell suspension (approx. 1 ml) into 15 ml falcon tube containing 5 ml of culture media (resuspend 2–4 times using 2 ml pipette and pipette the cell suspension, drop by drop, into cold culture media; for 1 l of culture media, mix 890 ml of DMEM with 100 ml of freshly thawed FBS and 10 ml of 1× P/S; filter through 0.22 μm filter, aliquot, and store at 4 °C). Close the tube and centrifuge at 1000 RPM (168 × *g*), 5 min, RT.
2. Transfer the tube back to flow hood, discard the supernatant, and resuspend in 2 ml of pre-warmed culture media. Count the cells using single-use hemocytometer (such as cell chip from Bioswisstec), optical-based cell counter (such as Countess II from Thermo Fisher Scientific), or electric field multichannel cell counting system (CASY by OLS). Resuspend the cells into larger volume of pre-warmed culture media so that you plate around 5 million viable cells per 20 ml of media per 1 cell culture for 15 cm dish or T-160 flask. Place the culture vessels with cells into humidified cell culture incubator (5% CO_2, 37 °C) and expand as necessary (change culture media three times a week).
3. When cells reach approx. 80–90% confluency, wash the cells using pre-warmed PBS, then passage the cells using 0.05%

Trypsin-EDTA (8–10 ml for 15 cm dishes, T-160 flasks), then block trypsin with the same volume of culture media (containing FBS), and centrifuge at 1000 RPM (168 × *g*), 5 min, RT. Count the cells and subculture them for experiment or freeze down for later use.

4. To freeze the cells, centrifuge at 1000 RPM (168 × *g*), 5 min, RT, and resuspend in freezing media containing 90% of culture media and 10% of sterile DMSO so that there will be approx. 6–7 million cells per ml per 1 cryotube (*see* **Note 15**). Pipette the cell suspension into cryotubes, close the caps, and let it stand for approx. 10 min at RT. Place the tubes into Nalgene Mr. Frosty freezing container, place into −80 °C freezer O/N, then transfer the tubes into liquid nitrogen cell freezer, and store until further use.
5. For one batch of lentiviral vector production, grow HEK293T cells in four 15 cm or T-160 flasks until they are approx. 90–100% confluent (do not overgrow). The day before the transfection, split the cells diluting 2.5-fold, so that four 15 cm plates (or T-160 flasks) will become ten 15 cm plates (or T-160 flasks). On the day of transfection of ten 15 cm plates, take 5 ml of Opti-MEM (Gibco-Thermo Fisher) and add each 100 μg of the transfer pHIV7 vector containing your gene of interest, LP1, LP2, and 50ug of pVMV-G plasmid (1:1:1:0.5 ratio), and mix to dissolve the plasmid DNAs well in the Opti-MEM. Add 900 μl of PEI solution into the DNA-Opti-MEM solution, mix and leave 15–20 min at RT in the BSL2 hood (*see* **Note 16**) (Fig. 2).
6. After 15–20 min incubation, add 14.1 ml of 10% FBS-DMEM pre-warmed at 37 °C, mix well, add 2 ml of solution dropwise to each culture plate, and mix gently.
7. Place the cells back to incubator and culture for 3 days.

3.4 Lentiviral Vector Preparation and Tittering

1. At the day 1, 2, and 3 post-transfection, collect the conditioned culture medium from producer cells (collect the conditioned medium and add new culture medium) (*see* **Note 17**). Keep the day 1 and day 2 media on ice until the day 3 collection is completed. Filter the collected conditioned medium through a 0.45 μm filter.
2. Concentrate the filtered medium by centrifugation at 7000 RPM (6000 × *g*) for 16 h at 4 °C with an appropriate centrifuge and rotor (for instance, Sorvall centrifuge with Sorvall GS-3 rotor).
3. Resuspend the resulting pellet in 10 mM Tris–HCl, pH 7.8, 1 mM $MgCl_2$, and 3% sucrose (*see* **Note 18**), place into Eppendorf tubes making 10–100 μl aliquots, and store in −80 °C.

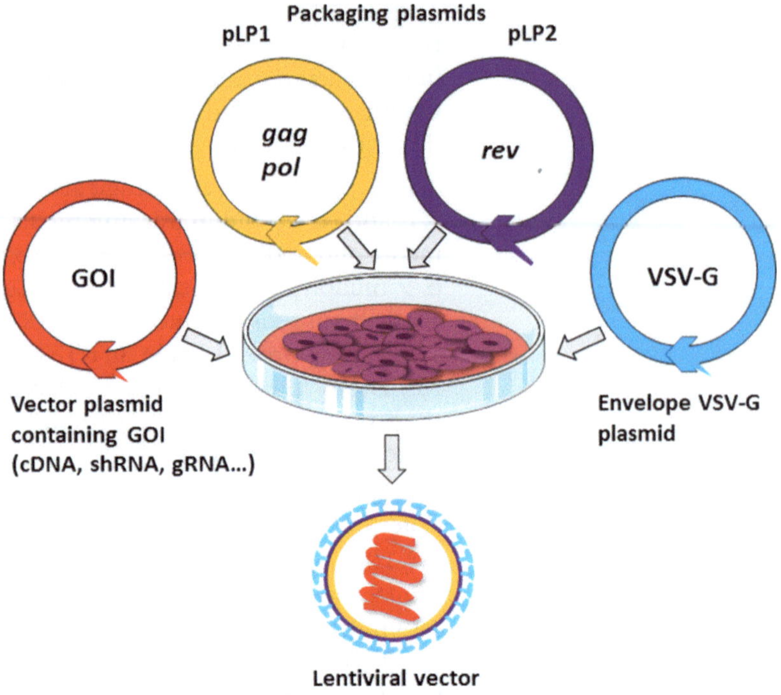

Fig. 2 Lentiviral production. Simplified overview of lentiviral production where all four plasmids of the third lentiviral packaging system (color-coded) are transfected into the producer HEK293T cells which leads to the expression of the encoded viral genes (along with the GOI as a part of the vector plasmid), subsequent self-assembly of the lentiviral vectors within the producer cells, and their release into the medium which is then collected

4. To assess the infectious titers of the newly produced lentiviral vector, use a standard vector with known titer and with the same backbone and promoter preferably driving some common reporter gene, such as EGFP. Plate the HEK293T cells in a 6-well plate and transduce them with different amounts of lentiviral vectors (both the standard and new vector) in the presence of Polybrene (Millipore, 4 μg/ml). Passage the transduced cells once every 4 days in 1/3 ratio so that 1/3 of the culture is passaged and 2/3 is trashed. At the day 14 post-transduction, wash the cells with PBS; detach with 0.05% trypsin; centrifuge at 1000 RPM (168 × *g*), 5 min, RT; wash with PBS; and centrifuge once more. Discard the supernatant and prepare cell DNAs according to the instructions in DNeasy Blood & Tissue Kit (Qiagen). To estimate the titers, perform real-time qPCR using a primer set selected from the WPRE sequence (same for both vectors). Compare the results between standard and new vectors and calculate the titer of the new vector accordingly (*see* **Note 19**).

3.4.1 Breeding Care and Egg Collection

1. Breeding colony. The breeding colony of the zebra finch is established by pairing adult males and females, depending on the number of eggs required for egg injection. Each breeding pair is housed in a small breeding cage (size = 61 × 51.3 × 36.8 cm). In each breeding cage, water, cuttlebone, gravel, dry seeds, and soft food are provided. The bottom paper of the cage is replaced every other day, and the cage is cleaned once a month (*see* **Note 20**). Each breeding pair produces 3–7 eggs per clutch, and the number of eggs being produced declines with consecutive breeding trials. We give each breeding pair a few months of rest after they lay three consecutive clutches of eggs.
2. Egg collection. The fresh-laid eggs are collected in the morning before the parents start incubating. Female finches lay one egg per day and continue to lay 3–7 eggs per clutch. Each fresh-laid egg is collected in the morning and is immediately replaced by a dummy egg; otherwise the female may stop laying additional eggs. The collected eggs can be temporarily stored in a low-temperature incubator or similar environment (~19 °C) to prevent embryo development. The fresh-laid eggs can be stored in this condition for up to a week. Each egg's hatching date is marked on an egg tray to identify the date of egg collection.

3.4.2 Egg Injection

1. Making micropipette injector. Pull glass micropipettes (Wiretrol-I, 1–5 μl) with a pipette puller (PE70). Using a fine stab knife to break the tip of the micropipette, bevel the tip with a pipette beveller, so the outer diameter of the tip is 20–30 μm.
2. Prepare egg holders by making silicone molds. Fill the silicone on the bottom of the culture dishes (35 mm × 10 mm). After an overnight of air-dry, place a finch egg in the silicon-filled culture dish (use eggs that have different sizes or shapes) and fill the silicone to the halves of the egg (*see* Fig. 3). Gently remove the egg after a few days of air-dry.
3. Loading of the viral solution. Before injection, mix 0.2 μl of the phenol red into an aliquot of lentiviral solution (2 μl). The phenol red allows us to better visualize the position of the viral solution in the micropipette and its movement during each injection.
4. Place a beveled micropipette on the top of the virus-phenol red mix (~2 μl), and let the micropipette slowly backfill the viral mix. After backfilling the viral mix into the glass pipette, add 3–4 μl of mineral oil backfilling into the pipette. Place a plunger into the glass pipette (comes with the micropipettes) and push the viral solution and mineral oil until the viral mix reaches near the tip of the micropipette, making sure there are no air bubbles. Place the micropipette and plunger into the holder of the micromanipulator (Fig. 3).

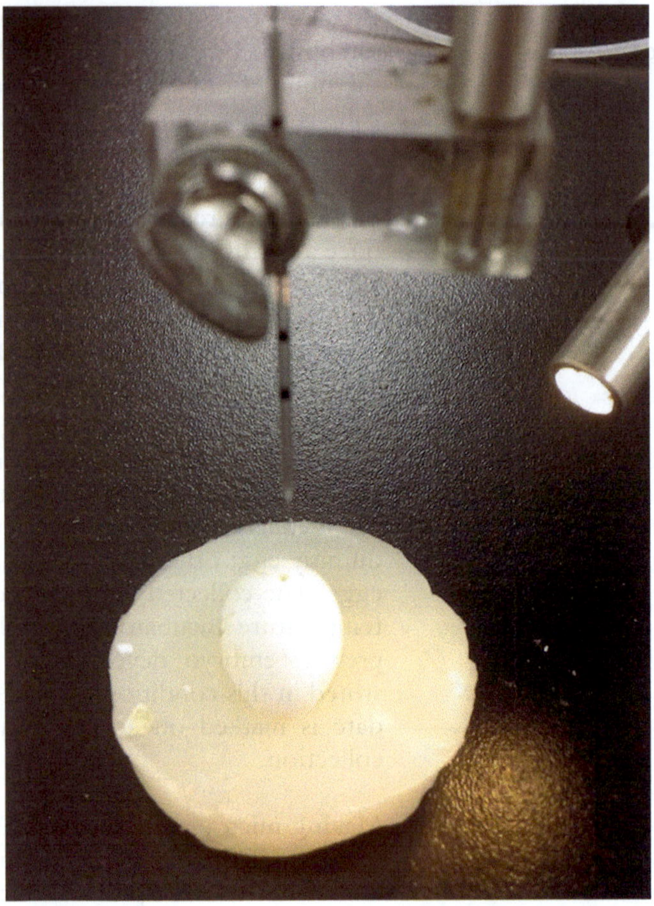

Fig. 3 Injection apparatus

5. Pick up a fertilized egg and gently clean the eggshell with a sterile cotton swab lightly dipped with 70% ethanol. Place a fertilized egg in an egg holder (Fig. 3) and let the egg sit for approximately half an hour. This will allow a healthy embryo slowly move to the top of the egg. Place the egg and egg holder under a stereomicroscope; use a fiber optic illuminator to identify the precise position of embryo through the microscope (Fig. 3). Lightly mark the embryo location with a pencil.
6. Under a stereomicroscope, use a fine stab knife to cut open a small window of the eggshell that is around the marked embryo location. Keep the size of window small, but the opening should be big enough to visualize the entire embryo (Fig. 4a). Use fine forceps to gently remove the eggshell, shell debris, and the underlining membrane. The healthy embryo should be exposed underneath the opening window. Use a Pipetman (200 μl) to add clear albumen collected from another egg or DPBS to prevent the embryo from drying.

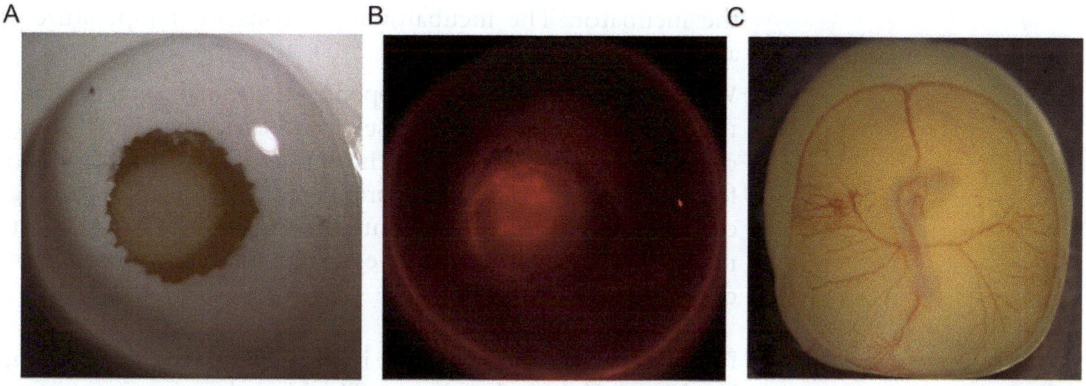

Fig. 4 Embryonic injection and development of the injected embryo. (**a**) A small window is cut open by a fine stab knife and the embryo can be seen underneath the opening window. (**b**) After viral injection, the injected sites can be identified by phenol red. (**c**) Injected eggs are incubated in an incubator, and at the fourth to fifth day of incubation, the eggs are transferred to the nests of foster parents

7. Adjust the location of the egg and egg holder so that both the injecting micropipette and the embryo can be seen under the microscope. Through the microscope, identify the center of the embryo (Fig. 4a). Slowly lower the micromanipulator until the tip of the micropipette is inside of the albumen, then release a small amount of viral solution to make sure the injecting micropipette is not clogged.
8. During each viral injection, quickly lower the micromanipulator so that the tip of micropipette penetrates the vitelline membrane that covers the embryo and pierce through the blastoderm (i.e., through the surface of the embryo). At the same time of penetration, quickly release a small amount of viral solution (30–50 nl) into the subgerminal cavity. The number of viral injections varies from 20 to 35 injections per embryo. The more injections we make in each embryo, the better chance the germ-line cells to be infected and labeled, but more injections also cause more damage to the embryo, and therefore the embryonic survival and hatching rate will be reduced. After injection, the injection sites might be floated with viral solution (Fig. 4b). Use a Pipetman to remove the floating viral solution and add DPBS solution or clear albumen to cover the embryo. Prevent any air bubble from generating during the pipetting.
9. After injection, gently shake the embryo away from the injecting window to prevent the embryo from sticking to the opening window. Cover the window with DPBS or clear albumen to prevent air-drying of the embryo, and then seal the window using a piece of eggshell that is freshly cut from an unfertilized egg. Let the sealed eggshell air-dry for 20 min. Use a pencil to label on the eggshell for identification, and transfer the eggs to

the incubator. The incubator has a constant temperature of approximately 37 °C.

10. When the embryos are developed to the fourth or fifth day in the incubator, the blood vessels can be clearly seen through candling (Fig. 4c). Transfer the injected eggs to the nest of foster parents. If the foster parents are not available, the eggs can be kept in the incubator until they hatch. The hatching rate might be slightly lower if the eggs remain in the incubator to develop (*see* **Note 21**).

3.4.3 Postinjection Care and Genotyping

1. Postinjection care is critically important for the survival and hatching of injected embryos. The ideal situation is to set up many breeding pairs of experienced zebra finches and use them as the foster parents. When the foster parents produce their own eggs and start incubating, place injected eggs into the foster parent's nest and replace with the foster's own eggs (*see* **Note 22**). Parental care from experienced and caring foster parents is important for the survival of embryos and hatchlings. The hatching rate of injected embryos can be determined by a number of other variables (*see* **Note 23**).
2. Genotyping. If the transgenic finch's transgene is fused with a fluorescent marker (e.g., GFP), one can examine the success of viral delivery by placing a hatchling bird under a fluorescent stereomicroscope and identify the expression of fluorescence. If the injected transgene does not have markers or the marker has a very weak expression that cannot be easily detected by a fluorescent microscope, immunohistochemistry or PCR will be used to identify the expression of transgene.
3. For PCR genotyping, the blood sample is collected from the brachial vein on the wing of hatchlings. Use a 30 gauge size syringe needle to poke a brachial vein, and then place a heparinized capillary tube to draw blood samples (4 μl). Extract DNA using DNeasy Blood & Tissue Kit (Qiagen) and perform PCR to identify the expression of GFP or transgene. Levels of GFP expression vary significantly among injected individuals. This genotyping method is combined with observation of behavioral phenotypes to select potential mosaics to breed.

4 Notes

1. Alternatively, ready-to-use commercial LB and LB agar preparations (powder, liquid LB, poured plates, etc.) are available.
2. Use water bath set to 55 °C so that you will prevent solidification of the agarose and so that you don't have to monitor the agarose so closely before adding the antibiotics.

3. Working concentrations of most antibiotics could be found using the following link: http://www.sigmaaldrich.com/life-science/core-bioreagents/learning-center/antibiotic-selection.html
4. Most of the plates could be stored for up to 2 months; however, ampicillin plates tend to be less stable. Always plate control-competent (non-transformed) cells when using ampicillin plates older than 6–8 weeks. Alternatively, use carbenicillin which is more resistant to degradation than ampicillin.
5. It is a good idea to resuspend the DNA pellets after the EtOH wash in sterile conditions of laminar flow.
6. Star activity may be a result of prolonged digestions. Check manufacturer's instructions for each restriction enzyme before pipetting the reaction.
7. Strong bands stained with SYBR® Gold that are a result of digestion of around 4 μg of plasmid could easily by seen in daylight. Cut the bands of digested DNA in a way so that there will be minimal inclusion of agar that does not contain DNA. Put the gel bands into the tube and weight.
8. Most ligations would work well at RT for 5–10 min. For blunt or single-base overhangs ligations, however, run the ligation up to 15 min RT or O/N at 16 °C.
9. For ligations, always transform bacteria with vector only to estimate the rate of re-circulated vectors that could be present in the ligation-transformed plates.
10. Use One Shot 50 μl competent cells or if you have produced competent cells yourself and have larger aliquots, thaw the aliquot on ice, remove 50 μl per transformation, and snap-freeze the rest that can be used for another transformation.
11. When cloning very long (close to 10 kb) or genes with long repetitive sequences into pHIV7, culture the NEB stable at 30 °C to prevent recombination.
12. Minipreps could either be grown in 12 ml Falcon tubes with round bottoms or in closed 50 ml Falcons (there is enough air for the culture to grow).
13. Preferably, use software, such as SnapGene.
14. It is not recommended to take cells from colleagues as these may come with potential mycoplasma or other contaminations.
15. This way, one should recover enough cells from one cryovial to plate onto one 15 cm dish or T-160 flask.
16. Polyethylenimine (PEI) (Polysciences Inc., PA): Make up stock solution at 1 mg/ml in ddH_2O, neutralize with HCI to pH 7.2, and filter sterilize (0.22 μm). Aliquot and store at −80 °C. A working stock can be kept at 4 °C.

17. Virus titer on the day 1 collection is not as high as those of at day 2 and day 3. If your centrifugation capacity is small, you may omit the day 1 collection.
18. This buffer formulation is an FDA-approved solution. We expect virus-protective effect with the 3% sucrose during the preservation.
19. This method is to quantitate the copy numbers of the provirus (vector sequence) integrated into the target cell chromosomes. LV vector preparations contain excess amount of plasmid DNAs that were used for LV production. These plasmid DNAs are carried into the target cell at the LV vector infection and interfere quantification of the infectious titer by qPCR. Culturing the infected cells for 2 weeks with a couple of passages eliminates the contaminated plasmid DNAs.

    WPAS primers:

    FW: 5′ GAGGAGTTGTGGCCCGTTGTCAGGCAACG 3′
    Rev.: 5′ CAGGCGAGCAGCCAAGGAAAGGACGATG 3′
    PCR product: 254 bp.
20. If the zebra finches are randomly paired, it usually takes 1–3 weeks for them to start laying eggs, depending on the compatibility of each breeding pair.
21. The injected eggs may hatch 1–2 days later than the normal incubation period of wild-type eggs (i.e., 14 days).
22. In our transgenic finches [18], injected embryos are weaker than wild-type embryos and may hatch 1–2 more days later than wild-type eggs. Sometimes the injected embryos were unable to open the eggshell and then died. Pay close attention to the injected eggs when they are about to hatch by inspecting the embryos 14–15 days of incubation period.
23. There are a number of factors that may affect the hatching rate of injected embryos:
    (a) Preinjection bird husbandry: The general health of the embryos may be important for the survival and hatching of injected embryos; it is essential to provide a good husbandry care of the finch colony.
    (b) The transgene itself or the quality of lentiviral construct can affect the embryonic development and hatching rate.
    (c) The size of the glass pipette tip for injection. Larger diameter of pipette tip is more likely to damage the embryos.
    (d) The amount of viral injection and the duration of injection in each embryo are important and a higher viral titer requires a smaller injection.
    (e) Postinjection husbandry care, documentation, and estimation of the expected hatching date of each injected egg.

References

1. Lipkind D, Marcus GF, Bemis DK, Sasahara K, Jacoby N, Takahasi M, Suzuki K, Feher O, Ravbar P, Okanoya K et al (2013) Stepwise acquisition of vocal combinatorial capacity in songbirds and human infants. Nature 498:104–108
2. Mooney R (2009) Neural mechanisms for learned birdsong. Learn Mem 16:655–669
3. Fee MS, Scharff C (2010) The songbird as a model for the generation and learning of complex sequential behaviors. ILAR J 51:362–377
4. Fee MS, Goldberg JH (2011) A hypothesis for basal ganglia-dependent reinforcement learning in the songbird. Neuroscience 198:152–170
5. Kao MH, Doupe AJ, Brainard MS (2005) Contributions of an avian basal ganglia-forebrain circuit to real-time modulation of song. Nature 433:638–643
6. Woolley SM (2012) Early experience shapes vocal neural coding and perception in songbirds. Dev Psychobiol 54:612–631
7. Theunissen FE, Shaevitz SS (2006) Auditory processing of vocal sounds in birds. Curr Opin Neurobiol 16:400–407
8. Knudsen DP, Gentner TQ (2010) Mechanisms of song perception in oscine birds. Brain Lang 115:59–68
9. Jarvis ED (2004) Learned birdsong and the neurobiology of human language. Ann N Y Acad Sci 1016:749–777
10. Doupe AJ, Kuhl PK (1999) Birdsong and human speech: common themes and mechanisms. Annu Rev Neurosci 22:567–631
11. Brainard MS, Doupe AJ (2013) Translating birdsong: songbirds as a model for basic and applied medical research. Annu Rev Neurosci 36:489–517
12. Konopka G, Roberts TF (2016) Animal models of speech and vocal communication deficits associated with psychiatric disorders. Biol Psychiatry 79:53–61
13. Fisher SE, Scharff C (2009) FOXP2 as a molecular window into speech and language. Trends Genet 25:166–177
14. Warren WC, Clayton DF, Ellegren H, Arnold AP, Hillier LW, Kunstner A, Searle S, White S, Vilella AJ, Fairley S et al (2010) The genome of a songbird. Nature 464:757–762
15. Zhang G, Jarvis ED, Gilbert MT (2014) Avian genomes. A flock of genomes. Introduction. Science 346:1308–1309
16. Velho TA, Lois C (2014) Generation of transgenic zebra finches with replication-deficient lentiviruses. Cold Spring Harb Protoc 2014:1284–1289
17. Abe K, Matsui S, Watanabe D (2015) Transgenic songbirds with suppressed or enhanced activity of CREB transcription factor. Proc Natl Acad Sci U S A 112:7599–7604
18. Liu WC, Kohn J, Szwed SK, Pariser E, Sepe S, Haripal B, Oshimori N, Marsala M, Miyanohara A, Lee R (2015) Human mutant huntingtin disrupts vocal learning in transgenic songbirds. Nat Neurosci 18:1617–1622
19. Agate RJ, Scott BB, Haripal B, Lois C, Nottebohm F (2009) Transgenic songbirds offer an opportunity to develop a genetic model for vocal learning. Proc Natl Acad Sci U S A 106:17963–17967
20. Dull T, Zufferey R, Kelly M, Mandel RJ, Nguyen M, Trono D, Naldini L (1998) A third-generation lentivirus vector with a conditional packaging system. J Virol 72:8463–8471
21. Yam PY, Li S, Wu J, Hu J, Zaia JA, Yee JK (2002) Design of HIV vectors for efficient gene delivery into human hematopoietic cells. Mol Ther 5:479–484
22. Yee JK, Moores JC, Jolly DJ, Wolff JA, Respess JG, Friedmann T (1987) Gene expression from transcriptionally disabled retroviral vectors. Proc Natl Acad Sci U S A 84:5197–5201
23. Yu SF, von Ruden T, Kantoff PW, Garber C, Seiberg M, Ruther U, Anderson WF, Wagner EF, Gilboa E (1986) Self-inactivating retroviral vectors designed for transfer of whole genes into mammalian cells. Proc Natl Acad Sci U S A 83:3194–3198
24. Zufferey R, Dull T, Mandel RJ, Bukovsky A, Quiroz D, Naldini L, Trono D (1998) Self-inactivating lentivirus vector for safe and efficient in vivo gene delivery. J Virol 72:9873–9880
25. Yee JK, Miyanohara A, LaPorte P, Bouic K, Burns JC, Friedmann T (1994) A general method for the generation of high-titer, pantropic retroviral vectors: highly efficient infection of primary hepatocytes. Proc Natl Acad Sci U S A 91:9564–9568

Chapter 10

In Ovo Electroporation Methods in Chick Embryos

Hidekiyo Harada, Minoru Omi, and Harukazu Nakamura

Abstract

To elucidate a gene function, in vivo analysis is indispensable. We can carry out gain and loss of function experiment of a gene of interest by electroporation in ovo and ex ovo culture system on early-stage and advanced-stage chick embryos, respectively. In this section, we introduce in/ex ovo electroporation methods for the development of the chick central nervous system and visual system investigation.

Key words In ovo electroporation, Ex ovo electroporation chick, Central nervous system, Visual system, Tol2 transposon, Transient misexpression, Long-term misexpression

1 Introduction

Chick embryos are one of the best model systems for the study of developmental biology, since their excellent accessibility enables us to carry out in ovo operations easily and quickly. Especially, in ovo electroporation has enabled for us to analyze gene function on wide range of organs at various developmental stages [1]. Here we introduce electroporation to the midbrain for transient misexpression and for long-term misexpression by transposon system. We also introduce temporal control of gene expression by combination of transposon and tetracycline-induced gene expression system. For electroporation to the advanced stage embryos, we introduce ex ovo electroporation in shell-less culture system.

2 Materials and Principles

2.1 Equipment

1. CUY21 (Bex Co. Ltd) (Subheadings 3.1–3.4)
2. CUY21EX (Bex Co. Ltd) (Subheading 3.5)
3. Dissection microscope

Guojun Sheng (ed.), *Avian and Reptilian Developmental Biology: Methods and Protocols*, Methods in Molecular Biology, vol. 1650, DOI 10.1007/978-1-4939-7216-6_10, © Springer Science+Business Media LLC 2017

4. Parallel electrodes for the mesencephalon (0.5 mm in diameter, 1.0 mm in length, 4.0 mm in distance; Unique Medical Imada, Natori, Japan) (Subheadings 3.1, 3.3, and 3.4)
5. Parallel electrodes for optic vesicle (0.3 mm in diameter, 1.0 mm in length, 2.0 mm in distance; Unique Medical Imada, Natori, Japan) (Subheadings 3.2 and 3.3)
6. Forceps-type electrode (3 mm square, LF646P3x3, Unique Medical Imada, Natori, Japan) (Subheading 3.5).
7. Glass capillary (capillary was a glass capillary puller; Narishige, Tokyo, Japan)
8. 5 ml syringe with 18 G needle
9. 1 ml syringe with 32 G needle
10. Cellophane tape
11. Scissors
12. Tungsten needle
13. 100 mm Petri dish

2.2 Reagents

1. 1× PBS (137 mM NaCl, 2.7 mM KCl, 10 mM Na_2HPO_4, 2 mM KH_2PO_4)
2. Black ink (Rotring Isograph and Variant Drawing Ink, black).
3. 0.5% fast green in 1× PBS
4. TE buffer (10 mM Tris-Cl pH 8.0, 1 mM disodium EDTA pH 8.0)

2.3 Animals

Fertilized chicken eggs incubated in a well-humidified chamber. For transfection to the brain vesicles in ovo, embryos at Hamburger and Hamilton (HH) stages 9–12 are preferable [2]. To reach HH stages 9–12, it takes 29–45 h. For ex ovo electroporation, embryos at HH stages 17–19 (E2.5–E3) are preferable.

2.4 DNA Constructs

2.4.1 Selecting Vectors for Electroporation Experiments

1. *Transient expression*

   pCAGGS or pMiwIII is suitable for electroporation for transient misexpression [3, 4]. These DNA vectors give a strong and ubiquitous expression and low toxicity for chick embryos. Since the plasmids in a cell are diluted at cell division, the expression level diminishes during development.
2. *Long-term expression*

   Recently, transposon-mediated gene transfer system has been developed for electroporation. Tol2 transposon, which was found in Medaka, could be integrated into genome by transposase activity [5–8]. Sato et al. [5] have generated transposase vector, pCAGGS-T2TP, and pT2K-CAGGS-EGFP vector for transgene expression, which enables long-term stable misexpression of transgene.

3. *Tetracycline-mediated gene-inducible system*
   Tetracycline-mediated gene-inducible system is a very powerful tool for analyzing gene function. pT2K-BI-TRE-EGFP is a bidirectional expression vector that contains transposon and tetracycline-responsive element (TRE) (Clontech) so that it assures stable and conditional expression of the gene of interest by tetracycline treatment. pT2K-CAGGS-tTA and pT2K-CAGGS-rtTA-M2 vectors could be used for Tet-Off and Tet-On system, respectively. tTA (*t*etracycline *t*rans*a*ctivator) constitutionally binds to TRE and activates the transgene. Tetracycline binds to tTA and provokes conformational change of tTA and dissociation of tTA from TRE, which silences transcription of the transgene. Thus it works as Tet-Off. Conversely tetracycline binding to rtTA (*r*everse *t*etracycline *t*rans*a*ctivator) enhances binding of the rtTA to TRE and activates transcription of the transgene. Thus it works as Tet-on [9]. If you co-transfect pT2K-BI-TRE-EGFP that contains transgene with pT2K-CAGGS-rtTA-M2 and pCAGGS-T2TP, you can activate transgene expression by Dox administration (Tet-On system). If you co-transfect pT2K-BI-TRE-EGFP with pT2K-CAGGS-tTA and pCAGGS-T2TP, you can deactivate transgene expression by Dox administration (Tet-Off system). Since transposase vector is co-electroporated, long-term regulation of the transgene expression is possible.

3 Method

By HH stage 12, five brain vesicles are visible: telencephalon, diencephalon, mesencephalon, metencephalon, and myelencephalon. Optic vesicles are visualized in the diencephalon around HH stage 9. These regions are easy to be recognized under a dissection microscope. It is easy to inject DNA plasmid vector solution into these regions by a micro-glass capillary. Since DNA plasmid vector is charged negatively, transgene expression is seen only at the anode side of the neural tube after electroporation (Fig. 1). Thus the cathode side could serve as the control. Protein expression could be detected in 2–3 h after electroporation.

3.1 Electroporation to the Midbrain

1. Sterilize eggshell by wiping 70% ethanol absorbed kimwipe.
2. Make a hole on the sharp end of the egg with a tip of a dissection scissors, and remove 2 ml of albumen by a syringe with 18 G needle.
3. Open the eggshell at the top of the egg by scissors. Tear the vitelline membrane by a tungsten needle under a dissection microscope (Fig. 1a).

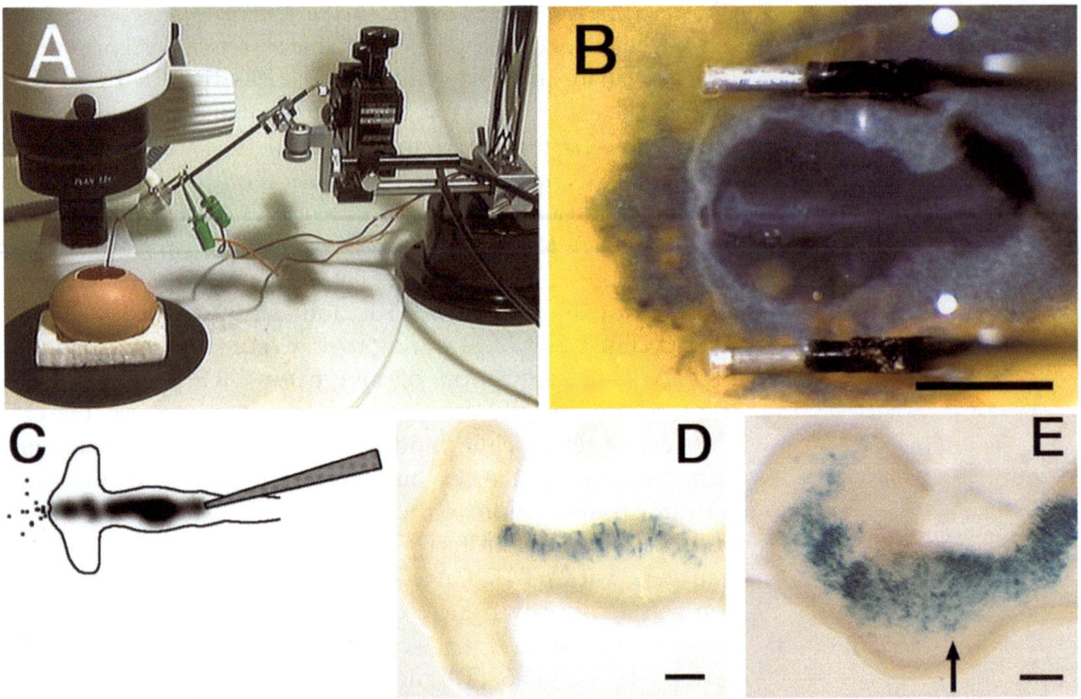

Fig. 1 In ovo electroporation to the midbrain. (**a**) Whole set of in ovo electroporation. A pair of electrodes held by a manipulator is inserted through a window cut through the shell. (**b**) A pair of electrodes is put on the vitelline membrane. The electrode is 0.5 mm in diameter, 1.0 mm in length, and 4.0 mm in distance. Ink injected beneath the embryo visualizes the embryo. (**c**) Plasmid DNA is injected from dorsal side of the brain vesicle by glass capillary. (**d**) LacZ expression at 3 h after electroporation. (LacZ expression is detectable 2 h after electroporation.) (**e**) LacZ expression at 24 h after electroporation

4. It is difficult to identify embryos at early developmental stages (around HH stage 10). Injection of ink (10× dilution by 1× PBS or 0.5% fast green solution) beneath the embryo by 32 G needle in 1 ml syringe will help you to visualize embryos under the dissection microscope (Fig. 1a, b).
5. Make a pore at the tip of the forebrain. Inject DNA plasmid vector solution (mixed with fast green, 0.05% final concentration) into the neural tube from the hindbrain enough amount to fill the entire mesencephalic vesicle (Fig. 1c). Since neural tube is closed after HH stage 8 and turn to 90° left side after HH stage 12, suitable stages for midbrain electroporation are between HH stages 8 and 12.
6. Place the parallel electrodes (0.5 mm in diameter, 1.0 mm in length, 4.0 mm in distance) across the mesencephalic vesicle (Fig. 1a). Pulse a 25 V rectangular pulse four times at 950 ms intervals.
7. Close the hole on the eggshell by tape (cellophane tape), and re-incubate the egg in 38 °C to reach the desired stage to observe.

Make sure the tape is completely closed to prevent infection and desiccation of the embryo. You can detect transfected gene expression from 2 h after electroporation (Fig. 1d, e).

3.2 Retinal Fiber Tracing by Tol2 Transposon-Mediated Gene Transfer

Recently, transposon-mediated gene transfer system has been developed for misexpression and widely applied to chick embryos by electroporation method for long-term misexpression. Tol2 transposon flanking region is integrated into genome by transposase activity. Sato et al. [5] have generated pT2K-CAGGS vector that enables us to long-term stable misexpression after in ovo electroporation.

With this method, we could trace retinal fibers by electroporating optic vesicles at early the stage of development [10–12]. By electroporating pT2K-CAGGS-GFP and pCAGGS-T2TP (transposase expression vector), we could get transfection of the temporal or nasal half of the optic vesicle, and as a result, we could get sharp border at the middle of the temporo-nasal axis of the optic tectum (Fig. 2c–f).

1. Follow the protocol described above from Subheadings 3.1 to 3.3.
2. Insert a glass capillary from the tip of the forebrain of the HH stages 10–12 embryos. When the tip of the glass capillary reached inside of the optic vesicle, inject a solution of a mixture of pCAGGS-T2TP and pT2K-CAGGS construct into the optic vesicle. Be careful not to leak the DNA solution outside of the optic vesicle.
3. Place the electrodes (2 mm distance) on the vitelline membrane to be rectangular to the nasal-temporal axis of the optic vesicle (Fig. 2a, b). If you target the nasal half of the eye, place the cathode just anterior of the optic vesicle and anode rectangular to the optic vesicle. To label the temporal half of the eye, the anode places just posterior to the caudal side of the optic vesicle, and cathode places anterior to the rostral side.
4. Give the rectangular pulses (three times, 13 V, 50 ms on, 950 ms off).
5. Seal the hole on the eggshell by tape (cellophane tape), and re-incubate the egg in 38 °C incubator to reach the desired stage for observation. Make sure the tape is completely closed to prevent infection and desiccation of the embryo. The retinotectal map is considered to be matured by E18 in chick.

3.3 Combination of Retinal Fiber Tracing and Midbrain Electroporation

If you combine retinal fiber tracing and transfection of the guidance molecule in the midbrain by electroporation, you can pursue the roles of these molecules in axon guidance.

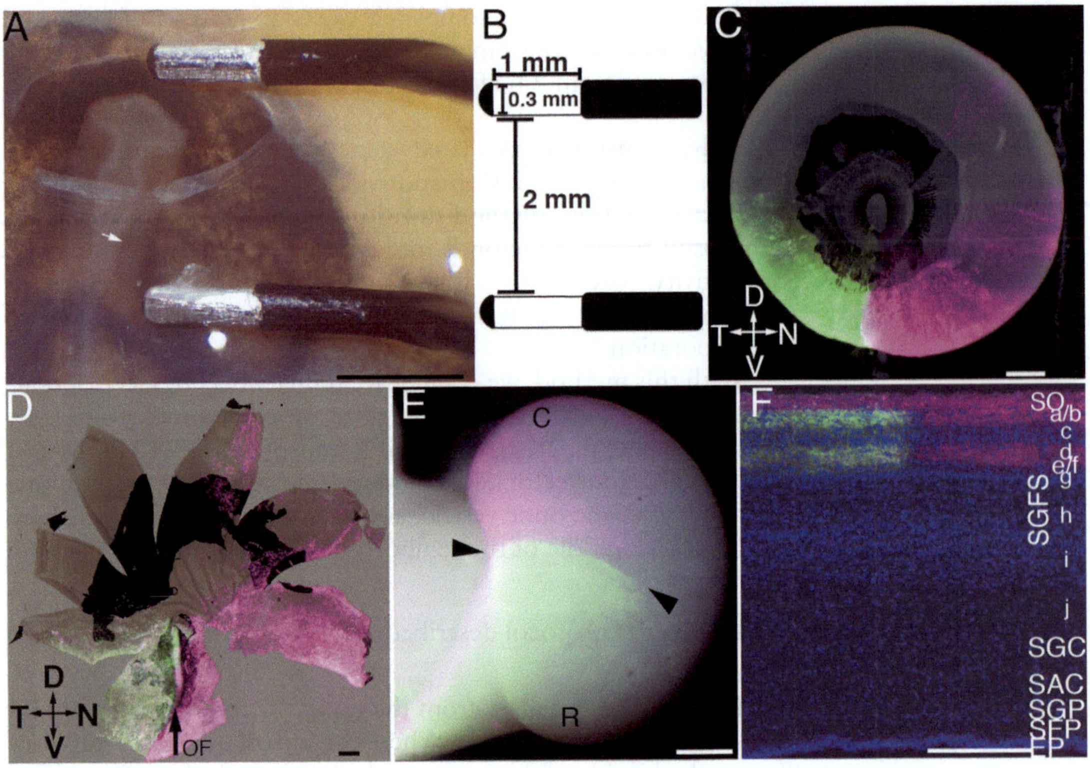

Fig. 2 In ovo electroporation to the optic vesicle. (**a**) A pair of electrodes is placed across the optic vesicle. (**b**) The electrodes are 0.3 mm in diameter, 1.0 mm in length, and 2.0 mm in distance. (**c**) Whole eye imaging of the eye that is transfected red and green fluorescent protein to the temporal and the nasal retina, respectively. (**d**) Flat-mount retina shows a sharp border at the optic fissure. (**e**) The sharp border (*arrowheads*) is seen in the optic tectum, reflecting retinotopic projection. (**f**) The section of the optic tectum shows that there is sharp border between temporal retinal ganglion cell (RGC) fiber terminal and nasal RGC fiber terminal

1. Electroporate plasmid such as pT2K-CAGGS-eGFP (or your desired gene as the tracer protein) together with transposase plasmid, pCAGGS-T2TP, to the optic vesicle by following Subheading 3.2 above.
2. After transfection to the optic vesicle, transfect the midbrain with the fluorescence protein of the different color. By using pT2K-CAGGS vector, sustainable expression during retinotopic map formation could be obtained.

3.4 Conditional Gene Expression by Tetracycline-Inducible System

Tetracycline-mediated gene-inducible system is also a powerful tool for the analysis of the gene function. Co-transfection of pT2K-BI-TRE-EGFP vector (Clontech), which contains gene of interest, pCAGGS-T2TP, and pT2K-CAGGS-rtTA-M2 or pT2K-CAGGS-tTA vectors enables temporal regulation of gene expression by application of doxycycline, Tet-On or Tet-Off, respectively. Here we introduce methods for Tet-On or Tet-Off system in midbrain development:

1. Follow the procedure in Subheading 3.1.
2. Induction or repression of the transgene could be controlled by application of a solution of doxycycline (Dox, an analog of tetracycline; 0.2 μg/μl in PBS). Inject 100 μl of Dox solution into the yolk sac every 24 h during desired expression period of transgene.

3.5 Electroporation to Chick Embryos at Advanced Stages (Ex Ovo Electroporation)

We have introduced in ovo electroporation to the young chick embryos. But it is difficult to electroporate chick embryos of advanced stages because the embryos after E4 are immersed in albumen in ovo, and it is hard to place electrodes close to the embryos. Shell-less culture system is suitable to overcome this difficulty [13, 14]. In the system, chick embryos grow in sterile Petri dish just as in ovo until E12, and embryos are accessible even at advanced stages. Over E12, embryos do not develop well because of the maldevelopment of the skeletal system. Combination of the shell-less culture and electroporation, which has been developed by Luo, Redies [15], enables us to transfer expression vectors into chick embryos at advanced stages. Here, we introduce this method modified at the step of pulse charge. This technique has been applied to analyses of gene function in developmental system of the brain [16, 17]:

1. Incubate fertilized chick eggs until E2.5–E3 (HH stages 17–19).
2. Sterilize eggshell by 70% ethanol.
3. Crack the eggshell on the sharp metal or acrylic edge at E2.5–E3 (Fig. 3a).

   Blood vessels do not develop well if you crack at E2. Note that vitelline membrane should not be broken (beyond E3, the yolk is fragile). Instead of the sharp metal edge, edge of 100-mm Petri dish can be used for cracking for simpler method.
4. Transfer whole content of the egg into the sterile Petri dish, 100 mm diameter and 2 cm high (Fig. 3b). Take care not to rotate the egg at cracking in order to keep the embryo on the top of the yolk. Cover the dish with its lid. Put the Petri dish into a humid incubator. Allow embryos to grow in the Petri dish until stages needed for electroporation (Fig. 3c).
5. Set the embryo in the Petri dish on the electroporation system.
6. Tear amniotic membrane by forceps, and drop PBS on the embryo.
7. Inject plasmid solution mixed with fast green into target of the brain vesicle, such as forebrain, the optic tectum and hindbrain, etc., with glass capillary.

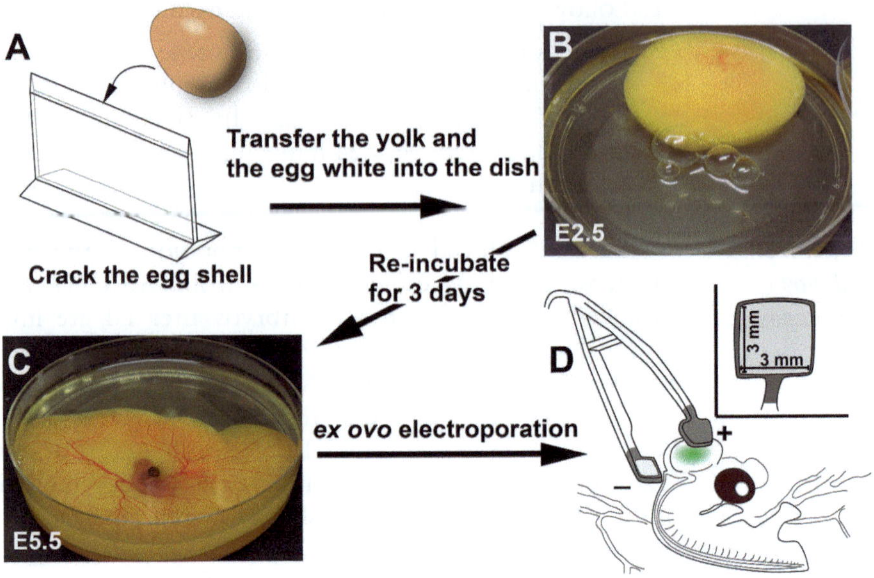

Fig. 3 Procedure for ex ovo electroporation. (**a**) Crack the eggshell on the sharp metal or acrylic edge at E2.5–E3. (**b**) Transfer the egg yolk and the egg white to the sterile 100 mm Petri dish. (**c**) Re-incubate until the desired stage (here the embryo is incubated for 3 days in vitro) at 38 °C in the humidified chamber. (**d**) Carry out ex ovo electroporation. Inject plasmid DNA mixture (*green*) to the optic tectum. Place the electrodes (3 mm square, LF646P3x3), and charge a single pre-pulse of 30 V, 1 ms, and then charge a rectangular pulse of 8 V, 5 ms/s four times

8. Place the forceps-type electrodes so that the target area is located between the electrodes (3 mm square, LF646P3x3, Unique Medical Imada, Japan) (Fig. 3d).
9. Charge a single pre-pulse of 30 V, 1 ms, and then charge a rectangular pulse of 8 V, 5 ms/s four times (Electroporator CUY21EX, Bex Co. Ltd).
10. Place the electroporated embryo in the Petri dish back into the humid incubator, and allow it to grow until the desired stage.

4 Notes

1. Ink is utilized for visualization of early chick embryos. We suggest utilizing Rotring or Pelikan drawing ink to obtain a good survival rate after in ovo electroporation.
2. When you carry out retinal fiber tracing using pT2K-CAGGS-GFP/RFP and pCAGGS-T2TP, fixation by 4% PFA/1× PBS diminishes GFP/RFP fluorescence intensity in the retinal fibers and increases autofluorescence. If you need to observe the retinal fiber trajectory on the whole-mount tectum, we recommend taking picture before fixation. When you take picture after fixation, whole-mount immunohistochemistry against

tracer protein enables us to visualize the retinal fiber trajectory, while autofluorescence background is higher than that before fixation.

3. If you add penicillin/streptomycin to 1× PBS (final concentration 1%), embryos' survival ratio after electroporation will be improved.

4. We have introduced electroporation method for gain of function in this section. Knockdown of the gene of interest is also possible by electroporating shRNA expression vector [18]; miRNA expression vectors for gene silencing were developed [18, 19]. Hou et al. [20] could conditionally knock down the gene of interest by combining shRNA and tetracycline-inducible system.

References

1. Funahashi J, Okafuji T, Ohuchi H, Noji S, Tanaka H, Nakamura H (1999) Role of Pax-5 in the regulation of a mid-hindbrain organizer's activity. Dev Growth Differ 41(1):59–72
2. Hamburger V, Hamilton HL (1992) A series of normal stages in the development of the chick embryo. 1951. Dev Dyn 195(4):231–272. doi:10.1002/aja.1001950404
3. Niwa H, Yamamura K, Miyazaki J (1991) Efficient selection for high-expression transfectants with a novel eukaryotic vector. Gene 108(2):193–199
4. Wakamatsu Y, Watanabe Y, Nakamura H, Kondoh H (1997) Regulation of the neural crest cell fate by N-myc: promotion of ventral migration and neuronal differentiation. Development 124(10):1953–1962
5. Sato Y, Kasai T, Nakagawa S, Tanabe K, Watanabe T, Kawakami K, Takahashi Y (2007) Stable integration and conditional expression of electroporated transgenes in chicken embryos. Dev Biol 305 (2):616-624. doi:10.1016/j.ydbio.2007.01.043, S0012-1606(07)00092-9 [pii]
6. Kawakami K, Takeda H, Kawakami N, Kobayashi M, Matsuda N, Mishina M (2004) A transposon-mediated gene trap approach identifies developmentally regulated genes in zebrafish. Dev Cell 7 (1):133-144. doi:10.1016/j.devcel.2004.06.005, S1534580704002059 [pii]
7. Kawakami K, Imanaka K, Itoh M, Taira M (2004) Excision of the Tol2 transposable element of the medaka fish Oryzias latipes in Xenopus laevis and Xenopus tropicalis. Gene 338 (1):93-98. doi:10.1016/j.gene.2004.05.013, S037811190400304X [pii]
8. Kawakami K, Noda T (2004) Transposition of the Tol2 element, an Ac-like element from the Japanese medaka fish Oryzias latipes, in mouse embryonic stem cells. Genetics 166 (2):895–899. 166/2/895 [pii]
9. Dubrulle J, Pourquie O (2009) Chapter 4: Temporal control of gene expression by combining electroporation and the tetracycline inducible systems in vertebrate embryos. In: Nakamura H (ed) Electroporation and sonoporation in developmental biology. Springer, Tokyo, pp 25–36
10. Harada H, Takahashi Y, Kawakami K, Ogura T, Nakamura H (2008) Tracing retinal fiber trajectory with a method of transposon-mediated genomic integration in chick embryo. Dev Growth Differ 50 (8):697-702. doi:10.1111/j.1440-169X.2008.01067.x, DGD1067 [pii]
11. Banerjee P, Harada H, Tassew NG, Charish J, Goldschneider D, Wallace VA, Sugita S, Mehlen P, Monnier PP (2016) Upsilon-secretase and LARG mediate distinct RGMa activities to control appropriate layer targeting within the optic tectum. Cell Death Differ 23 (3):442–453. doi:10.1038/cdd.2015.111
12. Omi M, Harada H, Nakamura H (2011) Identification of retinotectal projection pathway in the deep tectal laminae in the chick. J Comp Neurol 519(13):2615–2621. doi:10.1002/cne.22642
13. Nakamura H, O'Leary DD (1989) Inaccuracies in initial growth and arborization of chick retinotectal axons followed by course corrections and axon remodeling to develop topographic order. J Neurosci 9(11):3776–3795
14. Auerbach R, Kubai L, Knighton D, Folkman J (1974) A simple procedure for the long-term cultivation of chicken embryos. Dev Biol 41 (2):391–394
15. Luo J, Redies C (2005) Ex ovo electroporation for gene transfer into older chicken embryos.

Dev Dyn 233(4):1470–1477. doi:10.1002/dvdy.20454

16. Omi M, Harada H, Watanabe Y, Funahashi J, Nakamura H (2014) Role of En2 in the tectal laminar formation of chick embryos. Development 141(10):2131–2138. doi:10.1242/dev.102905
17. Watanabe Y, Sakuma C, Yaginuma H (2014) NRP1-mediated Sema3A signals coordinate laminar formation in the developing chick optic tectum. Development 141 (18):3572–3582. doi:10.1242/dev.110205
18. Katahira T, Nakamura H (2003) Gene silencing in chick embryos with a vector-based small interfering RNA system. Dev Growth Differ 45 (4):361–367
19. Das RM, Van Hateren NJ, Howell GR, Farrell ER, Bangs FK, Porteous VC, Manning EM, McGrew MJ, Ohyama K, Sacco MA, Halley PA, Sang HM, Storey KG, Placzek M, Tickle C, Nair VK, Wilson SA (2006) A robust system for RNA interference in the chicken using a modified microRNA operon. Dev Biol 294 (2):554–563. doi:10.1016/j.ydbio.2006.02.020
20. Hou X, Omi M, Harada H, Ishii S, Takahashi Y, Nakamura H (2011) Conditional knockdown of target gene expression by tetracycline regulated transcription of double strand RNA. Dev Growth Differ 53(1):69–75. doi:10.1111/j.1440-169X.2010.01229.x

Chapter 11

Genetic Manipulation of the Avian Urogenital System Using In Ovo Electroporation

Claire E. Hirst, Olivier Serralbo, Katie L. Ayers, Kelly N. Roeszler, and Craig A. Smith

Abstract

One of the advantages of the avian embryo as an experimental model is its in ovo development and hence accessibility for genetic manipulation. Electroporation has been used extensively in the past to study gene function in chicken and quail embryos. Readily accessible tissues such as the neural tube, somites, and limb bud, in particular, have been targeted. However, more inaccessible tissues, such as the embryonic urogenital system, have proven more challenging to study. Here, we describe the use of in ovo electroporation of TOL2 vectors or RCASBP avian viral vectors for the rapid functional analysis of genes involved in avian sex determination and urogenital development. In the context of the developing urogenital system, these vectors have inherent advantages and disadvantages, which will be considered here. Either vector can both be used for mis-expressing a gene and for targeting endogenous gene knockdown via expression of short hairpin RNAs (shRNAs). Both of these vectors integrate into the genome and are hence spread throughout developing tissues. Going forward, electroporation could be combined with CRISPR/Cas9 technology for targeted genome editing in the avian urogenital system.

Key words Sex determination, In ovo electroporation, Urogenital system, Embryonic chicken gonad, Testis, Ovary, Müllerian duct, DMRT1

1 Introduction

The avian embryo is an important experimental model, offering several advantages, such as in ovo development and hence accessibility for genetic manipulation [1–4]. Electroporation, the uptake of DNA by cells across an electric field, has been widely used and perfected by developmental biologists as a means of genetic manipulation, most notably in oviparous animal models, such as the chicken, frog, and fish [4–6]. In the avian model, this system works very well for studying developing somites, neural tube, limbs, and other accessible structures [3, 7–10]. A great advantage of this approach is that it is a rapid and inexpensive method of gene functional analysis. For organs developing deeper in the body, such

Guojun Sheng (ed.), *Avian and Reptilian Developmental Biology: Methods and Protocols*, Methods in Molecular Biology, vol. 1650, DOI 10.1007/978-1-4939-7216-6_11, © Springer Science+Business Media LLC 2017

as the urogenital system, effective gene delivery is more challenging. Here, we describe methods of functional analysis of genes involved in embryonic chicken urogenital development. Over the past several years, we have combined the use of either RCASBP viral vectors [7] or TOL2 transposable elements [11] with electroporation to manipulate gene activity in the embryonic chicken urogenital system (gonads and/or Müllerian ducts). This chapter describes the use of these approaches to study questions associated with sex determination and sexual differentiation in the avian model.

Sex determination in birds differs from that observed in eutherian mammals in several aspects. The sex chromosomes of avian species comprise a large Z chromosome, carrying up to 1000 genes, and a smaller W chromosome, which contains few genes and is largely heterochromatic. Furthermore, in birds, the male is the homogametic sex (ZZ), while the female is heterogametic (ZW). A gene/s carried on the sex chromosomes must play a role in initiating sexual differentiation of the embryonic gonads into ovaries or testes. Gonadal sex differentiation has long been studied in birds and in other organisms, as it offers a unique model system for studying the genetic regulation of sex determination and gonadal development. The embryonic gonads have a "developmental choice"—they become either testes or ovaries—and how that developmental decision is executed at the molecular level can inform our understanding of the genetic regulation of organogenesis. Birds lack *SRY*, the master testis determination gene located on the Y chromosome in mammals. Indeed, the underlying mechanism of sex determination in chickens and other birds is still not entirely clear. The most likely scenario is that sex is determined in avians by the Z-linked transcription factor gene, *DMRT1*, which controls testis formation via a dosage mechanism [12–14]. In the absence of global Z dosage compensation in birds, this gene is more highly expressed in males (ZZ), and its manipulation can alter gonadal development [12]. However, there is some evidence that a gene/s carried on the female-specific W chromosome may carry a dominant-acting ovary determinant (reviewed in [15]), yet a convincing candidate gene is yet to emerge [16, 17]. To complicate matters, recent evidence from gynandromorphic chickens (half male, half female) and cross-sex transplant experiments suggest that sex may be at least partly cell autonomous in birds, involving direct genetic effects in every body cell [18]. These effects may be a function of the lack of global dosage compensation in avians, with a different suite of Z-linked genes being involved in sexual differentiation in different tissues.

Manipulation of gene expression in the embryonic chicken urogenital system has been challenging, due to the location of the gonads deep within the embryo and their derivation from a restricted region of coelomic epithelium overlying the intermediate mesoderm of developing embryos. However, to functionally analyze novel candidate sex genes involved in embryonic chicken urogenital development, we have used RCASBP viral vectors or TOL2 transposable elements, through targeted electroporation into the presumptive urogenital system. In the chicken embryo, the first sign of the presumptive gonad is the formation of the genital ridge around embryonic day (E) 3.5–4 (Hamilton and Hamburger (HH) stage 19–24, Fig. 1) [19]. Cells proliferating from the coelomic epithelium, together with mesenchymal cells immigrating from the mesonephros, contribute to the gonad [20]. By embryonic day 4.5 (HH 26), the undifferentiated gonad is established. It comprises an outer epithelial layer and an underlying medulla, composed of cells arranged in loose cords and surrounding "interstitial" cells (Fig. 1). Having migrated via the bloodstream, germ cells populate the gonad in both the outer and inner layers. Morphological differentiation into the testis or ovary commences at day 6.0 (HH 29) (reviewed in [18]) (Fig. 1). The hallmark of testis formation is the condensation and differentiation of pre-Sertoli cells in the basement membrane-bound medullary cords. Indeed, it is here that the crucial male factors, DMRT1, HEMOGN, and SOX9, are expressed [13]. In females, the medullary cords express the FOXL2 transcription factor, and aromatase enzyme (which produces estrogen), and the medullary cord fragment. The outer cortex thickens and accumulates germ cells. This only occurs in the left gonad, as the right female gonad ultimately regresses (Fig. 1). Meanwhile, the paired Müllerian ducts (future oviducts) also develop from a thickening of the coelomic epithelium but on the opposite side of the mesonephric kidney to the gonad [21]. These form tubular structures that run along the length of the urogenital system, on the dorsal aspect of the mesonephric kidneys. An epithelial to mesenchyme transition generates cells beneath the coelomic epithelium, and an inner epithelial layer also forms [21]. The simple organization of this duct makes it an attractive target for the genetic dissection of Müllerian duct formation and development.

The exact type of genetic manipulation being performed on the avian urogenital system depends on the type of developmental question being asked. We typically use gene delivery vectors that integrate into the genome providing stable transgene expression. A gene delivery vehicle suitable for urogenital analysis is pTOL2-CAGGS. This is the TOL2 transposable element encompassing the CAGGS promoter that drives the expression of a gene of interest (GOI), or it may carry an eGFP reporter together with U6 or H1 polymerase type III promoters driving shRNA expression for RNAi. TOL2 stably integrates into the genome in the presence of coelectroporated transposase enzyme [11]. The system is highly robust, applicable to all vertebrates and has been used to mis-express genes and deliver shRNAs and can also be used in inducible systems [3, 22]. In chicken embryos, it can also be used

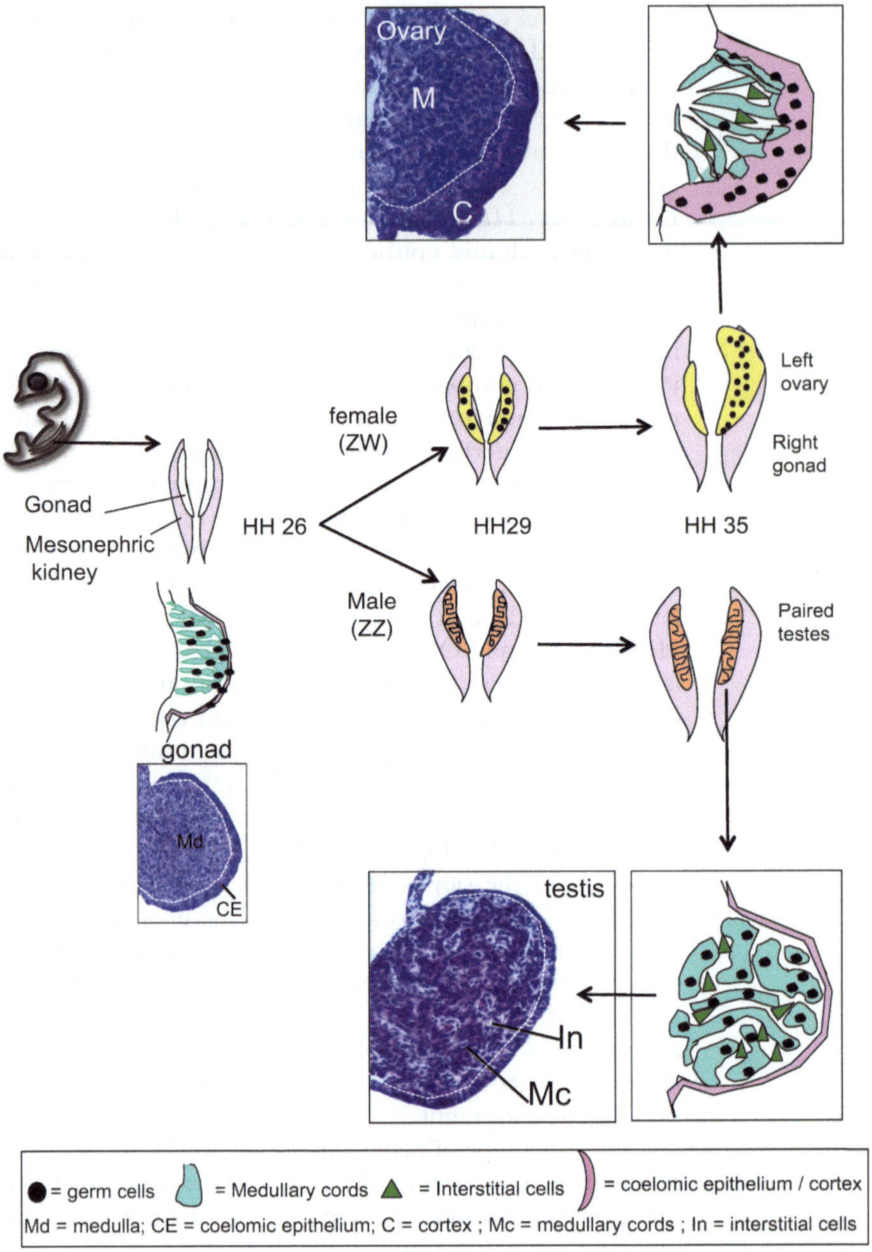

Fig. 1 Gonadal development in the chicken embryo. The gonads develop on the ventromedial aspect of the embryonic kidneys (mesonephric kidneys). By embryonic day 4.5 (HH 26), they comprise an outer epithelial layer and underlying cords of cells, the medullary cords. In ZZ males, Sertoli cells condense within the cords, which develop into the seminiferous cords. Germs cells become enclosed in these cords, and intersitial cells (steroidogenic Leydig cells, peritubular myoid, and other cells) differentiate around the cords. In the ZW female, the left gonad becomes an ovary, characterized by the fragmentation of the medullary cords and proliferation of the outer coelomic epithelia layer to a cortex, which encloses the meiotic germ cells

to assess the function of putative promoter or enhancers fused to GFP reporters in ovo [23]. In our hands, electroporation of TOL2 plus transposase into the coelomic epithelium at HH14 yields robust gene expression in the gonad and Müllerian duct when examined several days later (up to HH 34) (Fig. 2a). We have used this system to knock down expression of the male factor, *DMRT1* (TOL2 GFP-U6 *DMRT1* shRNA), in the developing chicken urogenital system [24]. An advantage of TOL2 is that it can accept large inserts (up to 11 kb), and expression from the CAGGS promoter is very strong. However, a disadvantage relates to the fact that electroporation in general only targets epithelia, whereas mesenchyme is largely refractory. Electroporation of TOL2 constructs into the coelomic epithelium at embryonic day 2.5 (HH 14) results in the robust transgene expression in the coelomic epithelium that covers to gonads (and Müllerian ducts) from 24 h after electroporation (Fig. 2b–d). This coelomic epithelium differentiates into the cortex of the gonads and also gives rise to the interstitial cells in embryonic chicken gonads [25]. Significantly, however, this vector does not target the medullary cords. In gonads electroporated at HH 14 and sectioned several days later at HH 34 (E8.5), TOL2-delivered eGFP is restricted to the interstitial cell population of the developing testis, and not the Sertoli cells, which derive from the medullary cords and are marked by SOX9 expression (Fig. 2e). This indicates that the key cell lineage for avian gonadal sex determination, the medullary cord cell, is not efficiently targeted using TOL2 electroporation. These cord cells must therefore derive not from the surface (coelomic) epithelia (that is well targeted by TOL2 electroporation), but from mesenchymal cells immigrating from the mesonephros, at least after HH 14 (the time of in ovo electroporation shown here). TOL2 is therefore best suited for studies of gonadal cortex development (in females), gonadal interstitial cells, and the Müllerian duct (Fig. 2a, c–e).

To target medullary cord cells in the gonads, we use the avian-specific viral vector, RCASBP (replication-competent *A*SLV long terminal repeat (LTR) with a splice acceptor, *B*ryan polymerase) [7, 26]. This viral vector infects chicken cells and is replication competent, such that it can spread both vertically to daughter cells and horizontally to neighboring cells. It integrates into the genome and can express a gene of interest off a viral LTR promoter as a spliced mRNA during its propagation in cells. Advantages of this system include the fact that the virus can spread following initial infection, delivering genes or shRNAs to a wide field of cells, including both epithelial and mesenchymal cells. That is, it is not restricted to epithelial cells and their direct derivatives. Either live virus can be injected directly into tissues of interest, or high-quality proviral DNA can be electroporated. For embryonic chicken gonads, we primarily use the latter option, as live virus does not appear to spread into the gonads as well. On the other hand, an advantage

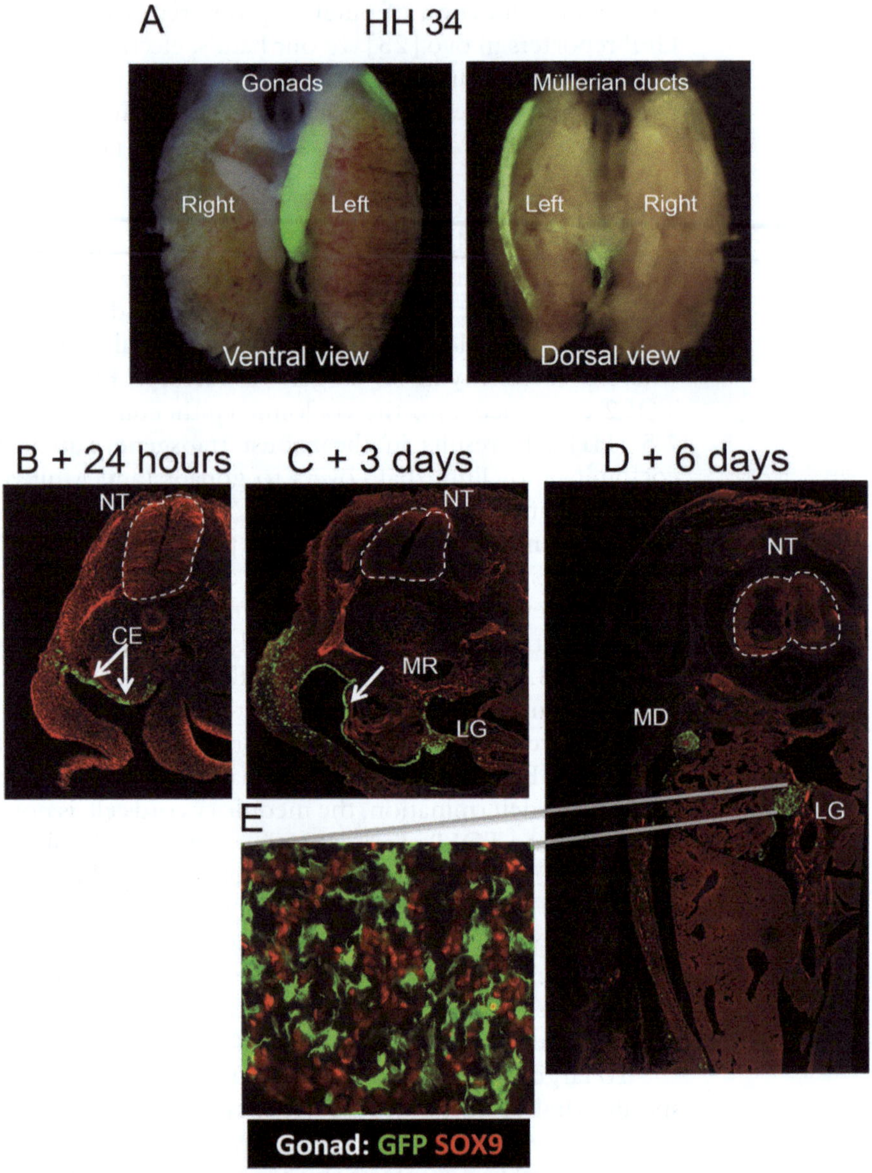

Fig. 2 Time course of eGFP expression within the electroporated embryonic chicken urogenital system. (**a**) Expression of eGFP in HH 34 chicken urogenital system, in the ventral view, unilateral delivery of pTOL2. CAGGS.eGFP into the left but not right gonads. In the dorsal view, HH 34 dorsal view, showing unilateral delivery of pTOL2.CAGGS.eGFP into the left but not right Müllerian duct (Modified from [23] with permission). (**b–d**) Transverse sections of electroporated embryos indicating eGFP-expressing cells, counterstained for acetylated tubulin (NT, neural tube). (**b**) After 24 h, the coelomic epithelium (CE) is labeled with eGFP. (**c**) After 3 days, expression is detected in the developing left gonad (LG) and Müllerian ridge (MR). (**d**) After 6 days, robust expression is seen in the left gonad (LG) and the Müllerian duct (MD). (**e**) High-power view of the gonad, counterstained with SOX9, which marks Sertoli cells in developing seminiferous cords, showing that the eGFP expression is confined to the interstitial cell population, which must therefore originate from the coelomic epithelium

of injecting live virus is a lower mortality rate than observed with electroporated embryos. Some disadvantages of RCASBP are (a) the maximum size of the insert is around 2.5 kb, (b) transgene expression is somewhat delayed (up to 20 h after initial infection), and (c) viral spread often cannot be contained within a small region. Nevertheless, using in ovo electroporation at HH 14, well prior to gonadogenesis, we can achieve good delivery of genes such as eGFP throughout the gonad using RCASBP [27]. Survival of embryos to E8.5 (HH 32, due to an electroporation-induced developmental lag) is around 40%. Recently, we have shown that unilateral electroporation of RCASBP carrying the candidate testis determinant, *DMRT1*, results in mis-expression of the gene in female gonads and ectopic expression of male markers, SOX9, HEMGN, and AMH [13]. Note that there are different strains of RCASBP, designated as A, B, D, and E. Researchers should empirically test which strain their embryos are susceptible to. Charles River SPF embryos are susceptible to RCASBP A and RCASBP B viral strains [26].

An alternative strategy is to inject live, purified RCASBP virus into the blastoderm, which then ensures widespread delivery of the GOI or an shRNA for gene knockdown. By this procedure, the live virus is propagated in chicken DF1 cells, harvested, and concentrated via ultracentrifugation, as described elsewhere [28]. Small volumes of high titer virus (1×10^8 infectious units/mL) are then injected into the subgerminal cavity of stage X blastoderms prior to gastrulation. Eggs are sealed and incubated until the desired stage. The virus becomes incorporated into the cells during gastrulation. We and others have described this direct blastoderm technique previously [29], and this rapid method can yield global gene mis-expression a few days after infection and at least embryo 60% survival to day 8.5 (HH 38) [29]. By this method, the transgene is ubiquitously expressed in the urogenital system but is also widely expressed throughout in the embryo. For knockdown, this may not matter, provided the gonadal gene to be targeted is not expressed in other embryonic sites, as is the case for *DMRT1* [12]. However, for overexpression, this global approach is often lethal to embryos, necessitating the use of RCASBP (or TOL2) vectors that are targeted into presumptive gonads via electroporation.

We now routinely use RCASBP or TOL2 that are electroporated specifically into the coelomic epithelium overlying intermediate mesoderm of the presumptive genital ridge (or the presumptive Müllerian ridge, if studying that organ). This results in variable (RCASBP) or very good (TOL2) delivery into the gonads and the Müllerian ducts, as shown in Fig. 2. The protocol below describes this methodology in more detail. Given its application to all vertebrates, the TOL2 vector could be used to study sex determination and gonadal development in reptilian as well as other avian embryos, although some reptile embryos appear to be laid at advanced stages compared to a chicken. The use of RCASBP is currently restricted to avians and even then only to specific lines of chickens that express the relevant receptors. However, some strains of quails may be susceptible to RCAS infection.

2 Materials

Electroporator: BTX Harvard ECM 830 Square Wave Electroporator or Intracel TSS20 Ovodyne electroporator.

Electrode wires: Tungsten wire for the black negative electrode and platinum wire for the red positive electrode, both with a 0.05 mm diameter. These are insulated with red and black nail polish, leaving a small (2–3 mm) area of exposed metal.

Fine needles made from glass capillaries: Borosilicate thin wall capillaries with an outer diameter of 1.2 mm and an internal diameter of 0.94 mm (Harvard GC120T-10). Fine needles are prepared on a Sutter P-2000 Laser-Based Micropipette Puller using the following settings: Heat 430, Fil 4, Vel 50, Del 225, and Pul 75. Snip the tip off the end of the pulled needles to produce find injection needles with an external bore size of approximately 70 μm.

2.1 Solutions

Prepare all solutions using ultrapure water (prepared by purifying deionized water, to attain a sensitivity of 18 MΩ cm at 25 °C) and analytical grade reagents. Prepare and store all reagents at room temperature (unless indicated otherwise).

1. 20% Fast Green: Dissolve 2 g of Fast Green in 10 mL of water, then filter, and sterilize.
2. 1% carboxymethylcellulose: Slowly dissolve 50 mg of carboxymethylcellulose powder in 5 mL of sterile water (*see* **Note 1**).
3. 50 mM $MgCl_2$: Dissolve 476 mg of $MgCl_2$ in 100 mL of water. Sterilize by autoclaving.
4. 10 × PBS: Dissolve 80 g of NaCl, 2.0 g of KCl, 14.4 g of Na_2HPO_4, and 2.4 g of KH_2PO_4 in 800 mL distilled H_2O. Adjust pH to 7.4. Adjust volume to 1 L. Sterilize by autoclaving.
5. Ringer solution: Dissolve 7.2 g of NaCl, 370 mg of KCl, and 335 mg of $CaCl_2 \cdot 6H_2O$ in water. Adjust the pH to 7.3–7.4 then adjust the final volume to 1 L. Sterilize by autoclaving. Dispense into 50 mL tubes. Store at 4 °C.
6. India ink: Sterile the ink by boiling (*see* **Note 2**), and then aliquot into 2 mL tubes. Store at −20 °C.
7. Penicillin/streptomycin: 5000 U/mL penicillin and 5 mg/mL streptomycin. Aliquot into 2 mL tubes. Store at −20 °C.
8. Tracking dye: The 3× tracking dye is made by adding 2 mL of 20% Fast Green, 5 mL of 1% carboxymethylcellulose, and

Table 1
Preparation and concentration of EP-DNA mixes

EP-DNA Mix	Volume (μL)	Concentration	Final concentration
pTOL2.CAGGS.GOI	5	3 μg/μL	1 μg/μL
pCAGGS.transposase	5	2 μg/μL	0.666 μg/μL
Tracking dye	5	3×	1×
RCASBP proviral DNA	5	3 μg/μL	1 μg/μL
Water	5	–	
Tracking dye	5	3×	1×

600 μL of 50 mM $MgCl_2$ to 3 mL of 10× PBS. Mix and dispense into 5 or 10 μL single use aliquots and store at −20 °C.

9. Plasmid DNA: Prepare high-quality endotoxin-free DNA of at least 2–3 μg/μL (use, e.g., QIAGEN or Macherey-Nagel Nucleospin Maxi kits). We use these DNA plasmids,

   (1) pTol2.CAGGS.GFP or pTOL2.CAGGS.GOI with pCAGGS.transposase (pCAGGS-T2-TP) [11] or (2) pRCASBP.EGFP or pRCASBP.GOI (from Cliff Tabin, Harvard, USA) [7].
10. Electroporation-DNA mix (EP-DNA mix): The EP-DNA mix is made up to a total volume of 15 μL, comprising 5 μL of 3× tracking dye and to 10 μL of DNA and water. The final concentration of each plasmid should be approximately 1 μg/μL; however, the transposase can be at a lower concentration. Examples of the EP-DNA mix for pTOL2-CAGGS-GOI with pCAGG-transposase and high-quality RCASBP proviral DNA are shown in Table 1.
11. Ringer + penicillin/streptomycin solution: Add 2 mL of penicillin/streptomycin to 50 mL of Ringer's solution.
12. India ink + Ringer's solution for visualizing embryos. Dilute 2 mL of Indian ink in 10 mL of the Ringer's + penicillin/streptomycin solution.

3 Method

1. For robust uptake of DNA by the urogenital system, embryos are electroporated at E2.5 (Hamburger Hamilton stages 14–16) prior to formation of the genital ridge. Place fresh fertile eggs on their sides and incubate at 37.8 °C and 70% humidity for 60 h. For TOL2 electroporations, any source of fertile eggs should

suffice. For RCAS electroporations, only eggs from susceptible strains of poultry can be used (*see* **Note 3**).

2. Place that egg on its side in a suitable sized holder (e.g., a Scott bottle lid padded with some tissues). Remove 3 mL of albumin from the blunt end of the egg using an 18 G needle and a 5 mL syringe (*see* **Note 4**).
3. Carefully cut a 1 cm circle of eggshell from the topside of the egg to reveal underlying embryo (*see* **Note 5**).
4. Place the egg under a dissecting microscope with dual movable fiber-optic light sources. **Steps 7–11** are all performed with the aid of the dissecting microscope.
5. Gently pipette 1 mL of Ringer's + penicillin/streptomycin solution on top of the embryo.
6. Inject Indian ink + Ringer under the embryo to improve contrast. Inject approximately 0.5 mL of India ink + Ringer's just beneath the embryo, using a 1 mL syringe with 26 G needle that is bent at 90°. Inject Indian ink/Ringer's just outside of the "blood circle," angling the bent needle under the embryo.
7. Gently tear away the overlying vitelline membrane (*see* **Note 6**), using a 1 mL insulin syringe with a 26 G needle that was tapped on the bench to make a small hook at the end.
8. Manually backfill the pulled capillary needle with 15 μL EP-DNA mix, using an Eppendorf Microloader Tip. Backfilling is faster and easier than front-filling by negative pressure, largely due to the viscosity of the mix. Mount the needle onto a small piece of tubing that has a rubber tear. Test that the mix is being expelled by manually injecting some into a small dish of Ringer. Using the rubber tear, manually inject DNA/EP mix into the coelomic cavity on the left side of the embryo, adjacent to somite 21 (*see* **Note 7**), which at this stage is just posterior to the vitelline vessels (Fig. 3a). Insert the capillary at the caudal region of the embryo, gently pushing the capillary up along the length of the embryo. Inject the DNA as the capillary is withdrawn: at HH 14 the EP-DNA mix should fill the entire coelomic cavity, using approximately 0.5 μL per embryo (Fig. 3a, b).
9. We typically electroporate the left gonad only, the right gonad serving as a negative control (and also because the right gonad becomes rudimentary in females). Position the electrodes parallel to each other on opposite sides of the embryo (Fig. 3c). The black (negative) electrode should be placed slightly below the level of the embryo on the left side of the embryo, while the red positive electrode is then placed slightly above the level of embryo on the right side; this should result in the dorsomedial

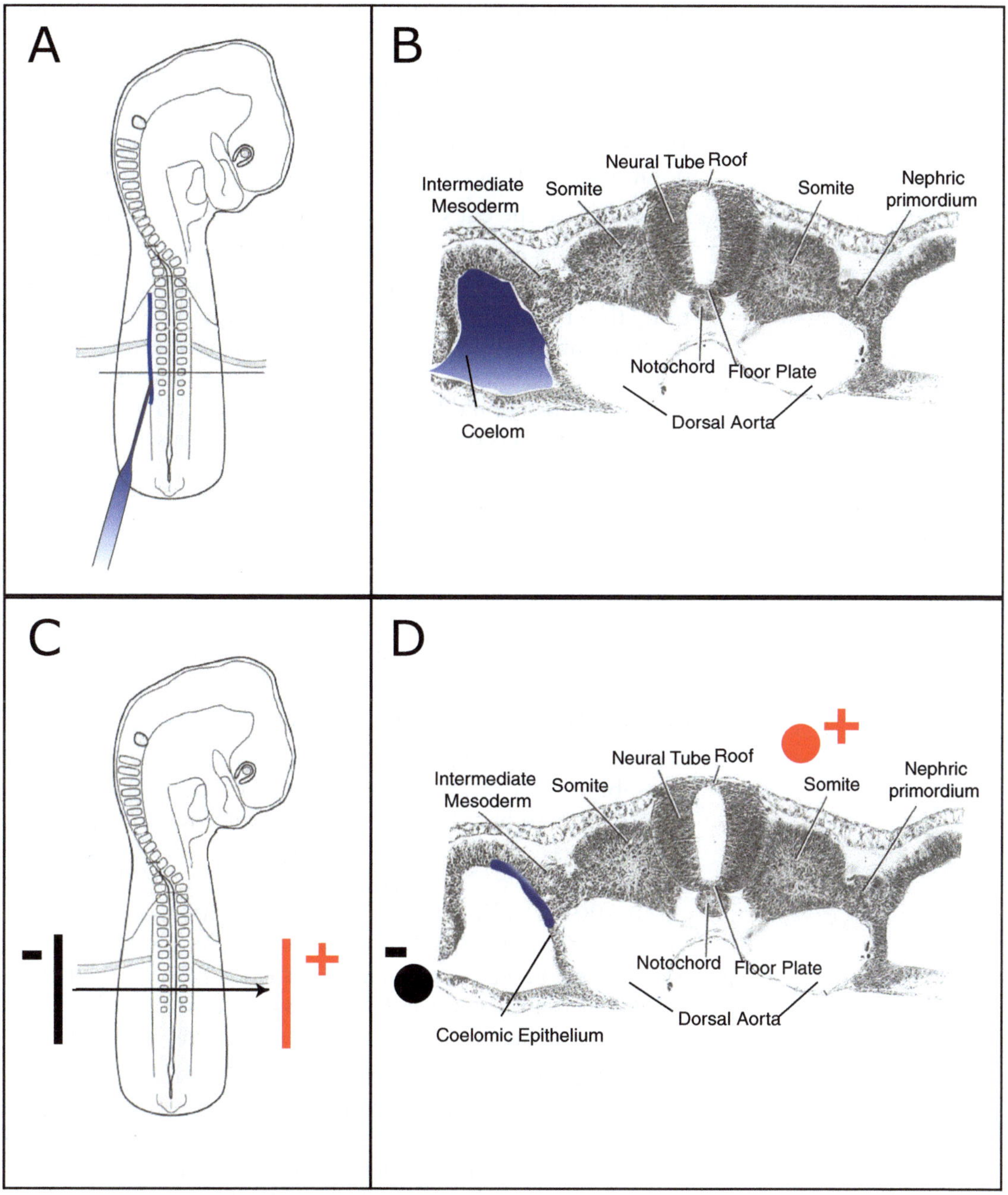

Fig. 3 Electroporation of the embryonic chicken urogenital system. (**a**) An HH stage 14–15 embryo indicating the site of the EP-DNA mix injection into the left coelom. (**b**) The EP-DNA mix fills the entire coelomic cavity. (**c**) Electrodes are placed parallel, on rather side of embryo, adjacent to the vitelline artery and somite 21. Electroporation of the coelomic epithelium in the vicinity of somite 21 at this stage of development should result in the targeting of the presumptive genital ridge. (**d**) The negative (*black*) electrode is placed slightly beneath the embryo, while the positive (*red*) electrode is placed slightly above the embryo. This results in the application of the current in a dorsomedial direction, which results in electroporation of the coelomic epithelium (figures modified from the Atlas of Chicken Development, Bellairs and Osmond)

movement of DNA into the upper-right or dorsomedial aspect of the coelomic epithelium (Fig. 3d; *see* **Note 8**).

10. Apply the voltage (*see* **Note 9**). For HH 14–16 chicken embryos, we routinely apply two pulses of 30–40 V with a 10 ms width, with 150 ms between pulses. We use an electroporator with a foot pump.
11. Carefully seal the eggs with clear package tape (*see* **Note 10**).
12. Incubate eggs on their sides without rocking until embryonic day 8.5. Due to a developmental lag following the manipulation, this results in embryos at HH stage 32 (=E7.5). We harvest embryos at this point because this is after the onset of gonadal sex differentiation (E6), and hence markers of sexual differentiation should be expressed. Embryos can be incubated for longer periods of time, but mortality increases with time.

4 Notes

1. Carboxymethylcellulose can be quite hard to dissolve; if the powder is added all at once, it will form clumps that are very difficult to dissolve. To assist dissolving, add the powder gradually to the water and place the solution on a rotating mixer or rocking platform to assist the powder in dissolving.
2. It is very important to use a brand of India ink that is not toxic to embryos. We routinely use either Pelikan or Lefranc and Bourgeois brands of Indian ink. We sterilize the ink by boiling in a microwave. The ink is quite viscous and will easily boil over in the microwave if not closely observed. To avoid this, pour about 100 mL of ink into a 500 mL capacity glass bottle, and microwave until the ink just begins to bubble, and then let cool for a few seconds, before mixing by swirling the bottle. Repeat boiling in the microwave 4–5 times letting the ink reach the boiling point each time. Let the ink cool before aliquotting.
3. The susceptibility of the embryo to RCAS infections is dependent on two factors, the presence of the appropriate receptor for uptake of the virus and whether the laying hen has previously been exposed to other members of the avian sarcoma leukosis virus subgroup A. The RCAS ASLV family of viruses encodes five envelope types, A, B, C, D, and E—which recognize three cell-surface receptors: A, C, and B/D/E. These are expressed in different strains of chickens, as such not all embryos are susceptible to all RCAS subtypes. Additionally, exposure of the laying hens to other members of the avian sarcoma leukosis virus subgroup A will preclude infection with the RCAS subtypes. As such we obtain specific pathogen-free (SPF) eggs from an SPF service provider.

4. Removing a small amount of albumin from the egg prior to cutting a hole in the eggshell prevents albumin from leaking out while you are cutting the hole. This also lowers the level of the embryo in the shell which helps prevent accidently nicking the embryo or yolk membranes while cutting the hole. The albumin does not need to be replaced if you are only planning on incubating the embryos to E9.5; however, the viability and growth of older embryos are improved by replacing the albumin before sealing the egg.
5. An adhesive tape can be stuck to the top of the eggshell after making an initial hole, to prevent pieces of eggshell from dropping onto the underlying embryo. Pieces of eggshell must be removed from the egg as they can prevent proper embryo growth and cause embryo mortality.
6. Only tear the vitelline membrane that is directly overlying the embryo. It is not necessary to tear all the way to the edge of the blood circle. Only expose enough of the embryo and surrounding membranes to allow injection of the DNA into the embryo and to be able to place the electrodes next to the embryo.
7. The genital ridge forms on the medial surfaces of the coelomic epithelium at approximately day 3.5 of embryonic development (HH stage 21). Electroporation of the region of the coelomic epithelium in the vicinity of somite 21 at this stage of development (HH 14) results in electroporation of the presumptive genital ridge.
8. When performing the electroporation, be careful to avoid any blood vessels when putting the negative (black) electrode in place. Indeed, it is important not to touch any part of the embryo with the electrodes; otherwise, this will result in burning of embryonic tissue which affects embryo growth and often leads to embryo lethality.
9. It is important to apply current within 60 s of DNA injection; otherwise the DNA begins to diffuse out of the coelomic cavity. The tracking dye is very viscous due to the carboxymethylcellulose, which slows down the diffusion of the DNA; however, the experiment will work best if the electroporation is performed as soon as possible after the injection of the DNA.
10. Some brands of packaging tape can be toxic to embryos and should be tested before use. Also it is important to only cover the hole, not the entire egg. The eggshell is permeable to gases, and if too much of the shell is covered, this can prevent adequate gas exchange, which can affect embryo growth.

References

1. Burt DW (2004) The chicken genome and the developmental biologist. Mech Dev 121:1129–1135
2. Nakamura H, Katahira T, Sato T et al (2004) Gain- and loss-of-function in chick embryos by electroporation. Mech Dev 121:1137–1143
3. Rios AC, Marcelle C, Serralbo O (2012) Gene loss-of-function and live imaging in chick embryos. Methods Mol Biol 839:105–117
4. Itasaki N, Bel-Vialar S, Krumlauf R (1999) "shocking" developments in chick embryology: electroporation and in ovo gene expression. Nat Cell Biol 1:E203–E207
5. Chernet BT, Levin M (2012) A versatile protocol for mRNA electroporation of *Xenopus laevis* embryos. Cold Spring Harb Protoc 2012:447–452
6. Teh C, Parinov S, Korzh V (2005) New ways to admire zebrafish: progress in functional genomics research methodology. BioTechniques 38:897–906
7. Logan M, Tabin C (1998) Targeted gene misexpression in chick limb buds using avian replication-competent retroviruses. Methods 14:407–420
8. Andermatt I, Wilson N, Stoeckli ET (2014) In ovo electroporation of miRNA-based-plasmids to investigate gene function in the developing neural tube. Methods Mol Biol 1101:353–368
9. Farley EK (2013) Gene transfer in developing chick embryos: in ovo electroporation. Methods Mol Biol 1018:141–150
10. Blank MC, Chizhikov V, Millen KJ (2007) In ovo electroporations of HH stage 10 chicken embryos. J Vis Exp 9:408
11. Sato Y, Kasai T, Nakagawa S et al (2007) Stable integration and conditional expression of electroporated transgenes in chicken embryos. Dev Biol 305:616–624
12. Smith C, Roeszler K, Ohnesorg T et al (2009) The avian Z-linked gene DMRT1 is required for male sex determination in the chicken. Nature 461:267–271
13. Lambeth LS, Raymond CS, Roeszler KN et al (2014) Over-expression of DMRT1 induces the male pathway in embryonic chicken gonads. Dev Biol 389:160–172
14. Zhang SO, Mathur S, Hattem G et al (2010) Sex-dimorphic gene expression and ineffective dosage compensation of Z-linked genes in gastrulating chicken embryos. BMC Genomics 11:13
15. Major AT, Smith CA (2016) Sex reversal in birds, sexual development. Sex Dev 10 (5–6):288–300
16. Ayers KL, Lambeth LS, Davidson NM et al (2015) Identification of candidate gonadal sex differentiation genes in the chicken embryo using RNA-seq. BMC Genomics 16:704
17. Ayers KL, Davidson NM, Demiyah D et al (2013) RNA sequencing reveals sexually dimorphic gene expression before gonadal differentiation in chicken and allows comprehensive annotation of the W-chromosome. Genome Biol 14:R26
18. Zhao D, McBride D, Nandi S et al (2010) Somatic sex identity is cell autonomous in the chicken. Nature 464:237–242
19. Hamburger V, Hamilton H (1951) A series of normal stages in the development of the chick embryo. J Morphol 88(1):49–92
20. Guioli S, Nandi S, Zhao D et al (2014) Gonadal asymmetry and sex determination in birds. Sex Dev 8:227–242
21. Guioli S, Sekido R, Lovell-Badge R (2007) The origin of the Mullerian duct in chick and mouse. Dev Biol 302:389–398
22. Takahashi Y, Watanabe T, Nakagawa S et al (2008) Transposon-mediated stable integration and tetracycline-inducible expression of electroporated transgenes in chicken embryos. Methods Cell Biol 87:271–280
23. Yokota Y, Saito D, Tadokoro R et al (2011) Genomically integrated transgenes are stably and conditionally expressed in neural crest cell-specific lineages. Dev Biol 353:382–395
24. Ayers KL, Cutting AD, Roeszler KN et al (2015) DMRT1 is required for Müllerian duct formation in the chicken embryo. Dev Biol 400:224–236
25. Sekido R, Lovell-Badge R (2007) Mechanisms of gonadal morphogenesis are not conserved between chick and mouse. Dev Biol 302:132–142
26. Hughes SH (2004) The RCAS system. Folia Biol (Praha) 50(3–4):107–119
27. Smith CA, Roeszler KN, Sinclair AH (2009) Robust and ubiquitous GFP expression in a single generation of chicken embryos using the avian retroviral vector, RCASBP. Differentiation 77:473–482
28. Ahronian LG, Lewis BC (2014) Generation of high-titer RCAS virus from DF1 chicken fibroblasts. Cold Spring Harb Protoc 2014:1161–1166
29. Smith CA, Roeszler KN, Sinclair AH (2009) Genetic evidence against a role for W-linked histidine triad nucleotide binding protein (HINTW) in avian sex determination. Int J Dev Biol 53:59–67

Chapter 12

Enhancer Analyses Using Chicken Embryo Electroporation

Masanori Uchikawa, Naoko Nishimura, Makiko Iwafuchi-Doi, and Hisato Kondoh

Abstract

Chicken embryo electroporation is a powerful tool used to identify and analyze enhancers involved in developmental gene regulation. In this chapter, the basic procedures and underlying principles of enhancer analysis using chicken embryo electroporation are described in the following steps: (1) identification of enhancers in a wide genomic region, (2) determination of the full enhancer region, (3) definition of the core enhancer regions, and (4) analysis of a functional transcription factor binding sequences in the core region.

Key words Enhancers, Electroporation, tkEGFP, Core region, Transcription factor binding

1 Introduction

Early-stage chicken embryos are easily accessible in ovo and in culture. They provide an excellent platform for analyzing tissue-specific and developmental stage-specific enhancers that are involved in developmental gene regulation. From a genomics perspective, genomic regulation in the chicken is generally conserved in mammals, reflecting the high conservation of regulatory sequences [1, 2]. However, the genomic size of noncoding regions in chickens is only one-third of that in mammals. In other words, the same repertoire of enhancers is contained within a much shorter genomic span in the chicken than that in mammals. This gives the chicken model a significant advantage over the mammal model in the quest for enhancer regions in the genome (Fig. 1).

Enhancer analysis using chicken embryos came of age when highly efficient DNA electroporation was established [3–5]. DNA can be efficiently electroporated into embryos in ovo and into isolated early-stage embryos in New's culture [1, 6]. For enhancer analysis, a reporter vector is an essential genetic tool. These vectors need to be silent without insertion of an enhancer and to be activated in specific tissues upon the insertion of specific enhancers.

Guojun Sheng (ed.), *Avian and Reptilian Developmental Biology: Methods and Protocols*, Methods in Molecular Biology, vol. 1650, DOI 10.1007/978-1-4939-7216-6_12, © Springer Science+Business Media LLC 2017

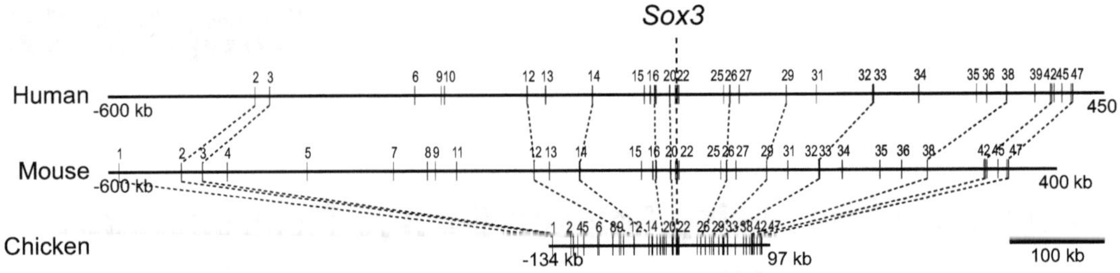

Fig. 1 Size differences in the genomic spans encompassing the Sox3 gene between mammals and chicken. Vertical bars represent the positions of conserved sequences >100 bp, indicating regional correspondences. (Reproduced from Nishimura et al. [8] Biology 1, 714–735, Fig. 1 with modifications)

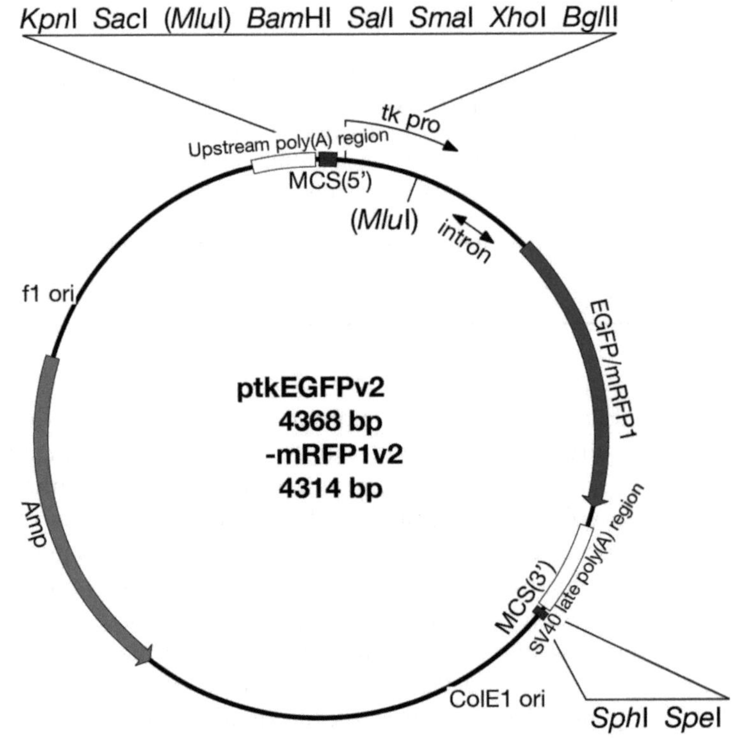

Fig. 2 Schematic representation of the ptkEGFP and ptkmRFP1 version 2 plasmid structures. (Reproduced from Uchikawa and Takemoto, [11] Chapter 7 Enhancer Analysis in Electroporation and Sonoporation in "Developmental Biology" (H. Nakamura, ed) pp. 55–71, Springer, Fig. 7.2a)

We demonstrated that tkEGFP or tkmRFP1 harboring the promoter of thymidine kinase gene satisfies these conditions [1, 2] (Fig. 2).

In this chapter, we describe the procedure and underlying principles of enhancer analysis using two typical examples: (1) the systematic identification of enhancers in a wide genomic region, and (2) detailed analysis of regulatory elements within an enhancer.

2 Equipment

1. PCR machine and other DNA preparation setups.
2. Incubators for fertilized eggs and for embryo culture at 38 °C.
3. Electroporation apparatus (e.g., CUY21, BEX; NEPA21, Nepagene; BTX T820, Harvard Apparatus).
4. Electroporation chamber for embryo culture (e.g., LF701P2, BEX; CUY701P2, Nepagene) or electrodes for in ovo electroporation (*see* the chapter by H. Nakamura).
5. A glass needle puller to prepare a pointed glass capillary.
6. Pointed glass capillary with rubber tubing and a mouthpiece to deliver DNA solution at the site of electroporation.
7. Dissecting microscope equipped with fluorescent optics to record GFP and RFP fluorescence.

3 Systematic Identification of Enhancers in a Wide Genomic Region

3.1 Preparation of Genomic Fragments and Their Introduction into tkEGFP/tkmRFP1 Vectors

Genomic DNA fragments shorter than 5 kb that cover the genomic region of interest with 1 kb overlaps should be prepared by PCR. The efficiency of electroporation steeply decreases when the plasmid size exceeds 10 kb. The vectors tkEGFP/tkmRFP1 are ~4.3 kb in size; therefore, the inserted DNA fragments need to be less than 5 kb. The 1 kb overlaps provide a safeguard for enhancers to be included in at least one of the two adjoining DNA fragments without interruption. Depending on whether the genomic fragments are inserted into the reporter vector via ligation at restriction sites, or via 3′ exonuclease-coupled PCR reactions, the primer sequences added to the ends of genomic fragments should be designed accordingly (*see* **Note 1**).

3.2 Preparation of Plasmid DNA for Electroporation

After the insertion of the genomic fragments into the vector plasmid, *E. coli* cells are transformed with the plasmids using ampicillin selection. Plasmid DNA is purified using an alkaline lysis-based method to obtain closed circular DNA. Pure plasmid DNA and pure closed circular DNA are crucial for high efficiency electroporation without tissue damage. An additional plasmid expressing a different fluorescent protein (e.g., pCMV-mRFP1 vs. tkEGFP) should also be prepared to control electroporation area and efficiency.

3.3 Preparation of Stage 4 (st. 4) Chicken Embryos for Electroporation and Culture

Incubate chicken fertilized eggs at 38 °C for 18–19 h until st. 4 of embryogenesis [7]. Crack eggs into a 10 cm diameter plastic dish and rotate the yolk so that the embryo is located at the center. Thoroughly strip the thick albumen off the embryonic vitelline membrane using scissors, forceps, and/or Kimwipes. Place a ring of filter paper around the embryo and cut the vitelline membrane along the periphery of the filter paper. Lift the embryo with the vitelline membrane attached to the filter ring by holding the ring with forceps. Afterward, rinse the adhering yolk material in phosphate-buffered saline (e.g., Hank's saline) (*see* **Note 2**).

3.4 Electroporation of Chicken Embryos at st. 4

The setup for electroporation and a diagram of tissue orientation during electroporation are shown in Fig. 3.

1. The electrode chamber has a square well at the center where the cathode for electroporation is placed. The chamber is covered with phosphate-buffered saline.
2. The rinsed embryos are placed at the center of an electroporation chamber with ventral side up (vitelline membrane at the bottom).
3. Prepare DNA mixture with tkEGFP constructs at 2 μg/μl and pCMV-mRFP1 at 0.2 μg/μl.

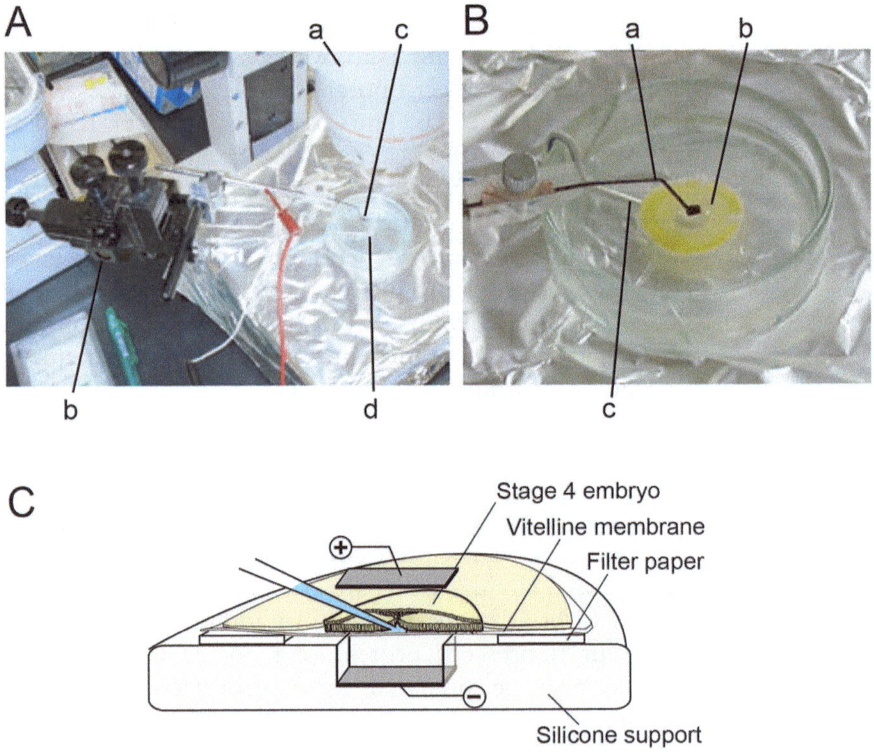

Fig. 3 Setup for the electroporation of st. 4 embryos (**a**, **b**) and the tissue orientation during electroporation (**c**). (**a**) (*a*) Dissecting microscope, (*b*) micromanipulator to hold the anodal electrode, (*c*) anodal electrode tip, and (*d*) cathodal electrode embedded in a silicone platform. (**b**) (*a*) Anodal lead, (*b*) an embryo on a filter ring support, and (*c*) cathodal lead. (Reproduced from Uchikawa et al. [6] Mech. Dev. 121, 1145–1158, Figs. 1B and 10, with modifications)

4. Suck DNA solution into the pointed end of the capillary. Insert the capillary tip into the space between the vitelline membrane and epiblast (dorsal) side of the embryo by piercing through the embryonic tissue. Once inserted, blow in a few μl of the DNA solution.
5. Quickly place the anode precisely 4 mm up from the cathode and apply 8–10 V electric pulses five times for 50 ms each, with an inter-pulse intermission of ~100 ms.
6. Place electroporated embryos on top of albumen-agar medium and incubate at 38 °C in a moist chamber (*see* **Note 3**).
7. Observe and photo-record fluorescent and bright-field images of embryos at intervals (*see* **Note 4**).

3.5 Example: Identification of Sox3 Enhancers

The 231 kb genomic region encompassing the chicken *Sox3* gene was tiled with 58 genomic fragments with overlaps >1 kb, and enhancer activity was tested by EGFP fluorescence [8]. The corresponding genomic region in mammals is >1 Mb (Fig. 1), emphasizing the advantage of using the chicken genome for enhancer analysis. Thirteen different enhancers, which are active in subdomains of the embryonic CNS or sensory primordial tissues before st. 11, were identified (Fig. 4) [8].

4 Trimming the Enhancer-Carrying Genomic Region to the Net Enhancer Sequence and the Core Sequence

4.1 Determination of the Enhancer Region

A full enhancer sequence usually spans a genomic region of hundreds of base pairs. This sequence can be identified either by stepwise trimming from either end of the original genomic fragment, or by focusing on a specific subsequence with strong phylogenic conservation. Using the first approach, deletion of the sequence from either end starts to reduce the enhancer activity once the deletion extends into the enhancer region. Using the second approach, if the isolated sequence contains the full enhancer sequence, the sequence will have an identical enhancer activity to the original genomic sequence.

4.2 Determination of the Core Region

1. The nature of the core region.

   An enhancer usually consists of a core region and adjoining auxiliary regions. The core region determines the specificity of the enhancer and activates the enhancer. The adjoining regions have broader specificities than the core, and they augment the core-dependent enhancer activity. The removal of the adjoining elements generally reduces the enhancer activity while maintaining the specificity, whereas the removal of the core region fully inactivates the enhancer. The core region on its own is often a very weak enhancer, and its activity may not be

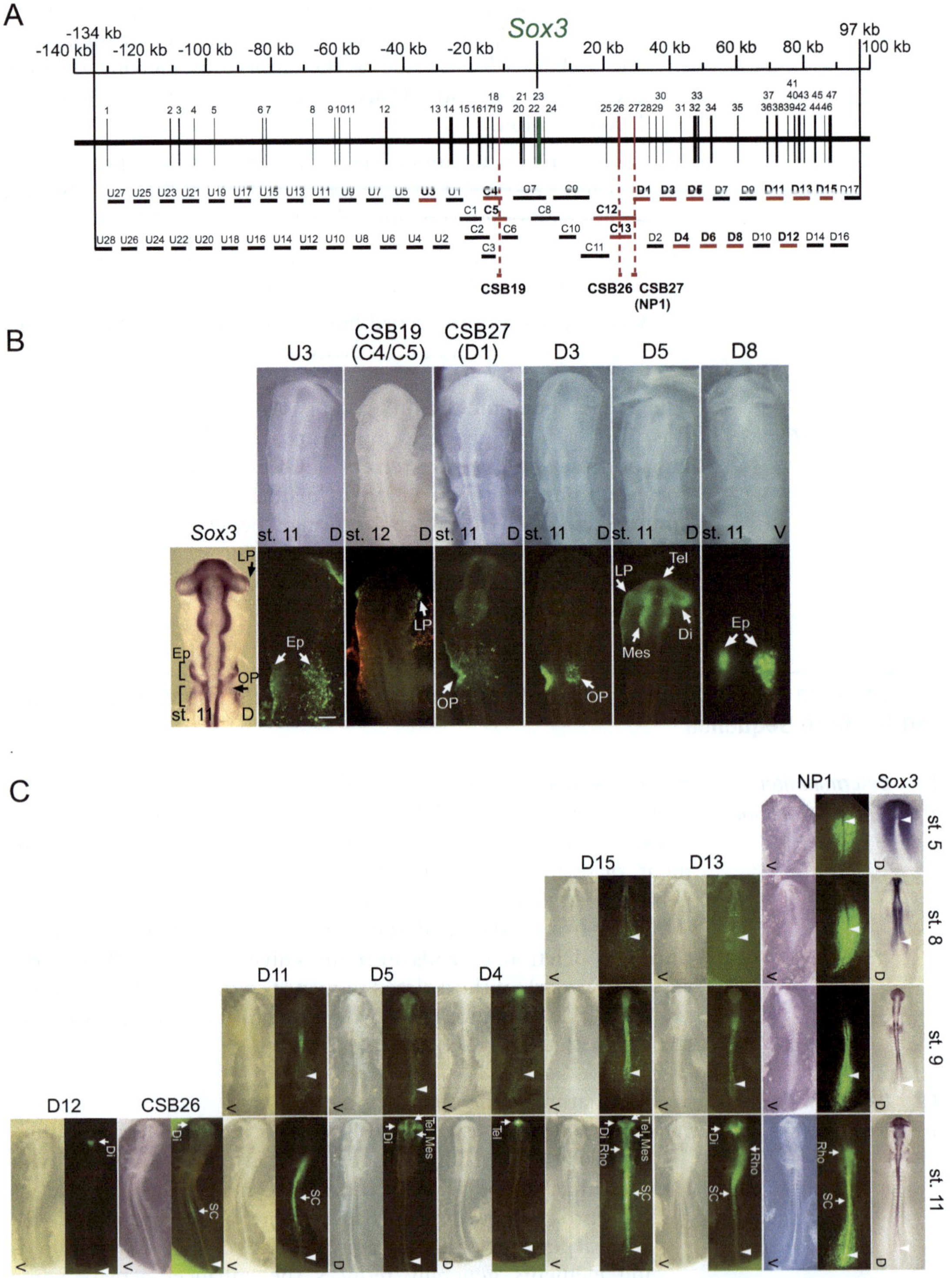

Fig. 4 (**a**) Genomic fragments in the 231 kb genomic sequence surrounding the Sox3 gene. Fragments with enhancer activity are highlighted in *red*. (**b**) Fluorescent images showing enhancer activities in the sensory primordia. *Top*: Bright-field images with indication of developmental stages. *V* ventral view, *D* dorsal view. *Bottom*: EGFP fluorescence images of the same field. The *bars* indicate 200 μm. *Di* diencephalon, *Ep*

detected. However, when the core region is multimerized into dimer to octamer depending on the enhancer, the multimeric core region exhibits its specific activity [2].

2. Multimerization of the core region.

   Multimerization can be performed systematically as follows: Insert the sequence at a multi-cloning site, flanked by a pair of restriction enzymes that share cohesive end sequences to tkEGFP (mRFP1) ver. 2, such as *Bam*HI and *Bgl*II or *Sal*I and *Xho*I. If the insert is flanked by *Bam*HI and *Bgl*II, digest the plasmid separately with *Bgl*II-*Sph*I and *Bam*HI-*Sph*I. Isolate restriction enzyme-digested DNA fragments that carry the enhancer sequence, and ligate them to reconstruct a plasmid sequence carrying a dimerized enhancer sequence joined by *Bgl*II-*Bam*HI site fusion. By repeating this procedure, tetramers and octamers of the enhancer sequence can be built in the tkEGFP (mRFP1) ver. 2 vector (Fig. 5).

3. Example: Identification of the 73 bp core region of the mouse *Sox2* N2 enhancer.

   The N2 enhancer is activated in precursor tissue of anterior neural plate [1]. The activity of the mouse N2 enhancer is identical to the chicken counterpart; therefore, mouse N2 enhancer was analyzed using chicken embryo electroporation (Fig. 6) [9].

5 Identification of Regulatory Elements Representing Transcription Factor Binding Sites Involved in Enhancer Specificity

An enhancer core region consists of a specific combination of multiple transcription factor binding sequences, which constitute the enhancer specificity. Analysis of these is highly informative in elucidating gene regulatory networks.

5.1 Introduction of Blockwise Base Substitution Mutations in the Core Sequence

1. Introduce transversion-type base alterations (G↔T and A↔C) in blocks of 3–10 bp that entirely cover the core region. Synthetic DNAs bearing the mutations may be used for this purpose. If some conspicuous transcription factor binding motif is

Fig. 4 (continued) epibranchial placode, *LP* lens placode, *Mes* mesencephalon, *OP* otic placode, *Rho* rhombencephalon, *Tel* telencephalon. (**c**) Fluorescent images showing enhancer activities in the CNS at different developmental satges in comparison with bright-field images. The rightmost images are Sox3 in situ hybridization patters. *V* ventral view, *D* dorsal view. The *white triangles* indicate the positions of Hensen's node. All photographs are shown on the same scale. The *bar* indicates 500 μm. *Di* diencephalon, *Mes* mesencephalon, *Rho* rhombencephalon, *SC* spinal cord, *Tel* telencephalon. (Reproduced from Nishimura et al. [8] Biology 1, 714–735, Figs. 2–4 with modifications)

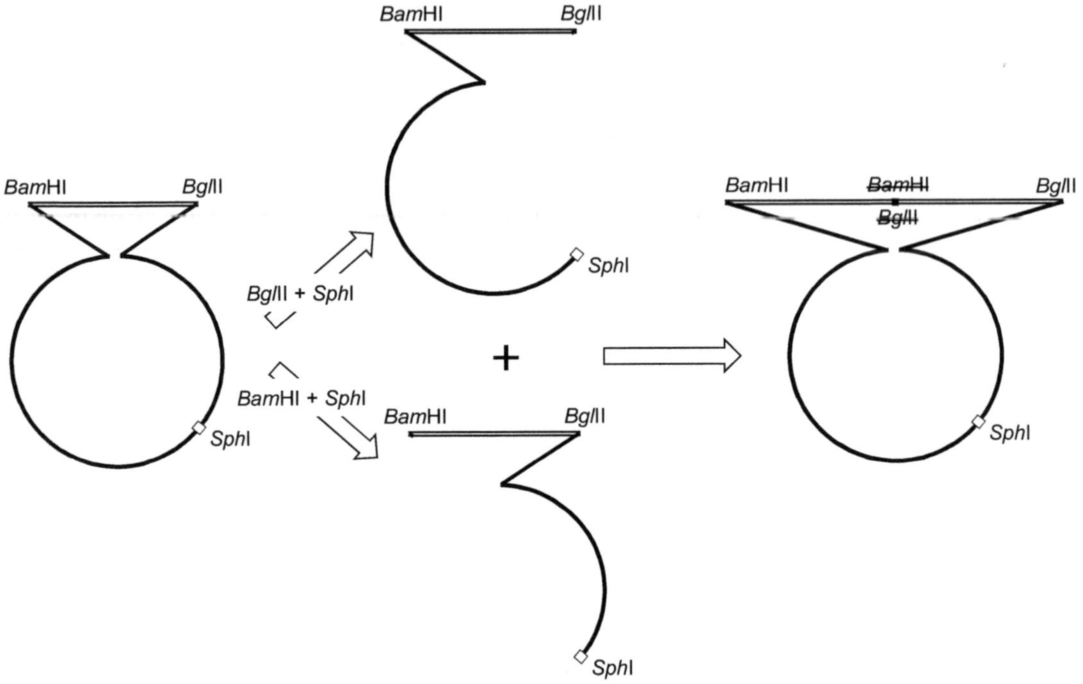

Fig. 5 Generation of multimeric enhancers on tkEGFP (mRFP1)

found in the core region, the mutations in the block may be targeted to the motif. It is important to check whether the base alterations, including non-mutated boundary sequences, create new transcription factor binding motifs. In such cases, alter the base sequences to avoid creating transcription-factor binding motifs.

2. Insert the mutated sequence into the tkEGFP vector, and multiply the mutated sequence as described above (Fig. 5).
3. Assess the mutational effects by co-electroporating embryos with the tkEGFP vector containing the mutated enhancer and tkmRFP1 containing the wild-type enhancer.

5.2 Example: Mutational Analysis of the Sox2 N2 Enhancer Core Region Identified Essential Transcription-Factor Binding Sequences (Fig. 7)

Effects of TF binding site mutations in the core sequence on N2 [176-bp] enhancer were examined. N2 core sequence of the mouse in comparison with that of chicken and zebrafish showed the conservation of potential TF binding sequences. Mutations were introduced in the TF binding sequences (Fig. 7, top). Mutation-bearing N2[176-bp]-tkEGFP were coelectroporated with the reference vector N2[176-bp]-tkmRFP1 into stage 4 embryos and enhancer activity assessed after 4 h at stage 5 (Fig. 7, bottom).

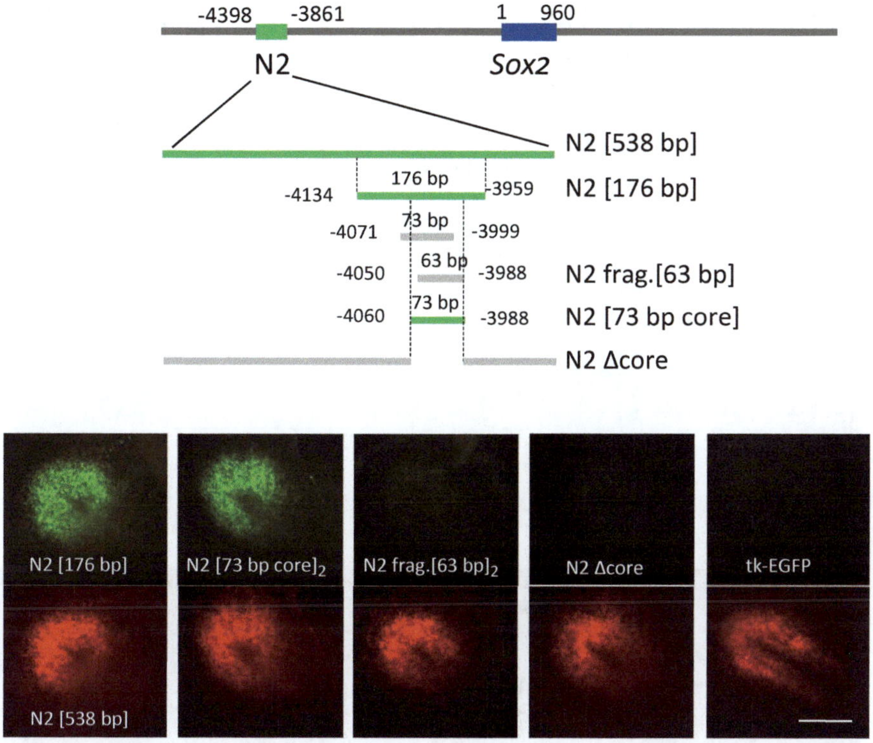

Fig. 6 *Top*: Identification of the 73 bp core region of the N2 enhancer, with active sequences in green and inactive sequences in *gray*. *Bottom*: Fluorescence data where full-length N2 in tkmRFP1 was co-electroporated as a control. DNA sequences shorter than 100 bp were analyzed using dimers. (The scheme on top was reproduced from Iwafuchi-Doi et al. [9] Dev. Biol. 352, 354–366 Fig. 2B)

6 Notes

1. Phylogenetic conservation of the noncoding sequence is useful for enhancer analysis. However, not all conserved noncoding sequences are enhancers and some species-specific enhancers are not conserved [1]. Therefore, in the initial survey of enhancers, the unbiased approach described in Subheading 3 is recommended. However, once the activity of an enhancer is detected in a genomic region, information concerning the conservation of sequences and sequence motifs may be fully utilized for finer analyses, as discussed in Subheadings 4 and 5.
2. In this chapter, the electroporation procedures using st. 4 embryos in New's culture are described (*see* **Note 5**). Depending on the developmental stages and tissues of interest, other electroporation methods can be employed. This is discussed elsewhere [10, 11], and relevant techniques are described in the chapter written by H. Nakamura. As embryos develop, the electroporation technique can only be applied to limited tissues. However, various embryonic chicken tissues can be easily

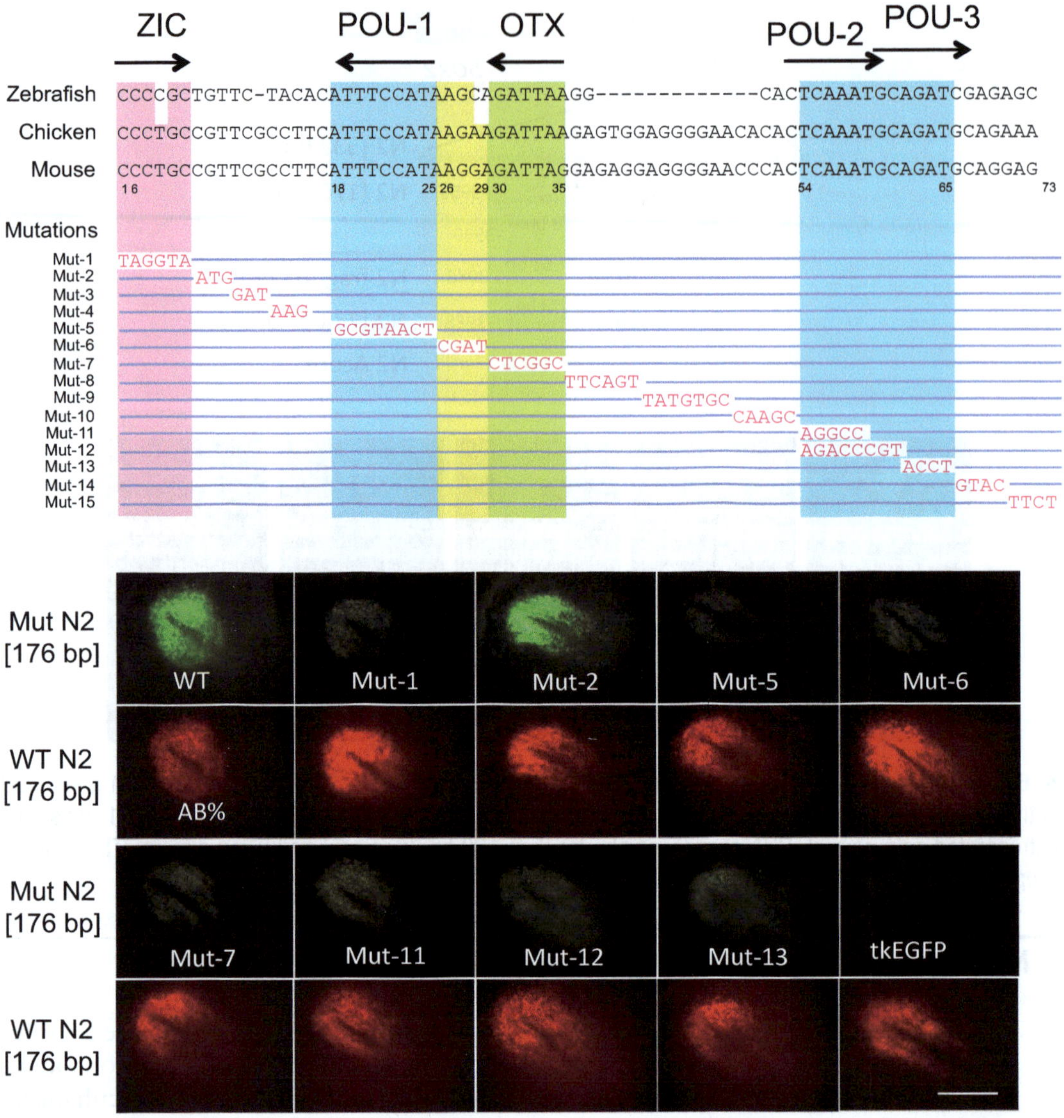

Fig. 7 The mutations shown on top were introduced into the core region sequence of N2[176-bp] in tkEGFP, and enhancer activity was compared with co-electroporated wild-type enhancer-bearing tkmRFP1. Mutations 3, 4, 8, 9, 10, 14, and 15 did not affect enhancer activity and produced results similar to Mut-2. Bar, 1 mm. (Combination of our unpublished data with that of Iwafuchi-Doi et al. [9] Dev. Biol. 352, 354–366, Fig. 3)

adapted to primary cultures, where transfection is efficient. Therefore, the transfection of primary cultures can replace electroporation of live embryos at later developmental stages. The basic principle of enhancer analysis remains the same when transfection of cultured cells is the method of choice [12].

3. The conditions for New's culture have not been detailed in this chapter because variations have been developed in different laboratories, following the first report of the modification of

New's culture using filter-paper support [13]. Any version of New's culture that has been successfully employed in a laboratory may be used. Our routine culture conditions have been reported previously [6].

4. Observation may be possible up to 48 h after electroporation, when embryos usually develop past st. 15. After st. 15, the circular filter paper can become an obstacle for normal embryonic development.
5. Some additional examples of enhancer analysis using the methods described in this chapter can be found in the following publications: [14–18].

References

1. Uchikawa M, Ishida Y, Takemoto T, Kamachi Y, Kondoh H (2003) Functional analysis of chicken Sox2 enhancers highlights an array of diverse regulatory elements that are conserved in mammals. Dev Cell 4:509–519
2. Kondoh H, Uchikawa M (2008) Dissection of chick genomic regulatory regions. Methods Cell Biol 87:313–336
3. Muramatsu T, Mizutani Y, Okumura J (1996) Live detection of firefly luciferase gene espression by bioluminescence in incubating chicken embryos. Anim Sci Technol Jpn 67:906–909
4. Funahashi J, Okafuji T, Ohuchi H, Noji S, Tanaka H, Nakamura H (1999) Role of Pax-5 in the regulation of a mid-hindbrain organizer's activity. Dev Growth Differ 41:59–72
5. Itasaki N, Bel-Vialar S, Krumlauf R (1999) 'Shocking' developments in chick embryology: electroporation and in ovo gene expression. Nat Cell Biol 1:E203–E207
6. Uchikawa M, Takemoto T, Kamachi Y, Kondoh H (2004) Efficient identification of regulatory sequences in the chicken genome by a powerful combination of embryo electroporation and genome comparison. Mech Dev 121:1145–1158
7. Hamburger V, Hamilton HL (1951) A series of normal stages in the development of the chick embryo. J Morphol 88:49–92
8. Nishimura N, Kamimura Y, Ishida Y, Takemoto T, Kondoh H, Uchikawa M (2012) A systematic survey and characterization of enhancers that regulate Sox3 in neuro-sensory development in comparison with Sox2 enhancers. Biology (Basel) 1:714–735
9. Iwafuchi-Doi M, Yoshida Y, Onichtchouk D, Leichsenring M, Driever W, Takemoto T, Uchikawa M, Kamachi Y, Kondoh H (2011) The Pou5fl/Pou3f-dependent but SoxB-independent regulation of conserved enhancer N2 initiates Sox2 expression during epiblast to neural plate stages in vertebrates. Dev Biol 352:354–366
10. Uchikawa M (2008) Enhancer analysis by chicken embryo electroporation with aid of genome comparison. Dev Growth Differ 50:467–474
11. Uchikawa M, Takemoto T (2009) Enhancer analysis. In: Nakamura H (ed) Electroporation and sonoporation in developmental biology. Springer, Tokyo, pp 55–71
12. Matsumata M, Uchikawa M, Kamachi Y, Kondoh H (2005) Multiple N-cadherin enhancers identified by systematic functional screening indicate its group B1 SOX-dependent regulation in neural and placodal development. Dev Biol 286:601–617
13. Chapman SC, Collignon J, Schoenwolf GC, Lumsden A (2001) Improved method for chick whole-embryo culture using a filter paper carrier. Dev Dyn 220:284–289
14. Takemoto T, Uchikawa M, Kamachi Y, Kondoh H (2006) Convergence of Wnt and FGF signals in the genesis of posterior neural plate through activation of the Sox2 enhancer N-1. Development 133:297–306
15. Inoue M, Kamachi Y, Matsunami H, Imada K, Uchikawa M, Kondoh H (2007) PAX6 and SOX2-dependent regulation of the Sox2 enhancer N-3 involved in embryonic visual system development. Genes Cells 12:1049–1061
16. Saigou Y, Kamimura Y, Inoue M, Kondoh H, Uchikawa M (2010) Regulation of Sox2 in the pre-placodal cephalic ectoderm and central nervous system by enhancer N-4. Dev Growth Differ 52:397–408
17. Okamoto R, Uchikawa M, Kondoh H (2015) Sixteen additional enhancers associated with

the chicken Sox2 locus outside the central 50-kb region. Dev Growth Differ 57:24–39

18. Uchikawa M, Kondoh H (2016) Regulation of *Sox2* via many enhancers of distinct specificities. In: Kondoh H, Lovell-Badge R (eds) Sox2: biology and role in development and disease. Academic Press/Elsevier, Boston, Oxford, San Diego, pp 107–129

Chapter 13

Transgene Introduction into the Chick Limb Bud by Electroporation

Shogo Ueda, Takayuki Suzuki, and Mikiko Tanaka

Abstract

Electroporation enables delivering bionanomolecules, such as DNAs, RNAs, siRNAs, and morpholinos, into chick embryos in a spatially and temporally restricted fashion. Recent advances in electroporation techniques allowed us to deliver transgenes into the restricted area of the limb bud and to analyze the function of the enhancers in the limb field. Here, we describe the introduction of transgenes by electroporation in the limb field and its application on enhancer analysis.

Key words Electroporation, Gene delivery, Enhancer analysis, Chick embryo, Limb buds

1 Introduction

The chicken embryo has provided an excellent model system for studying the developmental mechanisms of limb patterning. The ease of surgical manipulation by using the incubated eggs has led to the discovery of the apical ectodermal ridge, which patterns the limbs along the proximal-distal axis [1], and the polarizing region at the posterior margin of the limb bud, which specifies the posterior identity [2]. Chicken models also allowed us to develop unique systems to deliver transgenes in a spatially and temporally controlled manner, such as the retrovirus system [3] and the electroporation [4–7]. Electroporation-based procedures have been developed in chicken models as well as in various non-transgenic models for delivery of various biomolecules, such as DNAs, small interference RNAs, or antisense morpholino oligonucleotides.

Methods for gene delivery by electroporation into limb buds of chicken embryos have been established by Toshihiko Ogura's group [6, 7]. Recently, methods were also established that enabled delivery of transgenes into spatially restricted domains within the limb bud [7] and the analysis of enhancer function in the limbs [8] (Fig. 1d–f). In this chapter, we will describe the gene delivery

Guojun Sheng (ed.), *Avian and Reptilian Developmental Biology: Methods and Protocols*, Methods in Molecular Biology, vol. 1650, DOI 10.1007/978-1-4939-7216-6_13, © Springer Science+Business Media LLC 2017

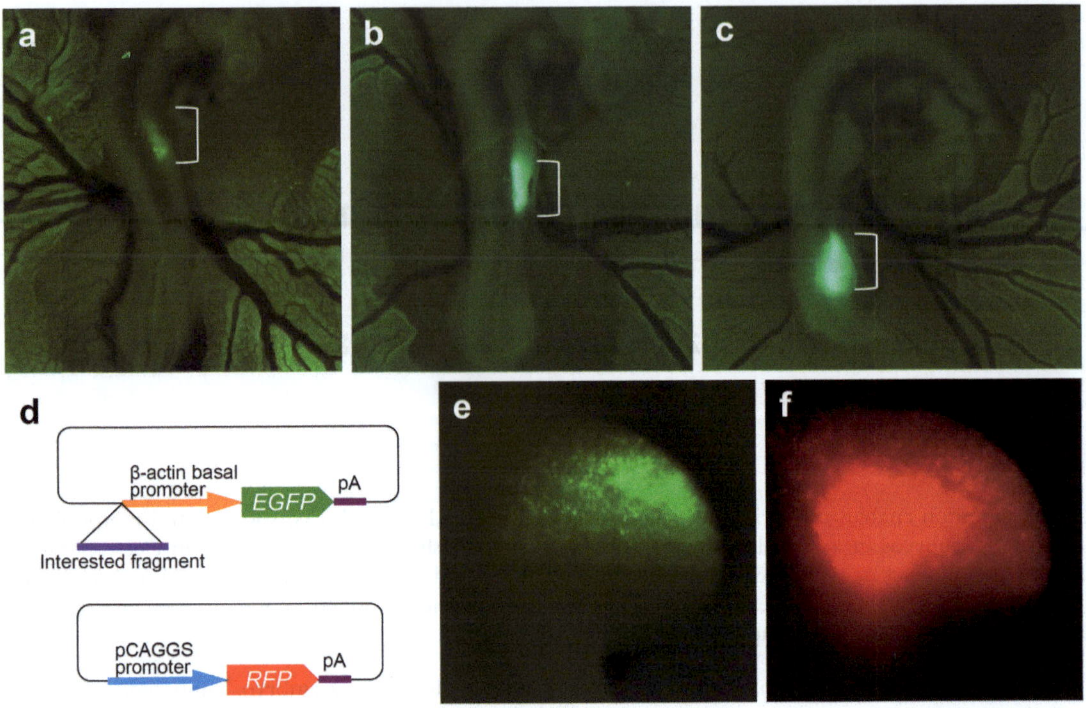

Fig. 1 Limb buds of chick embryos electroporated with DNA. (**a–c**) Expression of *EGFP* 24 h after electroporation of pCAGGS-*EGFP* into the anterior region of the wing field (**a**), the posterior region of the wing field (**b**), and the leg field (**c**). (**d**) The *EGFP* reporter vector (*top*) and the marker vector, pCAGGS-*RFP* (*bottom*) are electroporated into the chick embryos to assess the function of the inserted DNA fragment. (**e**, **f**) Expression of *EGFP* driven by the *Gli3* limb-specific enhancer (**e**). The *EGFP* reporter vector was co-electroporated with pCAGGS-*RFP* (**f**). (**e**, **f**, Courtesy of K. Onimaru [8])

technique into limb buds (based on the method established by [7]) as well as its application for enhancer analysis [8] by using the latest electroporator (*see* **Note 1**).

2 Materials

2.1 Instruments (Fig. 2a)

1. Tungsten needles: Sharpen 0.50 mm in diameter tungsten wire tips (Nilaco Co.) via electrolysis by immersing in 5 N NaOH and applying electric current at 15 V. Attach the sharpened needle to a needle holder.
2. Glass capillaries for DNA injection: Pull the 90 mm glass capillaries (G-1, Narishige) using a puller, break the tip, and attach to an aspirator tube (Drummond).
3. Bended glass pipette for Ink-PBS injection: Bend and pull out the Pasteur pipette sideways over a flame and cut the tip. Attach the pipette to a plastic tube (a 10 mm in diameter, a 100 cm in length), and attach a blue tip to another side of tube.

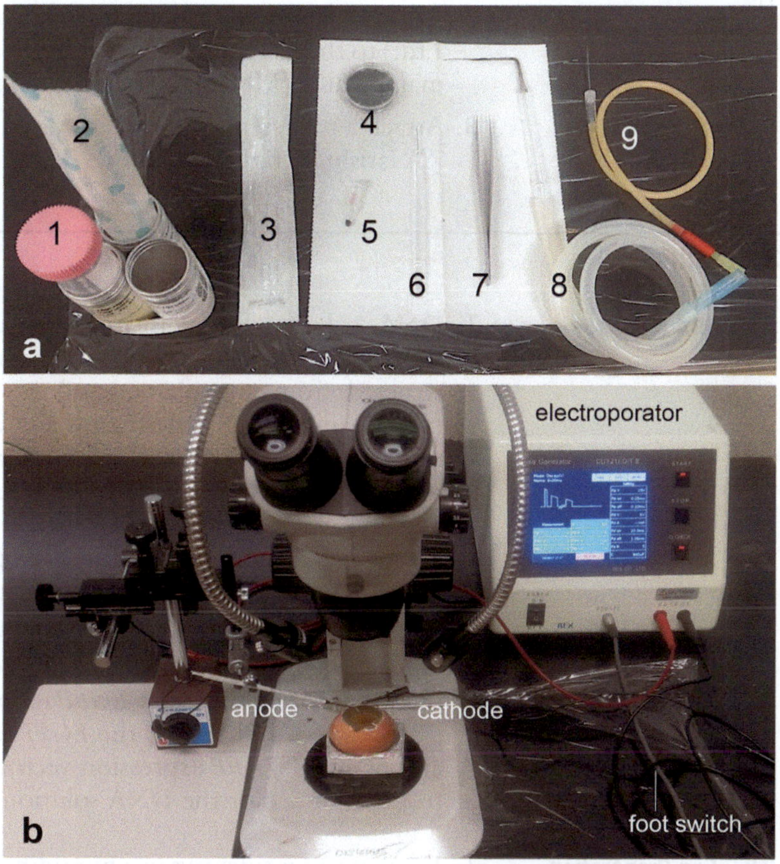

Fig. 2 Instruments and an electroporator setup. (**a**) Instruments and solutions required for electroporation. (*1*) Penicillin-streptomycin solution; (*2–3*) Disposable plastic pipettes for penicillin-streptomycin solution (2) and removing albumin (3); (*4*) Ink-PBS solution; (*5*) DNA solution; (*6*) Tungsten needle; (*7*) Forceps; (*8*), A bended glass pipette for Ink-PBS injection attached to a plastic tube; (*9*) Glass capillaries for DNA injection attached to an aspirator tube. (**b**) CUY21 EDITII electroporator, manipulator, and dissecting microscope are set up

4. Standard instruments for dissecting chick embryos: Forceps, plastic tape, disposable plastic pipettes.

2.2 Solutions (Fig. 2a)

1. Ink-PBS solution: 5 ~ 10% isograph liquid Ink (Rotring) in autoclaved phosphate-buffered saline (PBS).
2. Penicillin-streptomycin solution: 1% Penicillin-streptomycin (SIGMA) in autoclaved PBS.

2.3 Electroporator, Microscope, and Manipulator Setup (Fig. 2b)

1. A CUY21 EDIT II electroporator (BEX Co., Ltd.) (*see* **Note 2**) with foot switch, a stereo microscope, and a manipulator (M-152, Narishige) are set up as shown in Fig. 2.

2. Attach a platinum cathode electrode (custom order, BEX Co., Ltd.) to the electrode holder (H-7, Narishige) equipped on the manipulator (M-152, Narishige; left side in Fig. 2).
3. Attach a tungsten anode electrode to the electrode holder (H-7, Narishige; right side in Fig. 2).

3 Methods

3.1 DNA Solutions

1. DNA solution for misexpression of gene(s): Prepare the pCAGGS expression vector [9], an expression vector constructed from an enhancer of the human cytomegalovirus (CMV) immediate early region and a chick β-actin basal promoter, or pRCASBP vector, an avian replication competent retrovirus vector [10] with Maxi Prep (Qiagen; ~6 μg/μl), and mix with pCAGGS-*EGFP* or pCAGGS-*RFP* expression vector (2 ~ 5 μg/μl). Before electroporation, color the DNA solution with ~5% fast green (Nacalai Tesque).
2. DNA solution for enhancer analysis (Fig. 3): Subclone the enhancer sequence in front of a chicken β-actin basal promoter [11] that is followed by a *EGFP* reporter and polyA tail in the pBSSK+ vector [12]. Mix the *EGFP* reporter vector (~6 μg/μl) and pCAGGS-*RFP* expression vector (~2 μg/μl). Before electroporation, color the DNA solution with ~5% fast green.

3.2 Egg Preparation

1. Incubate eggs sideways in a humidified incubator at 38 °C until stage 12–16 (*see* **Note 1**).
2. Create a small window by forceps at the broader edge of the eggshell, and remove 3 ~ 5 ml of albumin with a disposable plastic Pasteur pipets.
3. Seal the small window with plastic tape.

3.3 Electroporation

1. Create a window on the eggshell, and drop 400 ~ 600 μl of penicillin-streptomycin solution on top of embryo.
2. Inject 100 ~ 200 μl of Ink-PBS solution underneath the embryo from outside of the extraembryonic blood vessels, using a bended glass pipette (Fig. 3a).
3. Draw 3 ~ 5 μl of DNA solution by a glass capillary attached with an aspirator tube, and inject into the embryonic coelom between the somatic and splanchnic mesoderm of the presumptive wing field (Fig. 3b, c).
4. Insert the tungsten cathode into the coelom using the small hole created by the glass capillary (Fig. 3b, d) (*see* **Note 3**).
5. Place the platinum anode over the presumptive wing field (Fig. 3b, e) (*see* **Note 4**). An anode must be placed at the

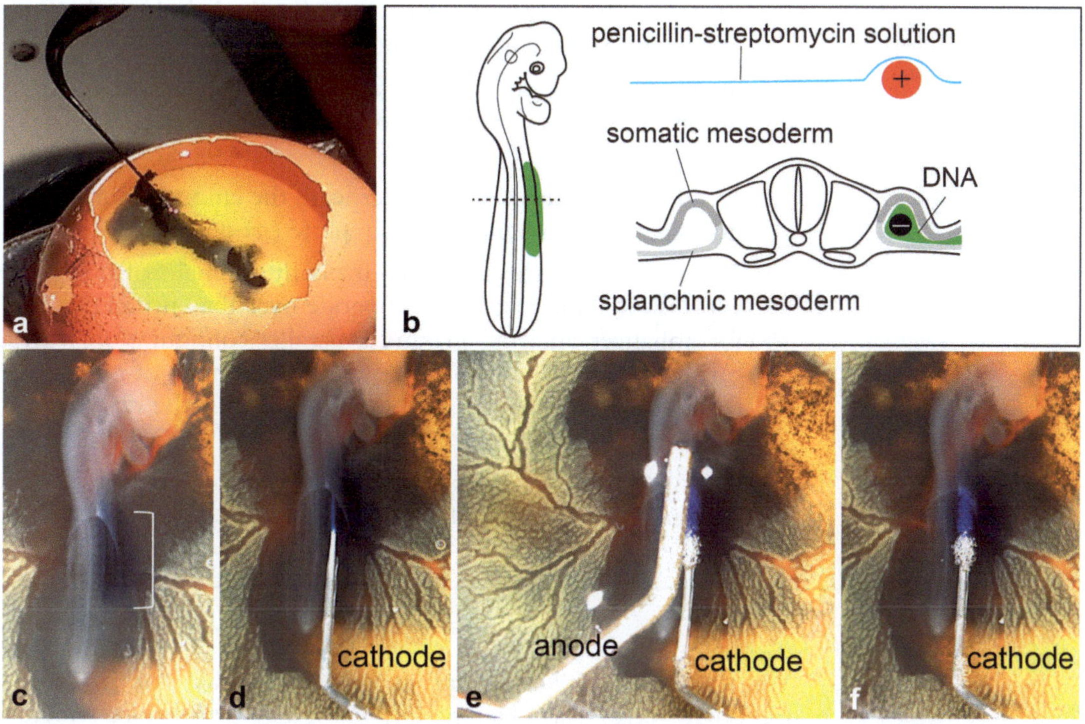

Fig. 3 Electroporation into wing field of stage 14 chick embryos. (**a**) Injection of Ink-PBS underneath the embryo by a bended glass Pasteur pipette. (**b**) A transverse section at the wing level of a stage 14 chick embryo. DNA solution is injected into the coelom between somatic and splanchnic mesoderm. A cathode (*minus*) is inserted into the coelom, and an anode (*plus*) is placed over the presumptive wing field. An anode should be placed at the same level of the surface of penicillin-streptomycin solution. (**c**) A stage 14 embryo injected with DNA solution mixed with fast green into the presumptive wing field (*a bracket*). (**d**) A cathode was inserted into the coelom. (**e**) An anode is placed over the presumptive wing field. (**f**) Immediately after successful electroporation. Small bubbles are observed around the cathode. Fast *green* turns into *purple*

same level of the surface of the penicillin-streptomycin solution (Fig. 3b). Avoid touching the surface of the embryo.

6. Apply electric pulses, consisted of one short pulse (25 V, 0.05 ms) and a 0.1 ms interval followed by five long pulses (8 V, 20.0 ms) with 1.0 ms intervals, by pressing the foot switch (Fig. 3e) (*see* **Note 5**).
7. Gently withdraw the electrodes from the egg and drop 400 ~ 600 μl of penicillin-streptomycin solution on the top of embryo. Seal the window with a plastic tape.

4 Notes

1. The latest model generates poration pulses, which make small holes in the cell surface, and subsequently generates driving pulses, which transfer molecules into the cell. This minimizes

damages and more efficiently introduces genes into the limb mesoderm than that of previous single pulse electroporator.

2. Stages 14 and 16 are the most appropriate stages for electroporation into the presumptive wing field and leg field, respectively.
3. A cathode can be inserted underneath the embryo from a small hole created by a bended glass pipette (*see* details in [7]).
4. The position of the anode controls the domain of transgene expression. Thus, we place an anode over the entire limb field for enhancer analysis, making a parallel configuration with a cathode (Fig. 3b, e). For restricting the domain of misexpression, we use a smaller anode and place it on the desired area within the limb field.
5. After successful electroporation, small bubbles appear around the cathode and color of fast green turns into purple (Fig. 3f).

Acknowledgment

We thank Toshihiko Ogura and Jun Miyazaki for providing pCAGGS vectors, and Hajime Ogino for an *EGFP* reporter vector. We also thank Koh Onimaru for providing figures. This work was supported in part by a Grant-in-Aid for Scientific Research (B) (25291086) to M.T.

References

1. Saunders JW Jr (1948) The proximo-distal sequence of origin of parts of the chick wing and the role of the ectoderm. J Exp Zool 108:363–404
2. Tickle C, Summerbell D, Wolpert L (1975) Positional signalling and specification of digits in chick limb morphogenesis. Nature 254 (5497):199–202
3. Logan M, Tabin C (1998) Targeted gene misexpression in chick limb buds using avian replication-competent retroviruses. Methods 14(4):407–420
4. Muramatsu T, Mizutani Y, Ohmori Y, Okumura J (1997) Comparison of three nonviral transfection methods for foreign gene expression in early chicken embryos in ovo. Biochem Biophys Res Commun 230(2):376–380
5. Yasugi S, Nakamura H (2000) Gene transfer into chicken embryos as an effective system of analysis in developmental biology. Dev Growth Differ 42(3):195–197
6. Ogura T (2002) In vivo electroporation: a new frontier for gene delivery and embryology. Differentiation 70(4–5):163–171
7. Suzuki T, Ogura T (2008) Congenic method in the chick limb buds by electroporation. Dev Growth Differ 50(6):459–465
8. Onimaru K, Kuraku S, Takagi W, Hyodo S, Sharpe J, Tanaka M (2015) A shift in anterior-posterior positional information underlies the fin-to-limb evolution. elife 4: e07048. doi:10.7554/eLife.07048Q2
9. Niwa H, Yamamura K, Miyazaki J (1991) Efficient selection for high-expression transfectants with a novel eukaryotic vector. Gene 108(2):193–199
10. Morgan BA, Fekete DM (1996) Manipulating gene expression with replication-competent retroviruses. Methods Cell Biol 51:185–218
11. Matsuo I, Kitamura M, Okazaki K, Yasuda K (1991) Binding of a factor to an enhancer element responsible for the tissue-specific expression of the chicken alpha A-crystallin gene. Development 113(2):539–550
12. Ogino H, Fisher M, Grainger RM (2008) Convergence of a head-field selector Otx2 and notch signaling: a mechanism for lens specification. Development 135(2):249–258

Part III

Stem Cells

Chapter 14

Chicken Induced Pluripotent Stem Cells: Establishment and Characterization

Aurelie Fuet and Bertrand Pain

Abstract

In mammals, the introduction of the OSKM (Oct4, Sox2, Klf4, and c-Myc) genes into somatic cells has allowed generating induced pluripotent stem (iPS) cells. So far, this process has been only clearly demonstrated in mammals. Here, using chicken as an avian model, we describe a set of protocols allowing the establishment, characterization, maintenance, differentiation, and injection of putative reprogrammed chicken Induced Pluripotent Stem (iPS) cells.

Key words Chicken, Induced pluripotent stem cells, Transfection

1 Introduction

The Chicken Embryonic Stem (cES) cells derived from chicken early blastoderm (stage IX-XII, EG & K) present typical features of ES pluripotent cells. Indeed, these cells are able to self renew for long-term culture, to differentiate in vitro and in vivo into various cell types and to contribute to chimeras when injected into recipient embryos as described in detail [1]. The pluripotency of these unique cells is maintained through the cOCT4 and cNANOG genes in the presence of other actors that were identified through a complete transcriptomic approach [2, 3]. In mammals, the introduction of the OSKM (Oct4, Sox2, Klf4 and c-Myc) genes into somatic cells has allowed us to generate induced pluripotent stem (iPS) cells [4, 5]. The chicken cDNAs for OCT4 (POUV), NANOG, SOX2, KLF4, c-MYC, and NANOG were delivered through a transposon efficient system to primary chicken embryonic fibroblasts (CEF). Even if some morphological changes were observed, the OSKM canonical combination was unable to provide long-term proliferating iPS-like cells in contrast to the OSKMN gene combination [6]. Only in those conditions, individual clones of reprogrammed cells were able to be long term cultured and

Guojun Sheng (ed.), *Avian and Reptilian Developmental Biology: Methods and Protocols*, Methods in Molecular Biology, vol. 1650, DOI 10.1007/978-1-4939-7216-6_14, © Springer Science+Business Media LLC 2017

established. The stem cell character of those reprogrammed cells was evaluated by using different markers such as the expression of alkaline phosphatase, of specific antibodies, of pluripotent associated genes and by their ability to differentiate both in vitro and in vivo.

2 Materials

2.1 Expression Vectors

1. Chicken cDNAs: The chicken cDNAs were obtained from the coding sequence of POUV (Gene ID: 427781), SOX2 (Gene ID: 396105), KLF4 (Gene ID: 770254), c-MYC (Gene ID: 420332), and NANOG (Gene ID: 100272166) by reverse transcription of chicken embryonic stem cells RNAs, inserted into pGET vector and sequenced.
2. Expressing vectors: pPB-POUV, pPB-SOX2, pPB-KLF4, pPB-cMYC, and pPB-NANOG vectors were obtained by inserting the corresponding cDNAs into the PB CAG-Transgene-PGK-Hygro expression vector [7]. This pPB vector backbone is derived from the Piggybac transposon [8] and carried the CAG promoter driving the cDNA expression and a hygromycine selection cassette under the control of the PGK promoter.
3. The pCAGPBase transposase expressing vector.
4. The pUC (or pBSK) plasmid used as a negative transfection control and the pPB-empty used as a positive selection plasmid.

2.2 Plasmid Preparation

1. Plasmids: all the plasmids were amplified in DH5a bacteria strain, grown in LB medium and prepared using HiPure column.

2.3 Eggs

1. Eggs: fertile freshly laid eggs were collected and stored at 16 °C for 2–5 days before being processed either for being incubated for CEF production or injected with reprogrammed cells.

2.4 Culture Media

1. Chicken Embryonic Fibroblast medium—CEF medium: in a 500 mL DMEM/F12 bottle, add 40 mL of non-heat inactivated ES tested batch of Fetal Bovine Serum (FBS), 10 mL decomplemented chicken serum, 50 mL TPB (Tryptose Phosphat broth) solution, 6 mL of 10,000 U/1000 U Penicillin/Stretomycin stock solution. Store the CEF medium at 4 °C for up to 2 weeks.
2. ES Complete Medium (ES CM): in a 500 mL DMEM/F12 bottle add 50 mL of ES tested batch of Fetal Bovine Serum (FBS), non-heat inactivated, 6 mL of NEAA (Non Essential Amino Acid) stock solution, 6 mL of 100 mM Sodium Pyruvate stock solution, 6 mL of 10,000 U/1000 U Penicillin/Stretomycin stock solution, 6 mL of 200 mM glutamine stock

solution, 1 mL of 50 mM β-Mercaptoethanol stock solution. Store the ES medium at 4 °C for up to 2 weeks.

3. ES Proliferative Medium (ES PM): add 5 ng/mL IGF1, 1 ng/mL SCF, 1 ng/mL IL6, 1 ng/mL sIL6 Rα, and 1000 U/mL LIF in the ES CM bottle. All the factors are recombinant either from human or mouse origin.
4. ES Selective Medium (ES SM): add 75 μg/mL Hygromycine in ES PM.
5. Freezing Medium: 80% FBS, 20% DMSO volume/volume. Keep on ice and prepare extemporaneously.
6. TPB: Dissolve and autoclave 29.5 g TPB per 1 L distilled water. Store at 4 °C.
7. 0.1% gelatin solution: Dissolve and autoclave 0.5 g bovine skin gelatin per 500 mL PBS 1×. Store at room temperature (RT). For the gelatin coating, add 0.1% gelatin solution at RT for at least 1 h using around 0.5 mL/cm^2 dish (i.e., 5 mL per 100 mm dish). Remove the solution before cell plating. Dishes, wells, slides can be coated in advance and kept dried and sterile for few days at RT once gelatin has been removed.
8. Pronase stock solution: dissolve 1 g of pronase powder in 80 mL of PBS by stirring vigorously to get a 10× stock solution. Filtrate on 25 μm and aliquot in 5 mL. Keep at −20 °c and at 4 °c for 1 week once thawed.

2.5 Disposable Sterile Plastic

1. 15 mL and 50 mL centrifuge tubes.
2. Cell culture-treated plates (12 wells, 6 wells), and 100 mm dish.
3. Micropipettes and sterile tips.
4. Pipettes (25, 10, 5 mL).
5. Bacteria Petri dishes (100 mm).
6. Eppendorf tubes (1.5 mL).
7. Cryovials.
8. Disposable Cell filters (40 μm mesh).
9. Kova slides.

2.6 Dissection Material

1. Scalpel blades.
2. Small cissors.
3. Small forceps.

2.7 Chemicals

1. PBS.
2. 70% Ethanol.
3. Trypan blue.

4. DMSO (dimethylsulfoxide). Store in the dark at RT.
5. Methanol.
6. Paraformaldehyde 32%.
7. Fixative solution: 4% paraformaldehyde in PBS. Keep at −20 °C.

2.8 Cell Culture Laboratory Environment

1. Bench coat.
2. Bench flux.
3. Low speed centrifuge (Cell culture).
4. High speed centrifuge (Molecular Biology).
5. Egg incubator.
6. CO_2 gas incubator.
7. Gloves.
8. Fluorescent microscope.
9. Inverted microscope.
10. Laboratory coat.
11. Nanodrop.
12. −20 °c freezers.
13. −80 °c freezers.
14. Liquid nitrogen tank.

3 Methods

3.1 CEF Preparation

Primary chicken embryonic fibroblasts were obtained from incubated embryos and prepared to get homogenous culture after 1–2 passages in a selective fibroblast medium containing TPB and serum (Fig. 1).

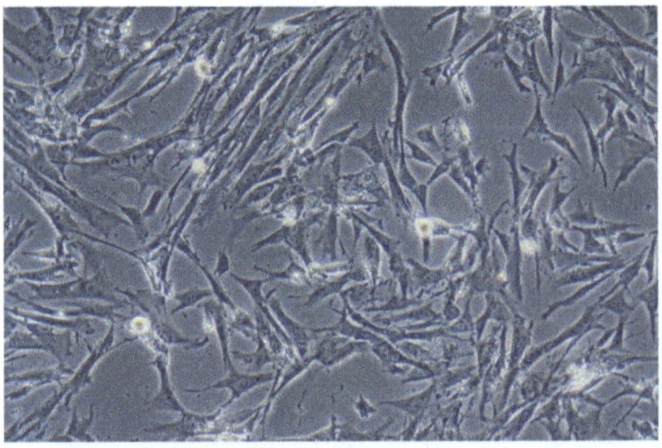

Fig. 1 Primary chicken Embryonic Fibroblasts (CEF). CEF were prepared as described and passaged every 4–5 days (100× magnification)

1. Wipe Fertile eggs with 70% ethanol and incubate them for 12 days at 37.5 °C, with rocking and 65% humidity, in an egg incubator.
2. At day 12, remove the eggs, wip them with 70% ethanol.
3. Under bench flux, break one egg and pour the egg content into a sterile bacterial Petri dish (*see* **Note 1**).
4. Take the embryo with dissecting forceps, cutting the embryological annexes with small surgical scissors.
5. Transfer the embryo in a new sterile bacterial Petri dish and wash it with PBS.
6. Transfer the embryo into a new bacterial Petri dish, hold it with dissecting forceps, and remove the head and the four members with a scalpel blade.
7. Eviscerate the carcass by a scalpel blade, still holding the carcass with dissecting forceps.
8. Transfer the carcass in a new bacterial Petri dish and mince it with scalpel blades to obtain a homogenous thick mash.
9. Transfer it with a 10 mL pipette into a 50 mL centrifuge tube.
10. Add 7.5 mL PBS and 2.5 mL of 10× trypsine stock solution, mix slowly, and incubate for 5 min at RT.
11. Decant the solution and take the supernatant.
12. Filter it on a 40 μm cell mesh on a 50 mL centrifuge tube containing 10 mL of CEF medium.
13. Repeat **step 10** and pool both filtered supernatants.
14. Centrifuge for 5 min at 400 × *g* at RT.
15. Resuspend the cell pellet in the 10 mL fresh fibroblast medium. Homogenize the cell suspension by gentle up and down pipetting.
16. Take a 50 μL aliquot.
17. Mix vol/vol with Trypan blue and count the cell suspension on Kova slide by taking care of not counting the remaining erythroid ellipsoid cells present in the preparation (*see* **Note 2**).
18. Adjust the cell concentration to 2×10^6 cells per mL.
19. Plate 2×10^6 cells per 100 mm dish in a 12 mL CEF medium.
20. Incubate at 38 °C under 10% CO_2 with 60% humidity for 2 days.
21. Change the medium, wash the cells with PBS, and add 12 mL fresh CEF medium.
22. For routine culture, change medium every 2 days.
23. Dissociate and pass the cells every 4–6 days.

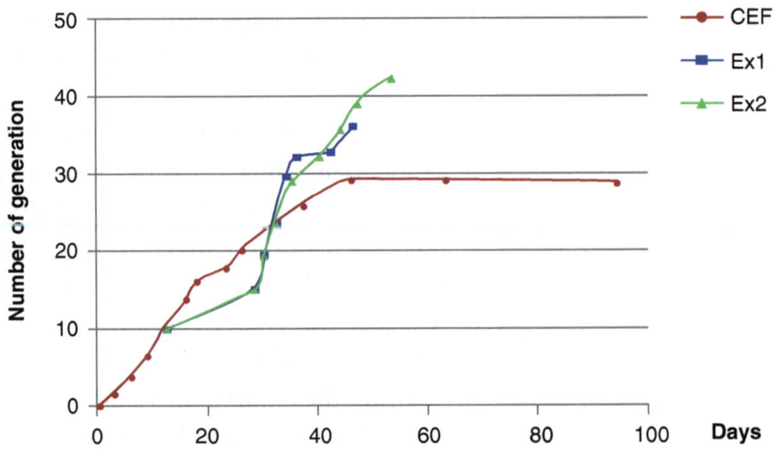

Fig. 2 Growth curve of CEF and the emergence of two examples of reprogrammed cells (Ex1 and Ex2). Cells were obtained as described, passaged and counted at each passage. The generation number (n) was calculated as $N = 2^n$ with N, the cumulated total of cell number

3.2 CEF Dissociation and Passage

CEF are easy to dissociate, propagate, and amplify for 8–12 passages before the cells enter into senescence as estimated by growth curve and rapid decrease of the proliferation rate (Fig. 2) (*see* **Note 3**).

1. Collect conditioned medium in a 50 mL tube.
2. Rinse the cells with 5 mL PBS twice.
3. Add 5 mL of Trypsine 1X in PBS and incubate the cells at 37 °C for 2–4 min until the cells detach.
4. Check the dissociation under the microscope.
5. Collect the cells in the 50 mL tube containing the conditioned medium.
6. Rinse the dish with 5 mL PBS and collect in the same tube.
7. Repeat **step 6** and homogenize the cell suspension by gentle up and down pipetting.
8. Count the cells as described.
9. Centrifuge for 5 min at 400 × *g* at RT.
10. For cell maintenance, resuspend the cell pellet at 10×10^6 per 1 mL of fresh medium.
11. Dilute the cell suspension to get 1×10^6 cells per one 100 mm dish in 12 mL fresh medium.

3.3 CEF Freezing

CEF are easy to freeze in large scale as cell amplification is rapid and easy. The present described procedure is using 10% of DMSO as cryoprotectant (*see* **Note 4**).

1. For freezing the cells, resuspend the cell pellet after dissociation and centrifugation at 10×10^6 cells per 1 mL of fresh medium.
2. Check the final volume of the cell suspension.
3. Add slowly, drop by drop, the same volume of ice-cold Freezing medium stirring gently between each drop but avoiding any bubbles.
4. Place the tube on ice.
5. Distribute 1 mL of cell suspension (5×10^6 cells) per cryovial placed on ice.
6. Close the vials tightly.
7. Allow the vials to be cooled at −80 °C in a cooling box overnight.
8. Transfer the frozen vials in liquid nitrogen for long-term storage (*see* **Note 5**).

3.4 CEF Thawing

CEF are primary cells easy to freeze and thaw as their recovery rate is near 90–95% in good conditions of freezing.

1. Remove the frozen vial from liquid nitrogen.
2. Thaw quickly the frozen vial in warm water until the cell suspension melts but not completely.
3. Open the vial and remove the cells suspension with 1 mL micropipette tip.
4. Place the cell suspension in a 50 mL centrifuge tube.
5. Rinse the vial with 1 mL of CEF medium.
6. Add drop by drop this 1 mL on the thawed cell suspension stirring gently between each drop.
7. Add 12 mL of fibroblast medium in the same way.
8. Centrifuge at $200 \times g$ for 10 min at RT.
9. Resuspend the cell pellet in 25 mL of CEF medium.
10. Plate two 100 mm dishes with 12 mL cell suspension each.
11. Incubate at 38.5 °C under CO_2 with 60% humidity for 2–3 days.
12. Rinse the cells with PBS 24 h after plating. Add fresh CEF medium.
13. Dissociate the cells as described once the dishes are full and prepare the plates for reprogramming transfection.

3.5 Plasmid Preparation

The quality of plasmid for reprogramming is crucial to allow an efficient transfection of the primary cells, even if CEF are cells transfected with high efficiency. The preparation uses commercially available columns and solutions.

1. Equilibrate the HiPure Midi column by adding 10 mL Equilibration Buffer.
2. Allow the solution to drain by gravity flow.
3. Harvest and centrifuge the bacteria at 4000 × *g* for 10 min at 4 °c.
4. Discard the supernatant.
5. Add 4 mL Resuspension Buffer with RNase A.
6. Resuspend and homogenize the cell pellet.
7. Add 4 mL Lysis Buffer.
8. Mix gently by inverting the capped tube until the mixture is homogeneous. Do not vortex. Incubate at room temperature for 5 min.
9. Add 4 mL Precipitation Buffer. Mix immediately by inverting the tube until the mixture is homogeneous. Do not vortex.
10. Centrifuge the lysate at >12,000 × *g* for 10 min at room temperature.
11. Load the supernatant onto the equilibrated column. Allow the solution in the column to drain by gravity flow.
12. Add 10 mL Wash Buffer to the column.
13. Discard the flow-through after Wash Buffer drains from the column. Repeat wash step once.
14. Place a sterile 15 mL centrifuge tube under the column.
15. Add 5 mL Elution Buffer to the column.
16. Allow the solution to drain by gravity flow. The elution tube contains the purified DNA.
17. Add 3.5 mL isopropanol to the eluate. Mix well.
18. Centrifuge at 12,000 × *g* for 30 min at 4 °c.
19. Discard the supernatant.
20. Wash the pellet in 3 mL 70% ethanol.
21. Centrifuge the tube at 12,000 × *g* for 5 min at 4 °C and discard the supernatant.
22. Air-dry the pellet for 10 min, and then resuspend the purified plasmid DNA in 100 μL TE Buffer (TE).
23. Check the plasmid quality and quantity with Nanodrop.
24. Adjust the concentration to 1 mg x mL^{-1}.
25. Store at −20 °C.

3.6 Fibroblast Reprogramming

CEF reprogramming appears to be a complex process: if morphological changes are relatively easy to observe after exogenous transgenes transduction, those are in many cases only transient. Indeed, the stabilization of the phenotype and the long-term establishment of reprogrammed cells are still randomly observed. The presented chosen gene combination—OSKMN—provides long-term proliferating reprogrammed cells.

1. Plate the dissociated CEF at 2×10^5 cells per well in 4 mL of CEF medium.
2. Prepare at least two wells for reprogramming, one well for positive selection control and one well for negative selection control.
3. Rinse the cells with PBS the next day and add 2 mL of CEF medium per well.
4. Prepare the reprogramming transfection mix in an Eppendorf tube for 1 well of a 6-well plate by adding in this order:
   - 300 μL OPTIMEM.
   - 0.8 μL of each plasmid corresponding to 0.8 μg of each of the five plasmid (pPB-POUV, pPB-SOX2, pPB-KLF4, pPB-c-MYC, pPB-NANOG).
   - 2 μL of pPBase corresponding to 2 μg of plasmid.
   - 18 μL of Fugene (*see* **Note 6**).
5. Prepare the selection positive control transfection mix for 1 well in an Eppendorf tube by adding in this order:
   - 300 μL OPTIMEM.
   - 4 μL of pPB-Empty plasmid corresponding to 4 μg.
   - 2 μL of pPBase corresponding to 2 μg of plasmid.
   - 18 μL of Fugene (*see* **Note 7**).
6. Prepare the negative control transfection mix for 1 well in an Eppendorf tube by adding in this order:
   - 300 μL OPTIMEM.
   - 4 μL of pUC(or pBSK) plasmid corresponding to 4 μg.
   - 2 μL of pPBase corresponding to 2 μg of plasmid.
   - 18 μL of Fugene (*see* **Note 8**).
7. Mix gently the tubes and incubate for 15 min at RT.
8. Add all the 324 μL of each mix in the well (*see* **Note 9**).
9. Rock gently the plate with the transfected wells and incubate at 38.5 °C under CO_2 with 60% humidity.
10. Rinse the cells with PBS 24 h after transfection and add 4 mL of fresh CEF medium.
11. Dissociate the cells with Pronase 1×, 2 days after transfection.
12. Plate all cells in two 100 mm dish with 12 mL of ES SM medium.
13. Rinse the cells with PBS 24 h after dissociation and add 12 mL of fresh ES SM medium.

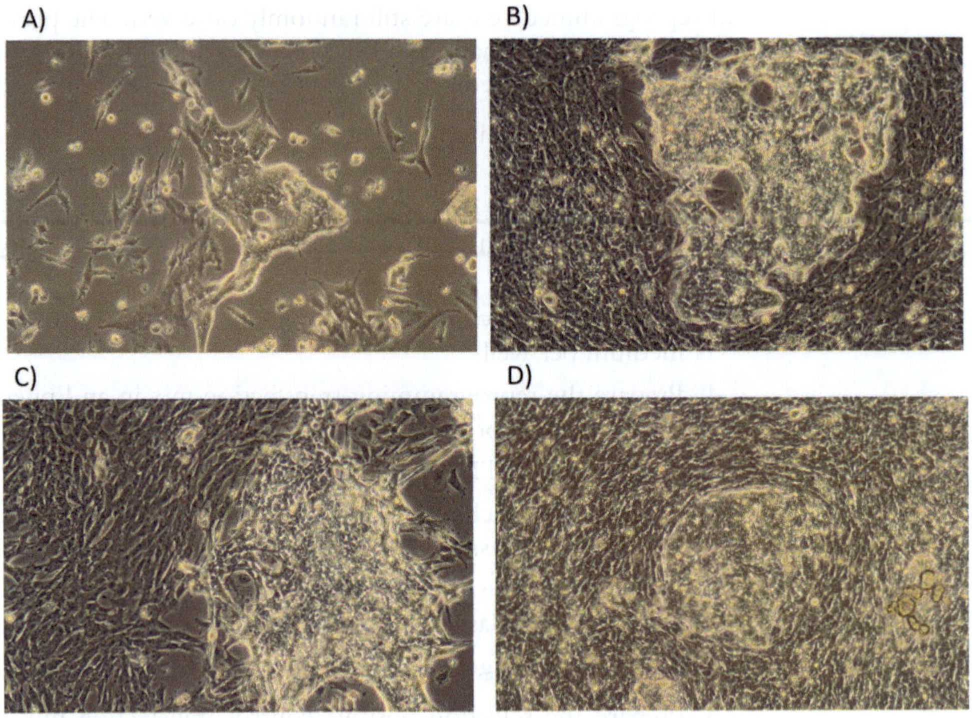

Fig. 3 Examples of emerging clones after CEF transduction and selection. Various morphologies are observed from ES-like' typical clones (**a**, **b**) and more mixed phenotypes (**c**, **d**) (100× magnification)

14. Change the ES SM medium every 2 days.
15. 7–8 days after the beginning of the selection, the negative control dishes should be empty in contrast to the positive control dishes with still proliferative CEF and the reprogrammed dishes that should present colonies with various morphologies surrounded by CEF (Fig. 3).
16. Pick individual clones between 8 and 10 days after the beginning of the selection according to their size and their most ES-like morphology.
17. Plate the picked clones in gelatin coated well of a 12-well plate in 1 mL of ES PM medium.
18. Change the wells every 2 days with 1–2 mL ES PM medium according to the cell density.
19. Screen the clones for the most ES-like morphology.
20. Dissociate the individual clones with good morphology with Pronase 1× when the well is full.
21. Plate the dissociated cells in a gelatin-coated well of a 6-well plate.
22. Change the medium every 2 days and dissociate the well when full.

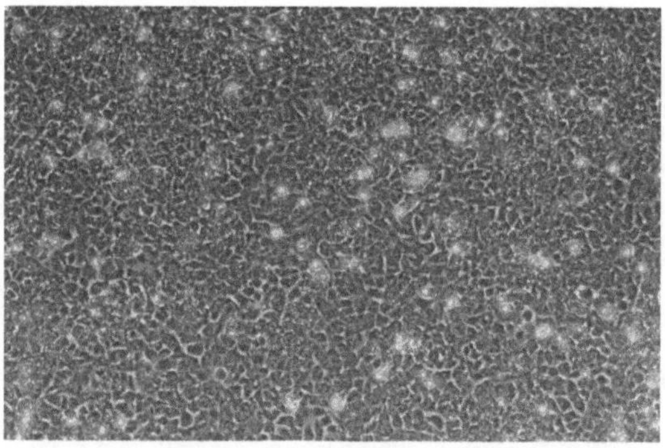

Fig. 4 Once passaged and stabilized for growth and phenotype, reprogrammed cells exhibit a homogenous morphology (100× magnification)

23. Start to analyze the cells only when the cells have been stabilized in both morphology and growth, usually observed 5–6 passages after the initial cloning (Fig. 4).
24. Freeze the cells at those early passages for stocks as previously described for the cES cells.
25. Maintain the cells for long-term establishment and further characterization.

3.7 Reprogrammed Cell Characterization

Reprogrammed chicken cells present some of the features found also in cES cells. Here, we described how to characterize the cells in similar and complementary ways as previously described for the cES cells [1].

The reprogrammed cells are positive for alkaline phosphatase (AP), exhibit an endogenous telomerase activity, present reactivity for SSEA-1 and EMA-1 antibodies, and form embryoid bodies. The cells expressed various molecular markers detected by real-time RT-PCR. The cells can be genetically labeled with a fluorescent reporter and injected into recipient embryo to evaluate their developmental contribution. The full characterization of the chicken reprogrammed cells, including the RNAseq data, is available in [6].

3.7.1 Alcaline Phosphatase Staining

1. Plate 2×10^5 cells in a well of a 6-well plate in ES PM medium.
2. Allow the cell to grow for 48 h by changing the medium 24 h after plating.
3. Wash the cells with PBS and fix the cells by adding 2 mL methanol for 15 min in a chemical extractor hood.
4. Remove the methanol (*see* **Note 8**) and allow the well to dry.
5. Prepare the diazonium salt solution by adding the ready to used kit solutions (AP Kit -Sigma ref 86R),
   - 100 μL of the FRV alkaline solution buffer in a 50 mL tube.
   - 100 μL of nitrite solution.

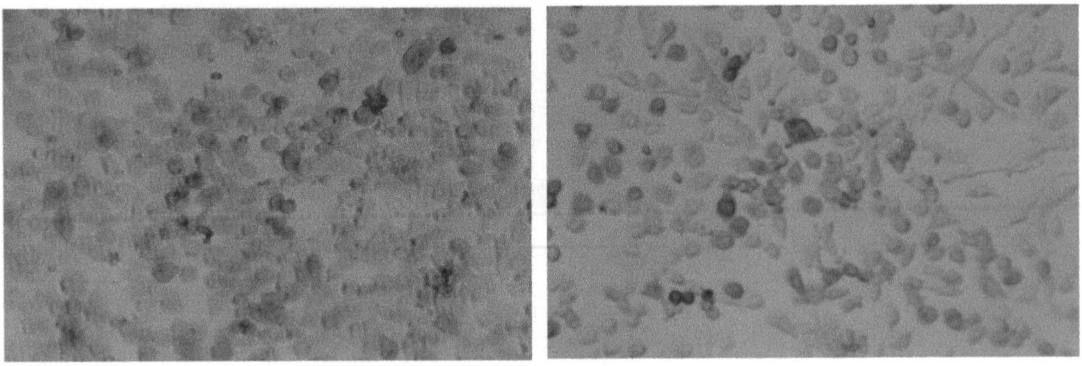

Fig. 5 Examples of AP positive reprogrammed cells from isolated clones (100× magnification)

- Mix and incubate for 2 min at RT.
- Add 4.7 mL distilled H_2O.
- Add 100 μL of Naphthol AS-BI alkaline solution.

6. Add this 5 mL mix solution directly to the dried well and incubate for 15 min.
7. The red color will develop for the positive cells (Fig. 5).
8. Remove the solution (*see* **Note 10**), wash with PBS, add 2 mL of PBS, and observe the cells under the inverted microscope.
9. cES cells are routinely used as positive control when the primary CEF are negative.

3.7.2 Telomerase Activity

The telomerase activity is detected and evaluated from the lysates of the reprogrammed cells using the Telo*TAGGG* telomerase PCR Elisa (Roche, ref 11854666910). The protocol is highly detailed and consists of three main steps: the preparation of the cell lysates, the TRAP (telomeric repeat amplification protocol) reaction, and the hybridization and ELISA step.

1. Dissociate the proliferating reprogrammed cells, count them, and harvest 2×10^5 cells in an Eppendorf tube.
2. Prepare cES cells as positive control and primary CEF, as negative control in similar conditions of 2×10^5 cells in Eppendorf tubes. The kit provides controls as well.
3. Prepare the cell lysates as indicated.
4. Follow the different steps of the detailed protocols.
5. Interpret the results and evaluate the telomerase activity in the reprogrammed cells comparatively to the cES cells as positive control and to the CEF as negative cells (Fig. 6).

3.7.3 Antibody Staining

Chicken ES cells are positive for both SSEA1 and EMA-1 antibodies. Once well established and phenotypically stabilized, the reprogrammed chicken cells exhibit also reactivity toward those

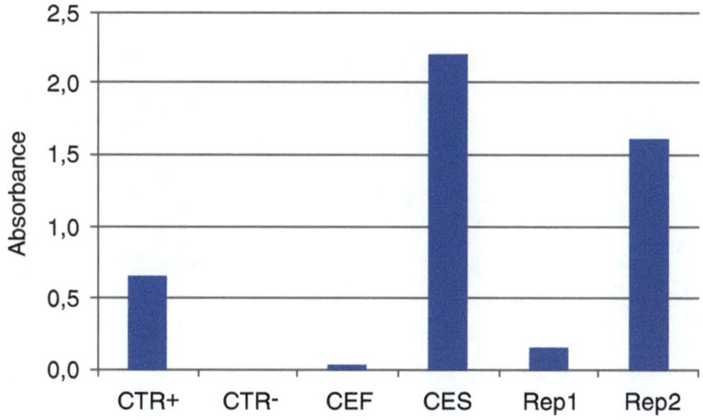

Fig. 6 Telomerase activity in chicken reprogrammed cells (Rep1 and Rep2) compared with cES, as positive cells and CEF as starting cells and negative ones. CTR+ is provided by the manufacturer and CTR– is obtained after the treatment of a positive sample with RNAse as recommended

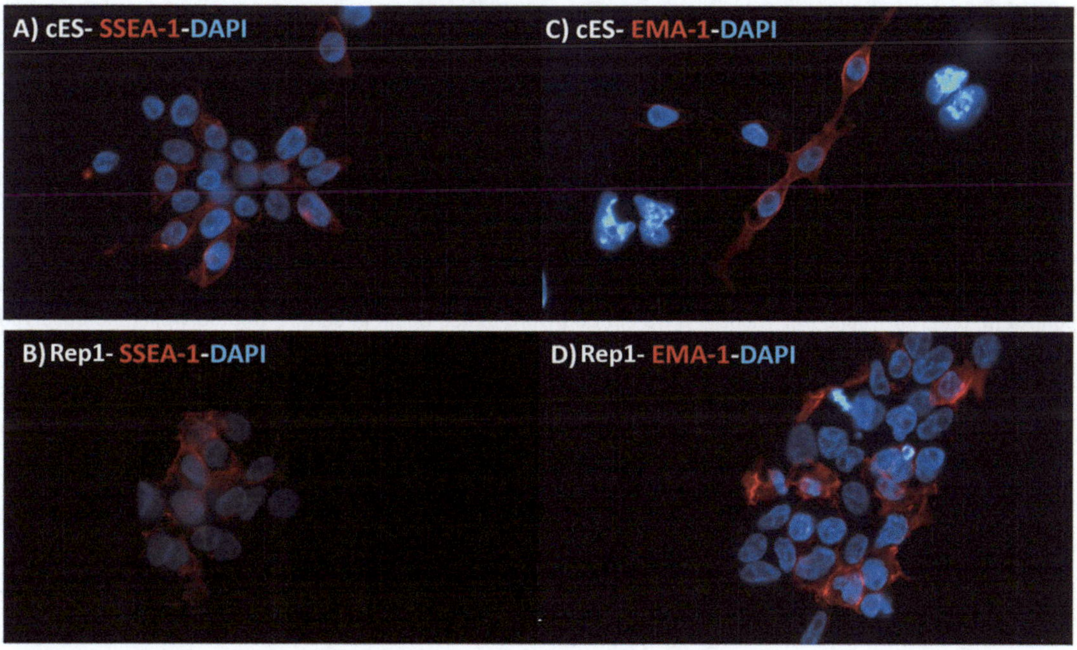

Fig. 7 Reprogrammed cells (Rep1 as an example) are positive for SSEA-1 and EMA-1 antibodies, ES cell-specific antibodies, as the cES cells taken as control are

antibodies. The detailed protocol has been reported previously for the cES cells [1] and will be followed in exactly similar conditions for the reprogrammed cells (Fig. 7).

3.7.4 Embryoïd Body Formation

Reprogrammed cells form embryoid bodies in similar conditions as previously described for the cES cells and as illustrated (Fig. 8).

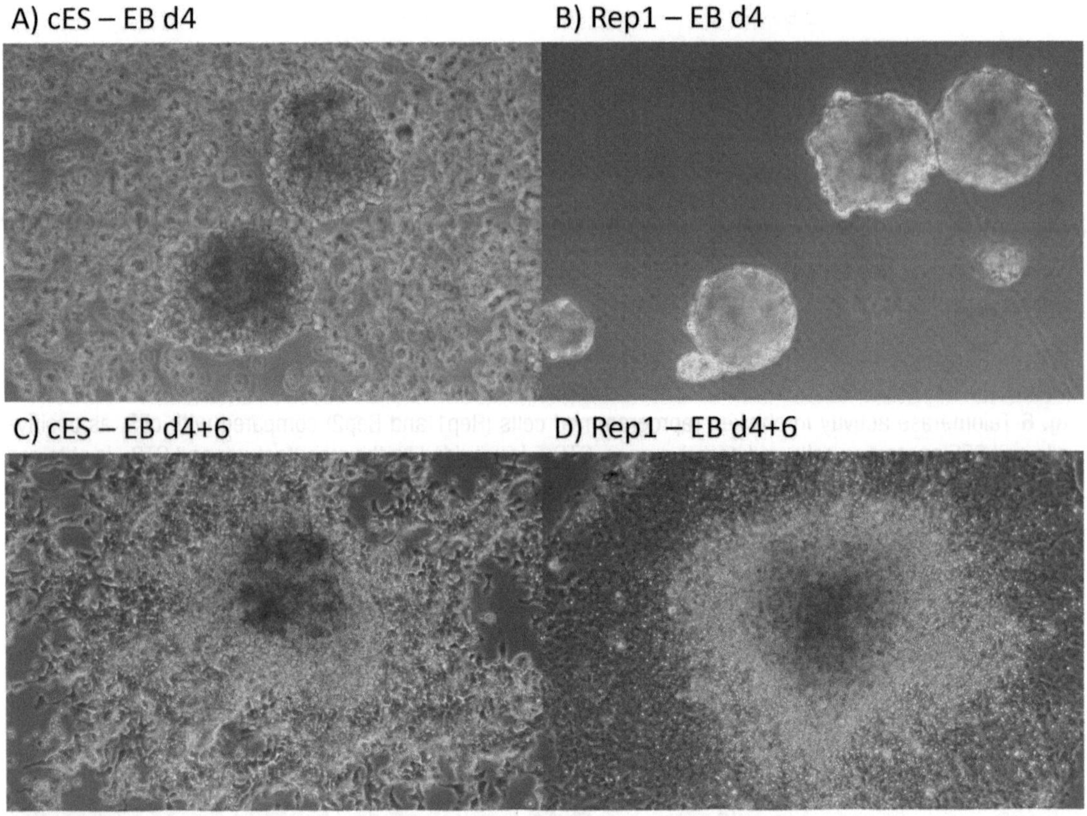

Fig. 8 Embryoid bodies (EBs) were obtained after 4 days by hanging drop method with cES cells as positive control (**a**) and Rep1 cells (**b**). Those 4 days old EBs were plated again in culture-treated wells and observed 6 days later for both cES (**c**) and Rep1 (**d**) derived cells, (40× magnification)

3.8 Detection of Molecular Markers

For determining the transcriptomic profile of the reprogrammed cells, total RNAs were extracted and submitted to real-time RT-PCR for some of the pluripotency-associated genes.

3.8.1 Reverse Transcription

1. To inactivate any trace of genomic DNA for the RT-PCR analysis, take 1 μg of total RNA, add up to 6.4 μL distilled H_2O, 0,8 μL DNAse buffer and 1 μL DNAse I (DNAseI Amp, Life).
2. Incubate for 15 min at 22 °C.
3. Inactivate the DNAseI by adding 0,8 μL of a 0.1 M EDTA solution and incubate for 10 min at 65 °C.
4. Add 10 μL of RT buffer and 1 μL of RT enzyme (High Capacity RNA-to-cDNA Kit, Life, 4387406) (*see* **Note 11**).
5. Incubate for 60 min at 37 °C and then for 5 min at 95 °C.
6. Stock and store the RT cDNAs at −20 °C.

3.8.2 Real-Time RT-PCR

1. Dilute to 10 μM the oligonucleotides (listed in Table 1) upon reception by mixing 10 μL of each of sens and antisens primer (stock at 100 μM) with 80 μL of distilled water.
2. Dilute 1/5 the RT cDNAs in distilled water.
3. Prepare the reaction mix by adding (1 target gene with 24 reactions) (*see* **Note 12**).
   - 132 μL of Fast Syber® Green Master mix.
   - 98 μL of distilled water.
   - 8 μL of the primer mix.
4. Prepare the PCR MicroAmp® Fast Optical 96-Well Reaction Plate by adding.
   - 1 μL of diluted RT cDNAs (**step 2**) in triplicate.
   - 9 μL of reaction mix (**step 3**) per well.
5. Add the adhesive film to the plate.
6. Centrifuge the plate.
7. Place the plate in the apparatus (such as StepOnePlus™ one) and launch an amplification program, typically consisting of denaturation steps at 95 °C and annealing and elongation steps at 60 °C.

Analyze the results according to the quantification comparative CT chosen method (ΔΔCT approach) and to chosen software (*see* **Note 13**).

8. Expression profiles of genes are evaluated regarding the RPS17 expression level in different samples run in triplicate and in the same set of experiments (Fig. 9a, b).

4 Notes

1. Depending on the purpose, several embryos can be mixed and pulled together to get a higher number of primary cells and constitute a stock. Alternatively, individual preparation can also be prepared.
2. An average of 12–15 $\times$ 10^6 cells per embryo is routinely obtained from a 12 days old embryo. This number is highly dependent on the chicken strain, on trypsin dissociation, and on filtration recovery.
3. The entry of CEF into senescence is highly dependent on the culture conditions (serum quality) as well as on the genetic background of the chicken strain. It occurs roughly when the harvested cells start to be fewer than the seeded cells after

Table 1
List of oligonucleotides used for real-time RT-PCR analysis

Gène	S	AS	NCBI Ref Seq
ALPL	ATTTCGCTCACGCCAACTAC	GGATGTAGTTCTGCTCATGCAC	NM_205360.1
CLDN3	GGGTGGTTTCGGTCAGCGGG	GATGCTGCACAGCCAGCCCA	NM_204202.1
KLF4	ATGCACAGGATGCTGCAACACG	TGGTGTGCGCCAGGATGAAGTC	XM_422416
NANOG	TGCACACCAGGCTTACAGCAGTG	TGCTGGGTGTTGCAGCTTGTTC	NM_001146142.1
OCT4 (POU5F3)	TGCAATGCAGAGCAAGTGCTGG	ACTGGGCTTCACACATTTGCGG	NM_001110178.1
RPS17	ACACCCGTCTGGGCAACGAC	CCCGCTGGATGCGCTTCATC	NM_204217.1
SALL4	GTCCACTGCGGACCCCAACG	GGTGGAGAAGGCACGGCCAC	NM_001080872.1
SOX2	ACTCGGCCGCGAACAACCAG	GCCCCGAGCCGTTTGCTGAT	NM_205188.1
TRIM71	CATCGTGGCTGACCGCAGCA	CGACGATCCGGCGTGAGACG	NM_001037275.1

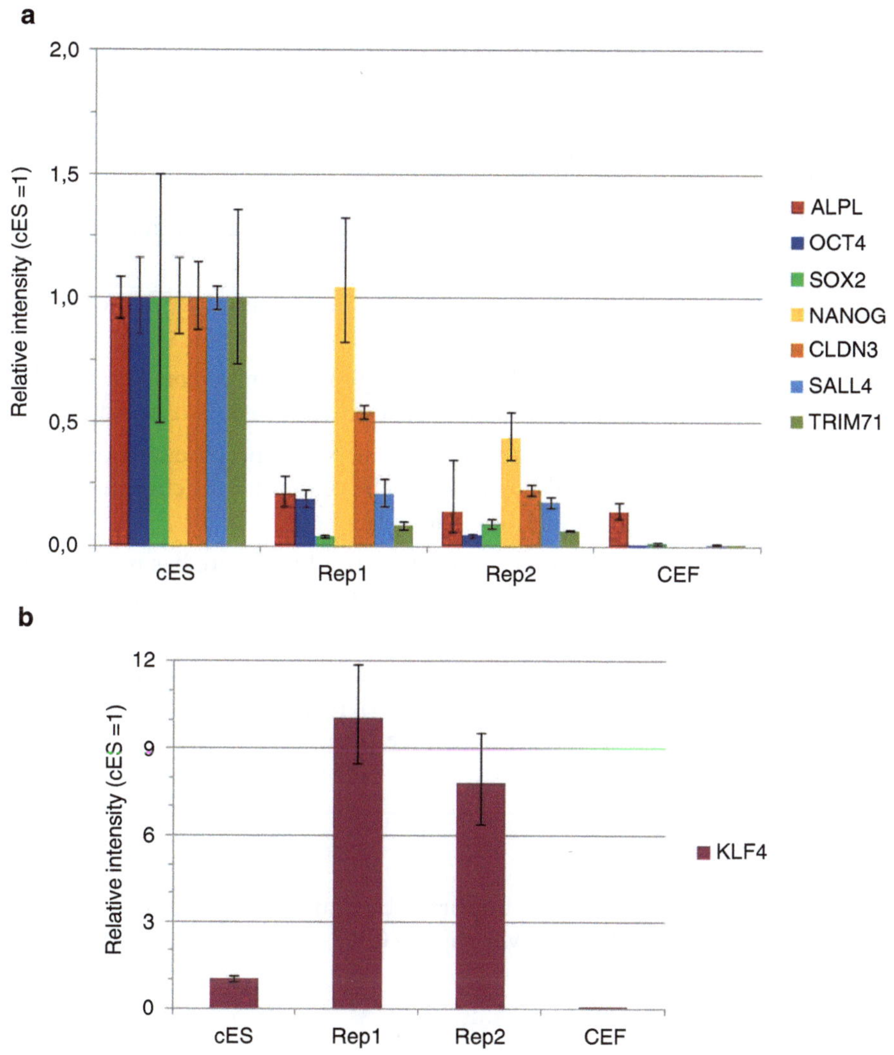

Fig. 9 (**a**) Expression profile of several pluripotency-associated genes in cES cells as positive control (level of expression taken at (1), in CEF as negative cells and in two independent reprogrammed cells (Rep1 and rep2). (**b**) The KLF4 expression level is particularly high in those specific clones. The value is the average (±SD) of a triplicate

dissociation. This corresponds to the plateau observed in the growth curve.

4. The percentage of cryoprotectant is submitted to variations depending on protocols, ranging from 10 to 5%. Additionally, some ready-to-use freezing solutions are also commercially available, some working even without serum. All of those solutions have to be tested by thawing the cells for evaluating their recovery before freezing all the cell stock in those new conditions.
5. Liquid nitrogen handling requires special precautions and is subject to specific regulations.

6. Different liposomes can be used including Lipofectamine (Life), FUGENE (Promega), etc… with an adaptation of the conditions according to the manufacturer's recommendation.
7. The final plate will have a negative control (pUC or pBSK plasmid), a positive transfection control (an empty plasmid), and at least one "reprogrammed" well.
8. Several wells can be transfected independently with individual mix to get independent clones.
9. Methanol is highly toxic and has to be handled under strict regulation for its use and waste.
10. Some of the used reagents exhibit toxicity and have to be handled under strict regulation for its use and waste.
11. Several systems are available to prepare and perform the real-time RT-PCR and can be preferentially chosen according to the different laboratory practices.
12. The number of wells in the PCR plate has to be optimized and the reference gene must be present in the analyzed series of samples to compare them.
13. Numerous QRT-PCR apparatus and analysis software are available. The chosen approach could provide differences in the way to present the relative data.

Acknowledgments

This work was supported by grants from INRA and from ANR with the CRB-ANIM project (CRB-ANIM-ANR-11-INBS-0003).

References

1. Aubel P, Pain B (2013) Chicken embryonic stem cells: establishment and characterization. In: Alberio R (ed) Epiblast stem cells: methods and protocols. Humana Press, Totowa, NJ
2. Lavial F, Acloque H, Bertocchini F, Macleod DJ, Boast S, Bachelard E, Montillet G, Thenot S, Sang HM, Stern CD, Samarut J, Pain B (2007) The Oct4 homologue PouV and Nanog regulate pluripotency in chicken embryonic stem cells. Development 134:3549–3563
3. Jean C, Oliveira NMM, Intarapat S, Fuet A, Mazoyer C, De Almeida I, Trevers K, Boast S, Aubel P, Bertocchini F, Stern CD, Pain B (2015) Transcriptome analysis of chicken ES, blastodermal and germ cells reveals that chick ES cells are equivalent to mouse ES cells rather than EpiSC. Stem Cell Res 14:54–67
4. Takahashi K, Yamanaka S (2006) Induction of pluripotent stem cells from mouse embryonic and adult fibroblast cultures by defined factors. Cell 126:663–676
5. Takahashi K, Yamanaka S (2016) A decade of transcription factor mediated reprogramming to pluripotency. Nat Rev Mol Cell Biol 17:183–193
6. Fuet A., Montillet G. Jean C., Aubel P., Kress C., Gervier S., Pain B. (2017). Somatic reprogramming in avian species remains elusive but provides new stem cell lines. Submitted.
7. Theunissen TW, Costa Y, Radzisheuskaya A, van Oosten AL, Lavial F, Pain B, Castro LF, Silva JC (2011) Reprogramming capacity of Nanog is functionally conserved in vertebrates and resides in a unique homeodomain. Development 138:4853–4865
8. Wang W, Lin C, Lu D, Ning Z, Cox T, Melvin D, Wang X, Bradley A, Liu P (2008) Chromosomal transposition of PiggyBac in mouse embryonic stem cells. Proc Natl Acad Sci U S A 105:9290–9295

Chapter 15

Isolation and Characterization of Chicken Primordial Germ Cells and Their Application in Transgenesis

Jae Yong Han and Bo Ram Lee

Abstract

Primordial germ cells (PGCs), precursors of functional gametes, offer great promise for the use of genetic resources and transgenesis in chickens. PGCs can be isolated from the developing embryo at diverse early stages and are subsequently expandable in vitro. In vitro proliferating chicken PGCs can facilitate the production of efficient germline chimeras and transgenic chickens. Here, we describe methods to isolate and characterize PGCs from chicken embryos and their application to transgenesis.

Key words Chicken, Primordial germ cells, Isolation, Characterization, Transgenesis

1 Introduction

The use of primordial germ cells (PGCs) to produce transgenic chickens has many advantages in animal biotechnology and biomodeling [1–3]. Because chicken embryos are oviparous, PGCs at an early stage are readily accessible and can be manipulated in vitro for practical applications, including the restoration of genetic material, genome editing, and transgenic research [4–9]. Significant efforts have been made to establish culture systems for chicken PGCs [10–12] from different embryonic origins [13], and it has been demonstrated that only chicken PGCs can be maintained indefinitely in vitro without losing their properties, compared to mammalian species [14]. Thus, chicken PGCs may be a promising source for transgenesis. Indeed, PGC-mediated transgenesis has garnered much attention in the fields of poultry biotechnology and developmental biology.

In chickens, PGCs first separate in the epiblast in Eyal-Giladi and Kochav (EGK) stage X embryos and then move down the hypoblast of the area pellucida, and subsequently to the germinal crescent and enter the blood stream [15–17]. After migration through the circulatory system, they arrive at the genital ridges

Guojun Sheng (ed.), *Avian and Reptilian Developmental Biology: Methods and Protocols*, Methods in Molecular Biology, vol. 1650, DOI 10.1007/978-1-4939-7216-6_15, © Springer Science+Business Media LLC 2017

[18]. These unique characteristics in germline development make it possible to use them to produce transgenic chickens via the injection of genetically manipulated PGCs into the blood vessels of recipient eggs [19]. Moreover, recent advances have demonstrated that germline-competent chicken PGC lines can facilitate the production of germline chimeras and transgenic chickens with higher germline transmission efficiency than embryonic stem cells (ESC) and spermatogonial stem cells (SSC) [14, 20–22].

Here, we describe methods to isolate and characterize chicken PGCs as well as fundamental tools required for transgenesis.

2 Materials

2.1 PGC Isolation

1. Microelectrode pipette puller (Shutter Instrument Co.).
2. Micro grinder (NARISHIGE).
3. Small-diameter (25 μm) glass micropipette (*see* **Note 1**).
4. Mouth-controlled pipette (Sigma-Aldrich).
5. 1× Hank's balanced salt solution (HBSS) without $CaCl_2$ or $MgCl_2$ (Hyclone, Logan, UT).
6. 1× Dulbecco's phosphate-buffered saline (PBS), without Ca^{2+} or Mg^{2+} (Hyclone).
7. Trypsin/ethylenediaminetetraacetic acid (EDTA) (Gibco).
8. Chicken embryos in Hamburger and Hamilton (HH) stages 14–17 (*see* **Note 2**).
9. Magnetic-activated cell sorting (MACS) buffer. 1× PBS supplemented with 0.5% bovine serum albumin (BSA) and 2 mM EDTA.
10. MiniMACS column (Miltenyi Biotec).
11. Stage-specific embryonic antigen 1 (SSEA-1) antibody (Santa Cruz Biotechnology, Santa Cruz, CA).
12. Anti-mouse IgM MicroBeads (Miltenyi Biotec).

2.2 PGC Culture

1. PBS (Hyclone).
2. HBSS (Hyclone).
3. PGC culture medium (50 mL). 3.75 mL fetal bovine serum (FBS, Hyclone), 1.25 mL chicken serum (Sigma-Aldrich, St. Louis, MO), 500 μL GlutaMAX-I Supplement (100×, Invitrogen), 500 μL nucleosides (100×, Millipore), 500 μL antibiotic/antimycotic (ABAM; 100×, Invitrogen), 500 μL insulin-transferrin-selenium supplement (100×, Gibco), 500 μL nonessential amino acids (NEAA; 100×, Invitrogen), 50 μL β-mercaptoethanol (1000×, Sigma-Aldrich), 10 μL of 50 ng/μL human basic fibroblast growth factor (bFGF, Sigma-Aldrich), and add

Table 1
Information of the primer sets used for RT-PCR analysis

Gene	Primers sequence	Product size (bp)
NANOG	F: CAGCAGACCTCTCCTTGACC R: AAGCCCTCATCCTCCACAGC	586
POUV	F: GCCAAGGACCTCAAGCACAA R: ATGTCACTGGGATGGGCAGA	511
CVH	F: GGGAAGATCAGTTTGGTGGA R: GACAAAGAAAGGCTGCAAGG	388
DAZL	F: CGTCAACAACCTGCCAAGGA R: TTCTTTGCTCCCCAGGAACC	540
KIT	F: GTGGGCAAGAAGTGGAAGCC R: GCAAACCAAGCATCTCATCCC	239
GAPDH	F: CACAGCCACACAGAAGACGG R: CCATCAAGTCCACAACACGG	443

Knockout Dulbecco's modified Eagle's medium (DMEM; Invitrogen) up to 50 mL. Store at 4 °C.

4. Accutase (Millipore).

2.3 PGC Characterization

2.3.1 RT-PCR

1. RNeasy mini kit (Qiagen).
2. Superscript III first-strand synthesis system (Invitrogen).
3. 10× PCR buffer with $MgCl_2$.
4. 2.5 mM dNTP mixture.
5. Taq DNA polymerase.
6. Germness-related and stemness-related primer sets (*see* Table 1).

2.3.2 Immunocytochemistry

1. Fixation solution: 4% formalin (Sigma-Aldrich) in 95% ethanol (Merck Millipore).
2. Permeabilization solution: 0.1% Triton X-100 (Sigma-Aldrich) in PBS.
3. Blocking solution: PBS containing 10% (v/v) normal goat serum and 1% BSA.
4. Primary antibodies: SSEA-1 (Santa Cruz Biotechnology).
5. Secondary antibodies: goat anti-mouse IgM-phycoerythrin (PE).
6. ProLong® Gold antifade reagent (with DAPI, or 4′,6-diamidino-2-phenylindole; Invitrogen).

2.3.3 Migration Capacity

1. Microelectrode pipette puller (Shutter Instrument Co.).
2. Micro grinder (NARISHIGE).
3. Small-diameter (25 μm) glass micropipette (Drummond Scientific).
4. HBSS (Hyclone).
5. PKH26 Red fluorescent cell linker kit (Sigma-Aldrich).
6. Recipient white leghorn (WL) chicken embryos at HH stages 14–17.
7. Mouth-controlled pipette (Sigma-Aldrich).
8. PBS, without Ca^{2+} or Mg^{2+} (Hyclone).
9. 100× ABAM (Invitrogen).
10. Hydrogen peroxide (Sigma-Aldrich).
11. Forceps.
12. Stereomicroscope with illuminator (Olympus).
13. Hot-melt glue sticks and glue gun.
14. Parafilm.
15. Accutase (Millipore).
16. Inverted fluorescence microscope (Leica Microsystems).

2.4 Germline Chimera Production and Transgenesis

2.4.1 Gene Transfer into Chicken PGC

1. Lipofectamine 2000 (Invitrogen).
2. Opti-MEM medium (Invitrogen).
3. Transfection medium (PGC culture media without ABAM; 50 mL): 3.75 mL FBS (Hyclone), 1.25 mL chicken serum (Sigma-Aldrich), 500 μL GlutaMAX-I supplement (100×, Invitrogen), 500 μL nucleosides (100×, Millipore), 500 μL insulin-transferrin-selenium supplement (100×, Gibco), 500 μL NEAA (100×, Invitrogen), 50 μL β-mercaptoethanol (1000×, Sigma-Aldrich), and 10 μL of 50 ng/μL bFGF (Sigma-Aldrich), and add Knockout DMEM (Invitrogen) up to 50 mL. Store at 4 °C.
4. PBS, without Ca^{2+} or Mg^{2+}.
5. HBSS, without Ca^{2+} or Mg^{2+}.
6. Prepare plasmid DNA at 1 μg/μL.

2.4.2 In Vitro Proliferation and Selection of Transgenic PGCs

1. HBSS, without Ca^{2+} or Mg^{2+} (Hyclone).
2. Hemocytometer.
3. Trypan blue, 0.4% (Invitrogen).
4. Inverted fluorescence microscope (Leica Microsystems).
5. Geneticin® Selective Antibiotic (Gibco).

2.4.3 PGC Transplantation into Recipient Eggs

1. Small-diameter (25 μm) glass micropipette (made with micro-electrode pipette puller (Shutter Instrument Co.) and a micro grinder (NARISHIGE).
2. Mouth-controlled pipette.
3. HBSS, without $CaCl_2$ or $MgCl_2$ (Hyclone).
4. Recipient chicken embryos at HH stages 14–17.
5. PBS, without Ca^{2+} or Mg^{2+} (Hyclone).
6. Pipette washing solution. 0.1% hydrogen peroxide (Sigma-Aldrich) in autoclaved distilled water.
7. Forceps.
8. Stereomicroscope with illuminator.
9. Hot-melt glue sticks and glue gun.
10. Parafilm.
11. Donor PGC prepared at 3×10^3 cells/μL in HBSS.

2.4.4 Testcross

1. Wild-type Korean Ogye (KO) female chicken.
2. 1 mL syringe.
3. PBS, without Ca^{2+} or Mg^{2+} (Hyclone).

2.4.5 Donor-Derived Progeny and Transgenic Chicken Validation

1. DNeasy Blood & Tissue kit (Qiagen).
2. Donor PGC-specific primer set.
3. Transgene-specific primer set.
4. PCR machine.
5. PCR reagents: PCR buffer, dNTP, Taq polymerase.
6. Agarose.
7. TAE buffer.
8. Electrophoresis machine.

3 Methods

Here, we describe the methods to produce transgenic chickens via PGC transplantation in detail. Figure 1 presents the overall procedures of PGC isolation and transgenesis in chickens.

3.1 PGC Isolation

1. Incubate fresh Korean Ogye (KO) chicken eggs (EGK stage X) at 37 °C for 50–54 h (HH stage 14–17) for blood PGC isolation. Place incubated eggs horizontally on the egg plate and gently wipe eggshell using sanitized cotton with 70% ethanol.
2. Using sharpened forceps, cautiously crack the eggshell (<1 cm diameter) for the isolation of whole blood cells.

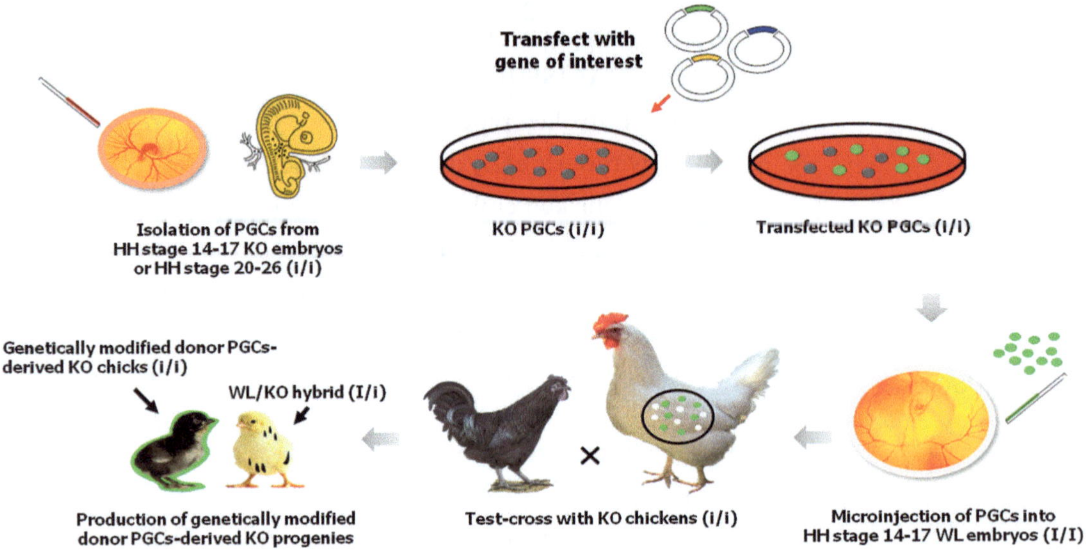

Fig. 1 Schematic representation of transgenic chicken production using PGCs. PGCs isolated from Korean Ogye (KO) (*i/i*) were cultivated and genetically modified. After genetically modifying the PGCs, they were transplanted into blood vessels of white leghorn (WL) embryos (*I/I*) in HH stages 14–17. Sexually mature germline chimeras (*I* and *i*) were crossed with KO (*i/i*). Feather color distinguished the genetically modified donor PGC-derived progeny (*i/i*) from the WL/KO hybrid (*I/i*)

3. Collect ~2–3 μL whole blood cells from the embryonic dorsal aorta using a 25 μm thinly ground glass needle and mouth pipette (*see* **Note 1**). Mix with 500 μL PBS.
4. Transfer whole-blood cells to a 1.5 mL sterile tube, centrifuge (250 × *g*, 5 min), and remove the supernatant.
5. For gonadal PGC isolation, incubate fresh chicken eggs at 37 °C for 5.5 days (HH 20–26).

   Extract embryonic gonads from embryos at HH stages 20–26 with sharpened forceps, and incubate with 500 μL of 0.05% trypsin/EDTA at 37 °C incubator for 5 min.
6. Add 50 μL FBS for inactivation, and centrifuge (250 × *g*, 5 min) and remove the supernatant.
7. Resuspend whole blood cells or dissociated gonadal cells in 1 mL PBS and apply anti-SSEA-1 antibody (Santa Cruz Biotechnology, SC-21702) at 1:200 dilution, and then incubate the mixture for 15 min at room temperature (RT).
8. Wash cells to remove unbound primary antibody by adding 5 mL MACS buffer (0.5% BSA and 2 mM EDTA in PBS, pH 7.2) per 10^7 total cells and centrifuge (250 × *g*, 5 min).
9. Aspirate supernatant completely and resuspend cell pellet in 80 μL MACS buffer per 10^7 total cells.

10. Add 20 μL rat anti-mouse IgM MicroBeads (Miltenyi Biotec. 130-047-301) per 10^7 total cells. For higher cell numbers, scale up buffer volume accordingly.
11. Incubate cells with antibody at 2–8 °C for 20 min.
12. Wash cells by adding 2 mL MACS buffer per 10^7 total cells and centrifuge (250 × *g*, 5 min).
13. Aspirate the supernatant completely and resuspend up to 10^8 total cells in 500 μL MACS buffer.
14. Place column in the magnetic field of a suitable MACS separator.
15. Prepare column by rinsing with 500 μL MACS buffer.
16. Apply cell suspension onto the column, and wash the column with 500 μL buffer three times. Add new buffer when the column reservoir is empty.
17. Remove column from the separator and place it on a suitable collection tube.
18. Add 1 mL MACS buffer to the column and immediately flush out the magnetically labeled cells by firmly pushing the plunger into the column.
19. Transfer the isolated cells containing PGCs to the pre-warmed PGC culture medium (1 mL PGC medium per 1 × 10^5 purified PGCs).

3.2 PGC Culture

1. Transfer the isolated cells containing PGCs to the pre-warmed PGC culture medium (1 mL PGC medium per 1 × 10^5 purified PGCs) and incubate at 37 °C.
2. After 7–14 days of growth, most of the primary cultured PGCs form colonies and lose aggregates of cell colonies.
3. The suspended PGC colonies can be withdrawn gently with medium after gentle pipetting and centrifuging (200 × *g*, 5 min).
4. Disaggregate the cell pellet with Accutase (1 mL Accutase solution per ~5 × 10^5 PGCs).
5. Centrifuge (200 × *g*, 5 min) and resuspend the cell pellet in a PGC culture medium.
6. Seed the suspended PGCs in a 12-well plate (1 × 10^5 cultured PGCs in 1 mL PGC culture medium per well of a 12-well plate).
7. PGCs can be routinely subcultured every 3–4 days.

3.3 Characterization of PGCs

Here, we introduce a series of experiments for investigating the characteristics of PGCs cultured in vitro.

3.3.1 RT-PCR Analysis

1. Collect approximately 1×10^6 PGCs for the preparation of RNA samples by centrifugation ($200 \times g$, 5 min).
2. Remove the supernatant and prepare total RNA from the cell pellet using the RNeasy Mini kit (Qiagen).
3. Prepare cDNA using the Superscript III first-strand synthesis system (Invitrogen) according to the manufacturer's protocol. Serially dilute the cDNA tenfold and equalize it quantitatively for PCR amplification as follows. PCR reactions (total volume, 20 μL) with 1 μL cDNA, 2 μL 10× PCR buffer (Solgent, Dajeon, Korea), 0.4 μL dNTPs (10 mM each), 2 pmol of each gene-specific primer, 0.25 U of Taq polymerase (Solgent), and H_2O up to 20 μL. PCR conditions as follows: denaturation at 94 °C for 5 min, followed by 35 cycles at 94 °C for 30 s, 60 °C for 30 s, and 72 °C for 1 min. Primer set information is provided in Table 1.
4. Analyze PCR products using a 1% agarose gel (*see* Fig. 2A) [10].

3.3.2 Immunocytochemistry

1. Disaggregate the PGC pellet with Accutase (Millipore) for 10 min at 37 °C and centrifuge ($200 \times g$, 5 min). Apply

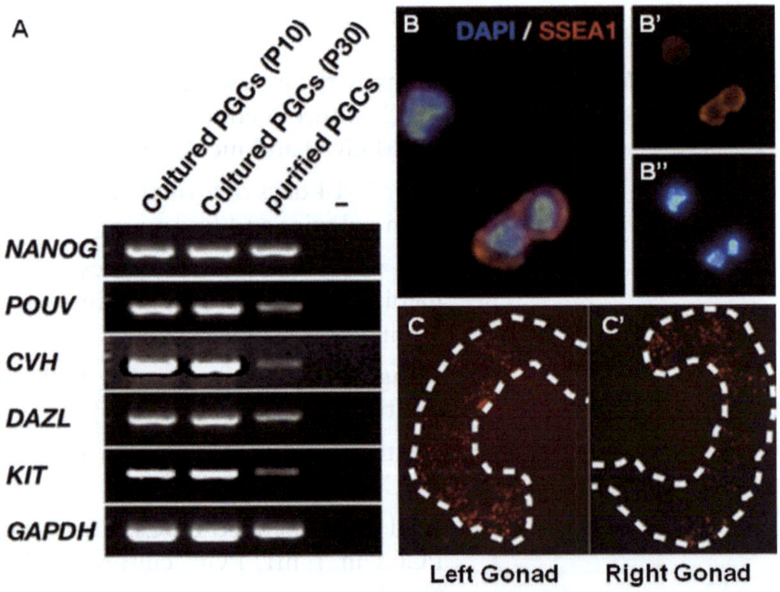

Fig. 2 Characterization of cultured PGCs. (**A**) RT-PCR analysis of *NANOG, POUV, CVH, DAZL*, and *KIT* in cultured PGCs and purified PGCs. (**B**) Immunocytochemical analysis of cultured PGCs against SSEA-1. (**C** and **C′**) Migration assay for cultured PGCs. PGCs were labeled with PKH26 and then injected into blood vessels of recipient embryos at stages 14–17. Left and right gonads were monitored using fluorescence microscopy

~1 × 10^4 of PGC suspension in 100 μL PBS on a slide glass and dry slowly on a 37 °C slide warmer.

2. Fix the cells with 500 μL 3.7% paraformaldehyde at RT for 10 min. Wash the slide in 1× PBS, three times.
3. Incubate with blocking solution (1× PBS containing 10% normal goat serum and 1% BSA) for 10 min at RT.
4. Add anti-SSEA-1 antibody at 1:200 titer in 1 mL blocking solution and incubate the cells for 1 h at RT under humid conditions in an air-tight container. After 1 h, wash the slide in 1× PBS, three times.
5. Add secondary antibody (Goat anti-mouse IgM-phycoerythrin (PE) at 1:500 dilution in 1 mL blocking solution) and incubate for 1 h at RT. After 1 h, wash the slide in 1× PBS, three times.
6. Mount the slides with ProLong® Gold antifade reagent (with DAPI, or 4′,6-diamidino-2-phenylindole) and analyze under a fluorescence microscope (*see* Fig. 2) [10].

3.3.3 Migration Capacity

1. Prepare an optimum number of cultured PGCs in the PGC culture medium.
2. Disaggregate a PGC pellet with Accutase (Millipore) for 10 min at 37 °C and collect single cells by centrifugation.
3. Mix cell pellets with 2 μL PKH26 (Sigma-Aldrich) and 500 μL diluent buffer and incubate for 5 min at RT without light exposure.
4. Inactivate the solution by adding 1 mL 10% FBS in DMEM and centrifuge (200 × *g*, 5 min).
5. Resuspend the pellet and wash in 1× PBS by gentle pipetting, and centrifuge (200 × *g*, 5 min).
6. Resuspend the pellet in an appropriate volume of PGC medium (usually 1,000–1,500 cells in 1 μL medium).
7. Incubate White Leghorn (WL) recipient eggs with their pointed end up for 53–56 h (at HH stage 14–17).
8. Make a small window at the pointed end of the recipient eggs using rotator tools (***Do not*** separate the cover from the egg).
9. Take 2 μL liquid containing an appropriate number of whole blood cells using a mouth pipette.
10. Inject into the upper portion of the dorsal aorta of the recipient embryo.
11. Seal the window twice with Parafilm using the glue gun (*see* **Note 3**). Incubate eggs further until HH stages 28–30.
12. Gonads from the recipient embryos can be retrieved, and then the number of fluorescent PGCs in the gonad can be counted under a fluorescence microscope (*see* Fig. 2) [10].

3.4 Gene Transfer into Chicken PGCs, Germline Chimera Production, and Transgenesis

3.4.1 Gene Transfer into Chicken PGCs

1. Mix 4 μg helper plasmid containing CAGG-PBase (pCyL43B) and 6 μg the *piggyBac* transposon (pCyL50) containing a reporter gene (GFP) or selectable marker (e.g., neomycin resistance gene) with 100 μL Opti-MEM (Gibco) and incubate for 5 min at RT [7]. For TALEN or CRISPR/Cas transfection, mix 2.5 μg CMV GFP expression plasmid vector, 2.5 μg each TALEN or Cas9 and guide RNA expression plasmid vectors with 100 μL Opti-MEM (Gibco) and incubate for 5 min at RT [8].
2. Mix 10 μL Lipofectamine 2000 reagent (Thermo Fisher-Invitrogen) with 100 μL Opti-MEM and incubate for 5 min at RT.
3. Mix the plasmids with Opti-MEM and Lipofectamine 2000 reagent with Opti-MEM and incubate for 20 min at RT.
4. During incubation, harvest cultured PGCs and centrifuge (200 × *g*, 5 min). Discard the supernatants.
5. Add 1 mL Accutase (Millipore) to the harvested PGCs and incubate for 10 min at 37 °C.
6. Determine the number of PGCs using a hemocytometer (Marienfeld) and seed 5 × 10^5 PGCs in a 12-well plate with 1 mL PGC culture medium without antibiotics.
7. Apply DNA-Lipofectamine complex to PGCs and incubate for 1 day at 37 °C in a CO_2 incubator.
8. Harvest the transfected PGCs and centrifuge (200 × *g*, 5 min). Remove the supernatants.
9. Wash the PGCs with 1 mL HBSS three times and suspend them with the PGC culture medium with antibiotics.

3.4.2 In Vitro Proliferation and Selection of Transgenic PGCs

1. Select the transfected PGCs in the PGC culture medium containing 100 μg/mL G418 (for the neomycin resistance gene) the day after transfection.
2. Monitor reporter gene expression using fluorescence microscopy.
3. Subculture the PGCs every 3–4 days. A complete selection period requires up to 3 weeks (*see* Fig. 3).
4. If there is no drug-selection marker in the transfected plasmid vectors, PGCs that express fluorescent protein can be sorted by FACS.

3.4.3 PGC Transplantation into Recipient Eggs

1. Incubate recipient eggs up to HH stages 14–17 at 37 °C in air with 60–70% relative humidity.
2. Harvest the transfected PGCs and centrifuge (200 × *g*, 5 min). Remove the supernatants.
3. Add 1 mL Accutase (Millipore) to the harvested PGCs and incubate for 10 min at 37 °C.

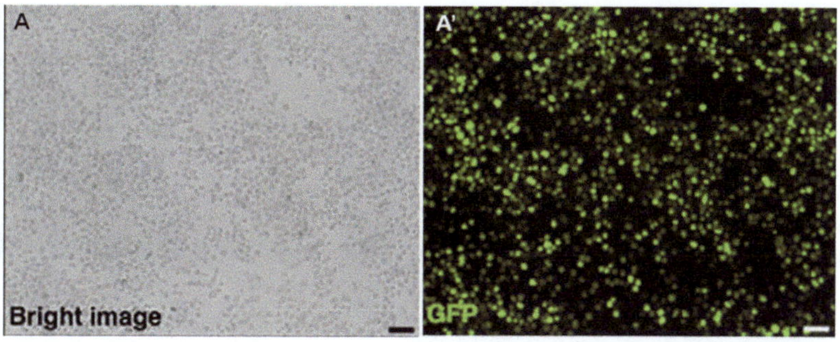

Fig. 3 Gene transfer to chicken PGCs and in vitro proliferation with G418 selection. (**A** and **A′**) PGCs were transfected with plasmid vectors containing GFP and a Neo resistance gene expression cassette after G418 selection. Scale bar = 50 μm. (Reproduced from ref. 7 with permission from The Federation of American Societies for Experimental Biology)

4. Centrifuge (200 × *g*, 5 min). Discard the supernatants.
5. Suspend the PGCs in HBSS (Hyclone).
6. Make a small window on the pointed end of the recipient egg and microinject a 2 μL aliquot containing more than 3,000 PGCs with a micropipette into the dorsal aorta of the recipient embryo.
7. Seal the egg window of the recipient embryo with parafilm using the glue gun, and incubate the egg with the pointed end down until hatching at 37 °C in air with 60–70% relative humidity.

3.4.4 Monitoring Genetically Modified PGCs in Embryonic Gonads

1. Incubate recipient eggs up to HH stages 28–30 at 37 °C in air with 60–70% relative humidity.
2. Dissect the gonad at embryonic day 6.
3. Monitor fluorescent protein expression in the embryonic gonads using fluorescence microscopy.

3.4.5 Testcross

1. Collect semen twice in 1 week from sexually mature recipients and wild-type (WT) chickens (*see* **Note 4**).
2. Introduce 50 μL semen from mature male recipients and WT roosters to WT laying hens and mature female recipients, respectively.
3. Collect eggs from WT laying hens and mature female recipients the day after artificial insemination, and incubate the egg with the pointed end down until hatching at 37 °C in air with 60–70% relative humidity.

3.4.6 Donor-Derived Progeny and Transgenic Chicken Validation

1. Donor-derived progeny can be distinguished by feather color or genomic DNA analysis. WL with a dominant pigmentation inhibitor gene (I/I) and KO with a recessive pigmentation

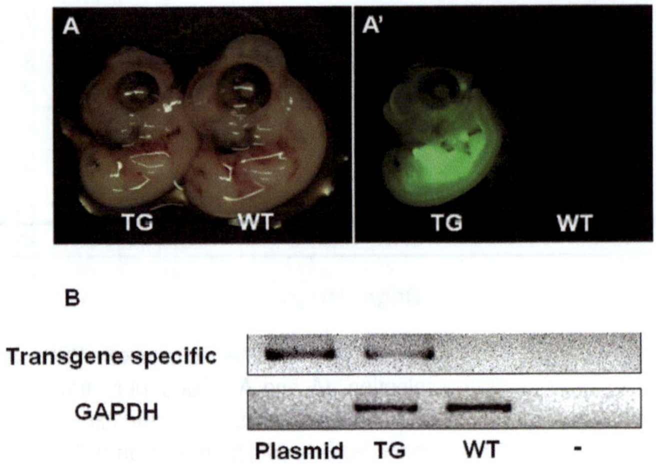

Fig. 4 Validation of donor PGCs-derived chicken. (**A** and **A′**) Monitoring GFP expression of transgenic chicken using fluorescence microscopy. (**B**) Genomic DNA analysis of transgenic chicken (TG) with wild-type (WT) using transgene-specific PCR conditions. (Reproduced from ref. 7 with permission from The Federation of American Societies for Experimental Biology)

inhibitor gene (i/i) can be used for the donor PGCs and the recipient embryos, respectively. Through test-cross analysis by mating with WL and KO females (I/I and i/i), donor-derived progenies can be identified by the phenotype of their offspring.

2. If the transgenic chickens express fluorescent protein in their bodies, they can be identified with a fluorescence excitation lamp with detection filters (BLS Ltd.) (*see* Fig. 4).
3. To perform genomic DNA analysis, extract genomic DNA from feather pulp of putative transgenic chicks' feather pulp according to manufacturer's protocol (QIAGEN DNeasy Blood & Tissue Kit).
4. Perform PCR using transfected plasmid vector-specific primers (*see* Fig. 4) [7].
5. To identify genome mutations, PCR products can be analyzed by sequencing.

4 Notes

1. For microinjection of PGCs, we generally use a mouth pipette and a glass micropipette of 20–25 μm diameter. The glass micropipette was made with a microelectrode pipette puller (Shutter Instrument Co.) and ground at 25 degrees using a micro grinder (NARISHIGE).

2. Fertilized and freshly laid eggs were brought to the lab within 1–3 h of oviposition for EGK stage X embryos. Developing embryos incubated at a relative humidity of 60–70% at 37.8 °C were staged according to the HH classification system. The incubation time up to HH 14–17 varied by chicken breed. For example, the Korean Ogye of HH stages 14–17 needed 50–54 h whereas the white leghorn embryo needed 52–56 h. Thus, researchers should determine the optimal incubation time for their own chicken breed.
3. A small window (<1 cm diameter) in recipient eggs was made with sharpened forceps. It is good to keep the window as small as possible. After the microinjection of donor PGCs, the window was sealed twice with Parafilm using hot-melt glue and a glue gun. Carefully seal the crack and window without microbial contamination. It is important to maintain clean conditions during the experiment. Gently wipe the eggshell using sanitized cotton with 70% ethanol before windowing the egg.
4. The testcross usually occurs between a male recipient of donor PGCs and a female wild-type chicken. For the easy identification of donor PGC-derived progeny, we generally use the dominance of feather color, because the WL chicken has a dominant pigmentation inhibitor gene (I/I), whereas KO has a recessive pigmentation inhibitor gene (i/i). Thus, donor-derived progeny can be distinguished by feather color. The donor KO PGC-derived sperm has a recessive gene (*i*) but the sperm of recipient WL has a dominant gene (*I*). Through test-cross analysis by mating with KO females (i/i), the donor-derived progeny can be identified by their phenotype (i/i).

Acknowledgments

This work was supported by a National Research Foundation of Korea (NRF) grant, funded by the Korean government (MSIP; No. 2015R1A3A2033826), and the "Cooperative Research Program for Agriculture Science & Technology Development (Project No. PJ0122362016)" of the Rural Development Administration, Korea.

References

1. Han JY (2009) Germ cells and transgenesis in chickens. Comp Immunol Microbiol Infect Dis 32(2):61–80. doi:10.1016/j.cimid.2007.11.010
2. Han JY, Lee HC, Park TS (2015) Germline-competent stem cell in avian species and its application. Asian J Androl 17(3):421–426. doi:10.4103/1008-682X.148073
3. Song G, Han JY (2011) Avian biomodels for use as pharmaceutical bioreactors and for studying human diseases. Ann N Y Acad Sci 1229:69–75. doi:10.1111/j.1749-6632.2011.06087.x
4. Han JY, Park TS, Hong YH, Jeong DK, Kim JN, Kim KD, Lim JM (2002) Production of germline chimeras by transfer of chicken gonadal primordial germ cells maintained

in vitro for an extended period. Theriogenology 58(8):1531–1539
5. Nakamura Y, Usui F, Miyahara D, Mori T, Ono T, Takeda K, Nirasawa K, Kagami H, Tagami T (2010) Efficient system for preservation and regeneration of genetic resources in chicken: concurrent storage of primordial germ cells and live animals from early embryos of a rare indigenous fowl (Gifujidori). Reprod Fertil Dev 22(8):1237–1246. doi:10.1071/RD10056
6. Kim JN, Park TS, Park SH, Park KJ, Kim TM, Lee SK, Lim JM, Han JY (2010) Migration and proliferation of intact and genetically modified primordial germ cells and the generation of a transgenic chicken. Biol Reprod 82(2):257–262. doi:10.1095/biolreprod.109.079723
7. Lee HJ, Lee HC, Kim YM, Hwang YS, Park YH, Park TS, Han JY (2016) Site-specific recombination in the chicken genome using Flipase recombinase-mediated cassette exchange. FASEB J 30(2):555–563. doi:10.1096/fj.15-274712
8. Park TS, Lee HJ, Kim KH, Kim JS, Han JY (2014) Targeted gene knockout in chickens mediated by TALENs. Proc Natl Acad Sci U S A 111(35):12716–12721. doi:10.1073/pnas.1410555111
9. van de Lavoir MC, Diamond JH, Leighton PA, Mather-Love C, Heyer BS, Bradshaw R, Kerchner A, Hooi LT, Gessaro TM, Swanberg SE, Delany ME, Etches RJ (2006) Germline transmission of genetically modified primordial germ cells. Nature 441(7094):766–769. doi:10.1038/nature04831
10. Choi JW, Kim S, Kim TM, Kim YM, Seo HW, Park TS, Jeong JW, Song G, Han JY (2010) Basic fibroblast growth factor activates MEK/ERK cell signaling pathway and stimulates the proliferation of chicken primordial germ cells. PLoS One 5(9):e12968. doi:10.1371/journal.pone.0012968
11. Macdonald J, Glover JD, Taylor L, Sang HM, McGrew MJ (2010) Characterisation and germline transmission of cultured avian primordial germ cells. PLoS One 5(11):e15518. doi:10.1371/journal.pone.0015518
12. Shiue YL, Tailiu JJ, Liou JF, Lu HT, Tai C, Shiau JW, Chen LR (2009) Establishment of the long-term in vitro culture system for chicken primordial germ cells. Reprod Domest Anim 44(1):55–61. doi:10.1111/j.1439-0531.2007.00990.x
13. Kagami H, Tagami T, Matsubara Y, Harumi T, Hanada H, Maruyama K, Sakurai M, Kuwana T, Naito M (1997) The developmental origin of primordial germ cells and the transmission of the donor-derived gametes in mixed-sex germline chimeras to the offspring in the chicken. Mol Reprod Dev 48(4):501–510. doi:10.1002/(SICI)1098-2795(199712)48:4<501::AID-MRD11>3.0.CO;2-W
14. Park TS, Han JY (2012) piggyBac transposition into primordial germ cells is an efficient tool for transgenesis in chickens. Proc Natl Acad Sci U S A 109(24):9337–9341. doi:10.1073/pnas.1203823109
15. Hamburger V, Hamilton HL (1951) A series of normal stages in the development of the chick embryo. J Morphol 88(1):49–92
16. Ando Y, Fujimoto T (1983) Ultrastructural evidence that chick primordial germ-cells leave the blood-vascular system prior to migrating to the gonadal anlagen. Dev Growth Differ 25(4):345–352
17. Ukeshima A, Yoshinaga K, Fujimoto T (1991) Scanning and transmission electronmicroscopic observations of chick primordial germ cells with special reference to the extravasation in their migration course. J Electron Microsc (Tokyo) 40(2):124–128
18. Meyer DB (1964) The migration of primordial germ cells in the chick embryo. Dev Biol 10:154–190
19. Eyal-Giladi H, Kochav S (1976) From cleavage to primitive streak formation: a complementary normal table and a new look at the first stages of the development of the chick. I. General morphology. Dev Biol 49(2):321–337
20. Park TS, Lee HG, Moon JK, Lee HJ, Yoon JW, Yun BNR, Kang SC, Kim J, Kim H, Han JY, Han BK (2015) Deposition of bioactive human epidermal growth factor in the egg white of transgenic hens using an oviduct-specific minisynthetic promoter. FASEB J 29(6):2386–2396. doi:10.1096/fj.14-264739
21. Yasuda Y, Tajima A, Fujimoto T, Kuwana T (1992) A method to obtain avian germ-line chimaeras using isolated primordial germ cells. J Reprod Fertil 96(2):521–528
22. Naito M, Tajima A, Yasuda Y, Kuwana T (1994) Production of germline chimeric chickens, with high transmission rate of donor-derived gametes, produced by transfer of primordial germ cells. Mol Reprod Dev 39(2):153–161. doi:10.1002/mrd.1080390206

Chapter 16

Handling of Gametes for In Vitro Insemination in Birds

Shusei Mizushima, Mei Matsuzaki, and Tomohiro Sasanami

Abstract

A characteristic biological property of avian gamete (e.g., extremely large egg and polyspermic fertilization) does not allow the direct observation of sperm-egg interactions in vitro, but recent research advances make it possible to manipulate the gamete in vitro. Here, we describe the techniques for the handling of gametes required for in vitro fertilization assay. In addition, we also introduce the procedures for sperm-perivitelline membrane assay, intracytoplasmic sperm injection, and ex ovo culture.

Key words In vitro insemination, Sperm-egg interactions, Perivitelline membrane, Intracytoplasmic sperm injection, Ex ovo culture, Birds

1 Introduction

Fertilization is an indispensable event for the production of the next generation in any animals that employ sexual reproduction. In order to study the mechanism of fertilization, in vitro systems that mimic the fertilization process are indispensable. There have been numerous studies showing that fertilization processes in many animals are able to proceed in vitro, in which isolated oocytes are incubated with ejaculated spermatozoa in a defined medium. In viviparous animals, including mammals, zygotes that develop to the blastocyst stage are able to be implanted into the uterus of a foster mother for the production of healthy offspring.

In comparison with those of mammalian species, avian oocytes are extremely large and this does not allow the direct observation of sperm-egg interactions in vitro. For instance, no available method for in vitro insemination exists to date, and researchers who engage in avian reproduction incubate the isolated perivitelline membrane (pvm), the homologous investments of the mammalian zona pellucida, with ejaculated sperm in vitro as an alternative model of in vitro fertilization [1–4]. In addition, evidence suggests that ovulated oocytes in birds quickly lose their fertilizability because the chalaza-layer, also referred to as the outer layer of the vitelline

Guojun Sheng (ed.), *Avian and Reptilian Developmental Biology: Methods and Protocols*, Methods in Molecular Biology, vol. 1650, DOI 10.1007/978-1-4939-7216-6_16,

membrane secreted from the infundibulum part of the oviduct, overlays the surface of the pvm immediately after ovulation and this layer prevents sperm penetration into the oocyte [5]. Thus, it is very difficult to isolate fertilizable eggs for studying the mechanism of fertilization in birds. Despite these difficulties, a few research groups, including ours, have investigated the mechanisms of avian fertilization using poultry birds, mainly chickens and quails.

Intracytoplasmic sperm injection (ICSI) has been very useful for studying the mechanisms of fertilization and is now an indispensable technique for generating mammalian offspring in laboratories and clinics [6–8]. In contrast, ICSI was not successful in birds using similar procedures as performed in mammalian species because of the natural polyspermic fertilization in birds, in which plural spermatozoa successively penetrate into one oocyte, and this is difficult to mimic in vitro [9–11]. However, we recently established an avian ICSI method in Japanese quail, in which a single spermatozoon was co-injected with sperm extract (SE), and this was able to enhance the full-term development of ICSI-generated quail zygotes all the way to hatching [12].

In this chapter, we will introduce the techniques for the handling of gametes required for in vitro analysis of sperm-pvm interactions in birds. We also would like to deal with the methods for avian ICSI, including isolation of the SE, sperm injection, and surrogate shell culture systems.

2 Materials

2.1 Isolation of Quail Spermatozoa

1. Microspatula.
2. Corrugated carton (514 × 364 × 591 mm).
3. Hanks balanced salt solution (HBSS). Weigh 0.14 g of $CaCl_2$ and dissolve in 10 mL H_2O. Weigh 0.2 g of $MgSO_4{\cdot}7H_2O$ and dissolve in 10 mL H_2O. Weigh 0.75 g of sodium bicarbonate and dissolve in 10 mL H_2O. Weigh 0.01 g phenol red Na and dissolve in 1 mL H_2O. Take 1 mL HBSS ×10 solution (H4385, Sigma, MO, USA) and transfer into a 15 mL plastic tube. Add 100 μL of $CaCl_2$ solution and $MgSO_4$ solution, 47.5 μL of sodium bicarbonate solution, and 10 μL of phenol red Na solution. Increase the volume to 10 mL with H_2O.
4. 1.5 mL micro test tube.

2.2 Isolation of an Ovum

1. Scissors.
2. Forceps.
3. Pentobarbital sodium (*see* **Note 1**).
4. Approx. 70 mL dish for chicken.

5. Plastic dish (35 × 18 mm; multidish 6-wells) for quail.
6. Dulbecco's modified Eagle's medium (DMEM D5796, Sigma, MO, USA).

2.3 Isolation of PVM

1. Scissors.
2. Forceps.
3. 0.9% (w/v) NaCl solution. Weigh 0.9 g of NaCl and dissolve in 1000 mL H_2O.
4. Petri dish (3 1/2 in.).
5. Phosphate-buffered saline (PBS). Weigh 0.8 g of sodium chloride, 0.02 g of potassium chloride transfer, 0.29 g of disodium hydrogen phosphate $12H_2O$, and 0.02 g of potassium dihydrogen phosphate, and transfer into a glass beaker. Add approximately 90 mL of sterile water, mix and adjust to pH 7.4. Make up to 100 mL with water and filter through a nylon membrane filter (0.45 μm).
6. HBSS.
7. Pasteur pipette (5 3/4 in.).

2.4 In Vitro Incubation of PVM and Sperm

1. Scissors.
2. Forceps.
3. Micro test tube.
4. HBSS.
5. Ice bath.
6. Glass slides.
7. 3.7% (v/v) formaldehyde solution. Mix 1 mL formaldehyde (Wako Pure Chemical Industries, Osaka, Japan) with 9 mL PBS.
8. Schiff's reagent (Wako Pure Chemical Industries).
9. Microscope; for example, BX51 (Olympus, Tokyo, Japan).

2.5 ICSI Devices

1. Inverted microscope with Hoffman module; for example, IX71 (Olympus) and Model EP 40 mm Condenser (Olympus).
2. Micromanipulator; for example, MMO-203/Three-dimensional oil hydraulic micromanipulator (Olympus).
3. Microinjector; for example, IM-9B, (Narishige Instrument, Tokyo, Japan) (*see* **Note 2**).
4. Stereomicroscope and universal stand; for example, SZ11 stereomicroscope, SZ-STU2 and SZ-STS (Olympus).
5. Fiber light source; for example, LG-PS2, (Olympus).

2.6 Preparation of Injection Pipette

1. Glass capillary (*see* **Note 3**).
2. 70% ethanol.
3. Autoclaved water.
4. Puller; for example, P-97/IVF (Sutter Instrument Co., CA, USA).
5. Microforge; for example, MF-900 (Narishige Instrument, Tokyo, Japan).
6. Beveler; for example, BV-10 (Sutter Instrument).
7. Syringe (5 mL).

2.7 Preparation of SE

1. Spatula.
2. PBS.
3. Centrifuge.
4. Cell counting chamber.
5. Liquid nitrogen.
6. Homogenizer.
7. Sonicator.

2.8 Preparation of Sperm Chamber

1. Petri dish (35 × 10 mm; cat. No. 35 1008; Falcon, Becton Dickinson Labware, NJ, USA).
2. 6% (w/v) polyvinylpyrrolidone (PVP; molecular weight 360 kDa): Weigh 0.06 g of polyvinylpyrrolidone and transfer into a 1.5 mL test tube. Add about 800 μL DMM, mix, and increase the volume to 1 mL with DMEM.
3. Mineral oil.
4. DMEM.

2.9 Ex Ovo Culture

The equipment and materials can be divided into two groups: those that are required for both the chicken and quail ova culture, and those that are needed additionally for quail or chicken.

2.10 Common Equipment and Materials for Chicken and Quail

1. 70% ethanol.
2. A laboratory egg incubator with an automatic turner (30°- and 90°-angle turning every 20 min); for example, P-008-B special model for embryo culture (Showa Furanki, Saitama, Japan).
3. An electric drill; for example, Mini-RooterMM100 No.28525 (PROXXON, Föhren Germany) and Cut-Off Wheel No. 409 (DREMEL, Robert Bosch Tool Co., Illinois, USA).
4. DMEM as a culture medium for culture **step 1**.
5. Thin albumen of hen's eggs for culture **steps 2** and **3**. Wipe an egg shell with 70% ethanol and crack open the egg into a 90 mm Petri dish and collect the thin albumen into a glass

beaker (*see* **Note 4**). Adjust to pH 7.4 by bubbling CO_2 with a CO_2 spray or an air gun attached to a pressure regulated CO_2 cylinder and add 10,000 IU penicillin and 50 mg streptomycin/L of albumen. Keep in a humidified CO_2 incubator at 41.5 °C with 5% CO_2 until use.

6. Spatula.
7. Circle template.

2.11 Additional Materials for Chicken

1. Approx. 70 mL plastic cups (60 mm diameter × 35 mm height).
2. Surrogate shell for the second culture step. Prepare a similar-sized chicken egg to that laid previously by the same chicken (approx. 50–60 g whole egg weight).
3. Surrogate shell for the third culture step. Prepare a chicken egg shell from an approx. 30 g heavier egg compared with the expected egg (at least more than 70 g, e.g., a double yolk egg).
4. Elastic bands: 16 mm diameter.
5. Plastic wrap: prepare two sheets of 70 × 70 mm for the second and third culture steps.
6. Plastic ring: prepare 7 and 15 mm tall polyvinyl chloride rings (36 and 42 mm inner and outer diameters, respectively) with four projections attached to the outside (Fig. 2).

2.12 Additional Materials for Quail

1. 20 mL plastic cups (32 mm tall, 35 and 30 mm upper and lower diameters, respectively).
2. Surrogate shell for the second culture step. Prepare a similar sized quail egg to that laid previously by the same quail (approx. 11–13 g whole egg weight).
3. Surrogate shell for the third culture step. Prepare a small-sized chicken egg (less than 45 g whole egg weight).
4. Elastic bands: 7 mm diameter.
5. Plastic wrap: prepare a sheet of 30 × 30 mm for the second culture step and 50 × 50 mm.
6. Plastic ring: prepare 7 and 15 mm tall polyvinyl chloride rings (20 and 26 mm inner and outer diameters, respectively) with four projections attached to the outside (Fig. 2).

3 Methods

3.1 Isolation of Quail Sperm

1. Transfer 1 mL of HBSS into a 1.5 mL micro test tube.
2. Hold the male quail face up and remove his cloacal gland secretion by squeezing the cloacal gland using the thumb and index finger.

3. Place a female and the male together in a corrugated carton and allow the male to copulate.
4. After mounting, catch the male face up and gently squeeze the cloacal gland using the thumb and index finger and collect the milky white spermatozoa that seep from the penis using a microspatula.
5. Suspend the spermatozoa in HBSS. Count the sperm density using the hemocytometer, dilute the sperm suspension to 1×10^7 cells/mL, and place in a humidified CO_2 incubator at 41.5 °C with 5% CO_2 until use.

3.2 Isolation of Ovulated Ovum

1. Sacrifice a hen by intravenous or intraperitouneal injection of pentobarbital sodium solution within 2 h after the preceding egg has been laid (*see* **Notes 1** and **5**).
2. Remove the abdominal feathers, laparotomize and find the ovum in the infundibulum or magnum. Cut out both the ends of the oviduct holding the ovum and place it in a Petri dish.
3. Insert scissors between the oviduct and the ovum, and carefully cut out the oviductal wall while holding the cut end of the oviduct using forceps. Drop the ovum gently into the 70 mL dish for chicken ova or the well of a 6-well plate for quail ova, into which DMEM has already been added.
4. Remove the thick albumen capsule with a spatula if present, rotate the ovum with the spatula so that the germinal disc is positioned at the top of the yolk, and clean the germinal disc with Kimwipes or gauze.
5. Add DMEM up to the ovum's equatorial level and place the plate or dish in a humidified CO_2 incubator at 41.5 °C with 5% CO_2 until use.

3.3 Isolation of PVM

1. Sacrifice a hen by intravenous or intraperitouneal injection of pentobarbital sodium solution (*see* **Notes 1** and **5**).
2. Remove the abdominal feathers, laparotomize and find the ovary. Remove the largest follicle and place it in a Petri dish.
3. Carefully cut the follicular wall and peel off the theca layer using forceps. After the removal of the theca layer, carefully peel off the granulosa layer and remove the adhering yolk materials using a stream of the water from a Pasteur pipette.
4. Transfer the isolated granulosa layer into a Petri dish filled with distilled water and wash it extensively using a stream of the water from a Pasteur pipette. This hypotonic treatment destroys the granulosa cells and the basal laminae separates from the pvm.
5. Transfer the isolated pvm into a Petri dish filled with PBS and store at 4 °C until use.

3.4 In Vitro Assay for Sperm-PVM Interaction

1. Cut the isolated pvm into pieces approximately 8 mm in diameter.
2. Transfer the pvm into a micro test tube containing 0.5 mL of sperm suspension at 1×10^7 sperm/mL in HBSS and incubate at 39 °C for 30 min (*see* **Note 6**).
3. After the 30 min incubation, the reaction is terminated by placing the tube on ice, and the pvm is washed three times with ice-cold PBS.
4. Transfer the pvm onto a glass slide, spread out under stereomicroscope. Fix the pvm by adding 3.7% (v/v) formaldehyde in PBS and allow it to stand for 10 min at room temperature.
5. After fixation for 10 min, wash the pvm with H_2O. Stain with Schiff's reagent by dropping on the reagent and allow it to stand for 5 min.
6. Wash the pvm with H_2O and allow it to dry.
7. Count the number of holes formed in the pvm in a × 40 field using a light microscope (BX 51). Select at least five areas randomly for the enumeration of perforations.

3.5 Setup of ICSI Devices

This section explains the setting up of equipment necessary for the preparation of an inverted microscope equipped with manipulators and injectors and a stereomicroscope, and the protocol for its use in injection. The protocol explained herein is basically in accordance with Hrabia et al. [9] and Mizushima et al. [12].

1. Set an inverted microscope on a bench. We currently use an Olympus IX70 inverted microscope at 200× or 400× magnifications for the aspiration of SE solution and sperm into an injection capillary with a Hoffman optical system, which produces pseudo three-dimensional images (Fig. 1a). In addition, the microscope is set on a vibration isolation table.
2. Mount the micromanipulator and its accessory to the stand of your inverted microscope with the controller on the bench (Fig. 1b). The micromanipulator allows fine and smooth three-dimensional movement required to position the injection needle.
3. Mount the injection pipette holder on the micromanipulator with the controller on the bench (Fig. 1c).
4. Set a stereomicroscope and its stand near the inverted stereomicroscope (Fig. 1d). A universal stand, SZ-STU2 and SZ-STS (Olympus), allows focusing of the ovum placed on the stage of the inverted microscope.

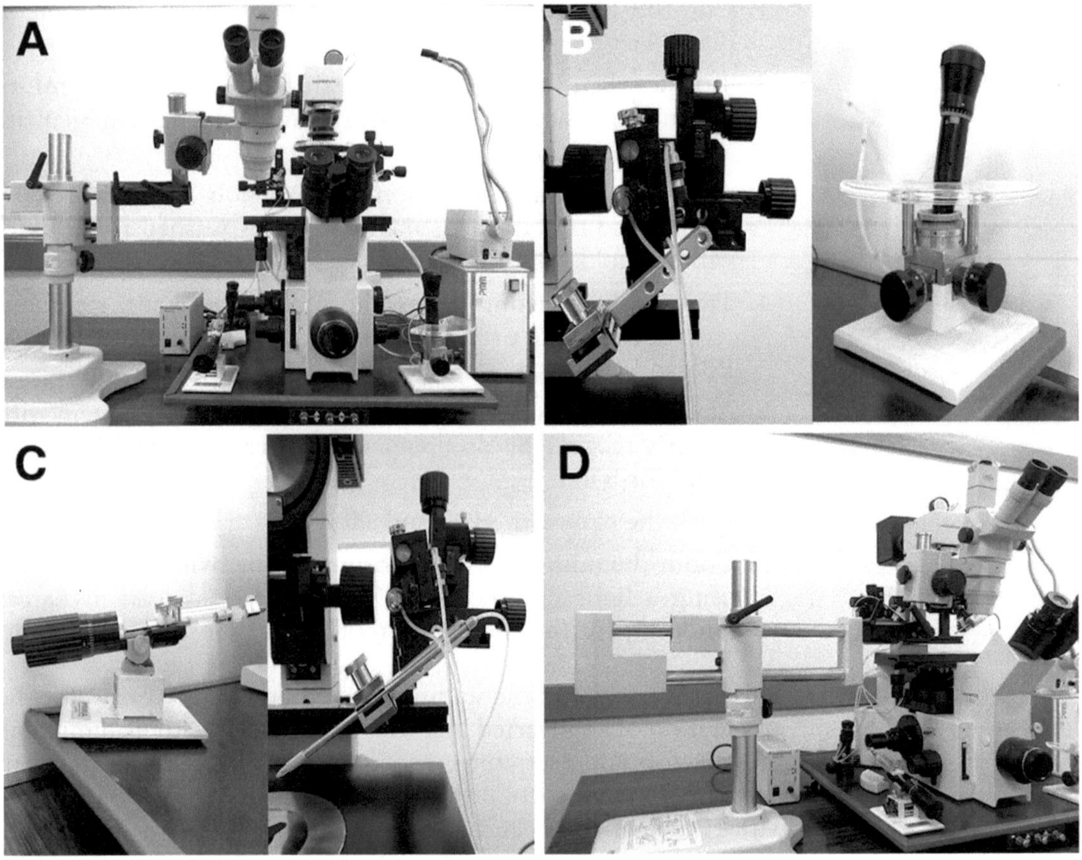

Fig. 1 Setup of the ICSI equipment. (**a**) Micromanipulation system for avian ICSI, (**b**) micromanipulator, (**c**) microinjector, (**d**) stereomicroscope and universal stand

3.6 Preparation of an Injection Pipette

1. Sterilize the glass capillaries at 120 °C for 30 min before use.
2. Load the glass capillary into a puller and adjust the HEAT, PULL, VELOCITY, TIME, and PRESSURE values in accordance with the manufacturer's instructions (*see* **Note** 7).
3. Press the PRESS key on the keypad.
4. Remove the capillaries from the puller bars.
5. Fix the pulled glass capillary into the pipette folder and cut the pipette to an outer diameter of 7–10 μm on a microforge.
6. Attach the capillary to a 5 mL syringe using silicon tubing (1 and 2 mm inner and outer diameters, respectively).
7. Adjust the angle plate (generally 25–30°) of the pipette folder on the micropipette beveler and attach the capillary to the pipette folder.
8. Switch the meter to an appropriate rotation rage and wash the abrasive surface using 70% ethanol.

9. Advance the capillary toward the abrasive surface until it makes a contact with 70% ethanol.
10. Furthermore, advance the abrasive plate under conditions of positive pressure in the capillary by pressing the syringe. The tip of the capillary is completely ground in approximately 1 min.
11. Back off the pipette folder and remove the capillary from the pipette folder.
12. Irrigate the capillary with water and subsequently 70% ethanol while monitoring the tip of the capillary. If the tip is open enough to emit the gas, you can see a continuous trail of bubbles from the tip.
13. Remove the capillary from the syringe connector.

3.7 Preparation of SE

1. Collect quail ejaculated semen in accordance with the method described in Subheading 3.1 and add to 1 mL of PBS. Centrifuge for 5 s in a standard table-top centrifuge to remove particles. Then collect the supernatant containing the sperm, and wash twice by centrifuge at 1000 × *g* at 4 °C for 5 min.
2. Calculate the sperm concentration using a counting chamber and adjust to 3×10^8 sperm/mL in PBS. Dispense 1 mL of sperm suspension into a new 1.5 mL microcentrifuge tube.
3. Freeze the tube for 10 min in liquid nitrogen and then thaw at room temperature.
4. Homogenize the sample for 15 s, two times on ice.
5. Immediately after homogenizing, sonicate the sample for 15 s at 50 W, five times on ice.
6. Centrifuge the lysed sample at 10,000 × *g* at 4 °C for 5 min to pellet the sperm debris (*see* **Note 8**).
7. Again, sonicate the sperm debris for 15 s at 50 W, four times on ice and centrifuge at 20,000 × *g* at 4 °C for 10 min to pellet the insoluble sperm debris.
8. Transfer only the supernatant to a new test tube and measure the protein concentration.
9. Adjust the protein concentration so that it is 2.0 mg/mL (*see* **Note 9**), divide into 10 μL aliquots, and store at −80 °C until use. After thawing, the extract is used only once.

3.8 Preparation of the Sperm Chamber

1. Collect the ejaculated semen in accordance with the method described in Subheading 3.1 and add it to 1 mL DMEM.
2. Centrifuge at 1000 × *g* at 4 °C for 5 min and remove the supernatant.
3. Add 1 mL of DMEM and centrifuge again using the same condition.

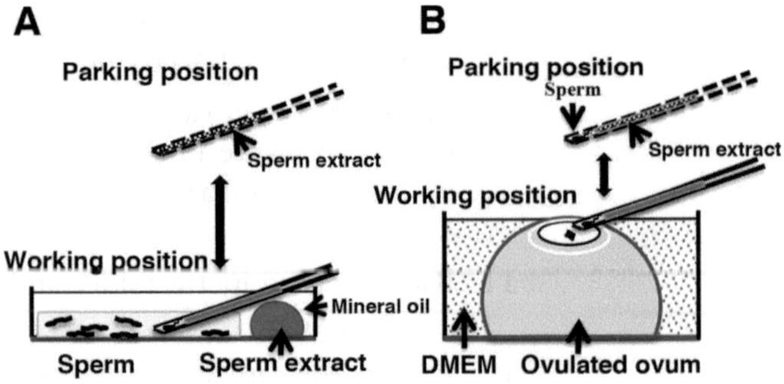

Fig. 2 Schematic drawing of handling of the injection capillary. Parking and working positions during sperm manipulation (**a**) and injection (**b**)

4. Remove the supernatant and dilute the sperm pellet with 6% PVP in DMEM so that a single sperm can be collected in a glass capillary.
5. Make a droplet of 10 μL of DMEM and 10 μL of 2 g/mL SE in a 35 mm Petri dish, and make a thin and even droplet of 20 μL of PVP-diluted sperm solution. Finally, cover with mineral oil and keep until injection at room temperature (*see* **Note 10**).

3.9 Storing Position of the Injection Capillary

The capillary can be moved easily in any direction (X−/Y−/Z-axis) by means of the manipulator. The working position is in the focal plane in the droplet, but the capillary can be placed above the droplet in a parking position so that they do not interfere with the sperm or the attachment of mineral oil to capillary wall as the dish is moved around the stage (Fig. 2a). The same applies to the ovum (Fig. 2b).

3.10 ICSI Procedure

1. It is essential to heat the DMEM to 41.5 °C prior to use.
2. Fix the injection capillary to the holder of micromanipulator at an angle of 30 °C and adjust the injection angle by rotating the capillary holder. The ground surface of the capillary must lie almost flat on the bottom of the dish in order to allow aspiration of the sperm in a controlled manner (Fig. 2a).
3. Introduce paraffin oil into the injection capillary to approx. 1 cm from the tip of pipetting folder and confirm no air bubbles are present in the oil.
4. Place the 35 mm Petri dish containing the sperm droplet onto the stage of the inverted microscope, position the capillary tip in the SE solution, and aspirate 1 μL of the solution into the capillary as gently as possible by rotating the knob of the microinjector.
5. Move the capillary from the working place to the parking place, and position the dish to bring the sperm into focus (Fig. 2a).

6. Select the sperm for injection, move the capillary to the working place, and aspirate the sperm, tail-first, into the same capillary (*see* **Note 11**).
7. Move the capillary from the working place to the parking place again, and set the dish containing the ovum onto the stage of the inverted microscope (Fig. 2b).
8. Set up the stereomicroscope and fiber lighting device, and focus on the germinal disc of the egg by rotating the focusing knob. Move the capillary until it contacts with the surface of the center of the germinal disc.
9. Press the capillary by controlling the micromanipulator toward 9 o'clock and prick the oolemma using the joystick of the micromanipulator (*see* **Note 12**).
10. Push out the sperm together with the SE solution into the cytoplasm (injection volume is approx. 1 nL for quail and approx. 5 nL for chicken, respectively (*see* **Notes 13** and **14**)).
11. Return the capillary to the parking place.

3.11 Ex Vivo Culture of Avian Embryo

Chicken embryo development is divided into three periods for the purpose of ex ovo culture: fertilization to blastoderm formation for 1 day, embryogenesis for 3 days, and embryonic growth for 18 days. The culture period for quail embryos is shorter than that of chicken embryos: embryogenesis for 2.5 days and embryonic growth for 14 days. The protocol explained herein is basically according to the method described by Ono [13] and Kato et al. [14] with some modification.

3.12 Culture Step 1: From Fertilization to Blastoderm

1. Place the 1-cell stage ovum without thick albumen capsule (ICSI-treated ovum) in a plastic cup and fill the cup with DMEM (*see* **Notes 15** and **16**).
2. Tightly seal the open surface of the cup by putting another cup on it, and eliminate the internal air space (Fig. 3a).
3. Incubate the culture-set for 24 h in a humidified CO_2 incubator at 41.5 °C with 5% CO_2.

3.13 Culture Step 2: Embryogenesis

1. Wipe the surrogate shell with 70% ethanol.
2. Place a circular template onto the narrow end of the surrogate shell and draw a line along the inner circumference of the template using a pencil (*see* **Note 17**).
3. Cut the shell along the line using the electric drill. Pour the thin albumen into a Petri dish and place the shell in the dish with the open face of the shell down to prevent drying until use.
4. Pour a little of the thin albumen into the surrogate shell to prevent damaging the yolk physically and subsequently transfer the cultured ovum using a spatula.

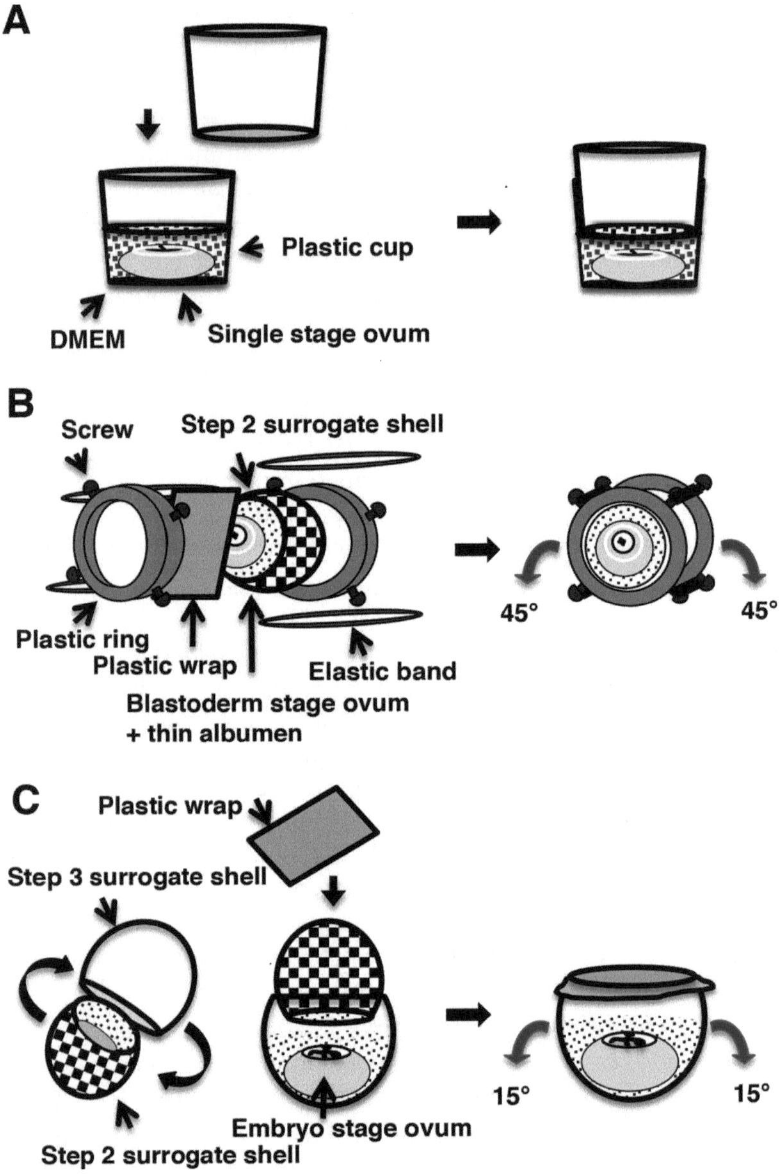

Fig. 3 Schematic drawing of ex vivo culture of quail and chicken ova. Culture step 1 (**a**), step 2 (**b**), and step 3 (**c**). Details of the culture systems and culture conditions are described in the text

5. Fill the thin albumen using a 5 mL syringe and remove the foam.
6. Seal with a sheet of plastic wrap, secured using a pair of plastic rings and elastic bands (Fig. 3b).
7. Place the culture-set in an incubator, with the long axis of the shell held horizontally, and culture the embryo at 37.5 °C and 70% relative humidity with rocking at a 90° angle at 20 min intervals.

3.14 Culture Step 3: Embryonic Growth

1. Wipe the surrogate shell with 70% ethanol.
2. Place a circular template onto the blunt end of the surrogate shell and draw a line along the inner circumference of the template using a pencil (*see* **Note 18**).
3. Cut the shell along the line using the electric drill. Use the narrow-end half without an air sac for culture. Pour the thin albumen into a Petri dish and place the shell into the dish with the open face of the shell down to prevent drying until use.
4. Remove the plastic rings, elastic bands, and plastic wrap from the **step 2** culture-set.
5. Transfer the ova including the thin albumen from the **step 2** surrogate shell into the **step 3** shell by overlaying the empty **step 3** surrogate shell (open face down) onto the **step 2** surrogate shell (open face up) and turning these shells upside down (Fig. 3c).
6. Seal with a sheet of plastic wrap using thin albumen as the glue.
7. Place the culture-set in an incubator with the cut end of the shell held upward and culture the embryo at 37.5 °C and 70% relative humidity with rocking at a 30° angle at 20-min intervals.
8. Stop rocking 1–2 day before the expected hatching day, Prick 5–10 holes in the plastic wrap with a pin (*see* **Note 19**).
9. Remove the punctured wrap and put a Petri dish on the open face of the shell. Hatching is considered when chick comes out of the surrogate shell.

4 Notes

1. You may only use a euthanasia method that is approved by your Animal Care and Use Protocol.
2. We recommend use after filling with paraffin oil.
3. We always use borosilicate glass capillary tubing (cat. no. B100-75-10; Sutter Instrument Co., Novato, CA) with inner and outer diameters of 0.75 and 1 mm, respectively, and a length of 10 cm.
4. Approx. 30 or 10 mL of thin albumen is required for ex vivo culture of one chicken or quail ova, respectively.
5. Ovulation can be considered to occur within 30 min after oviposition of the previous egg in the sequences [15].
6. For the observation of the sperm binding to the pvm, 2 μg/mL pertussis toxin can be added to the reaction mixture during the sperm-pvm incubation, which was reported previously to inhibit the sperm acrosome reaction in quail [16]. After

incubation, the nuclei of the adherent sperm can be stained with 4′,6-diamidino-2-phenylindole (DAPI) after fixation, and the numbers of sperm attached to the pvm can be observed using a fluorescence microscope (BX51).

7. The type and size of glass capillary pulled influences the preparation of the microtools and the result of the injection procedure, so that preliminary experiments are needed to establish these parameters. In particular, it depends on the type and shape of the heating element inside the puller. CAUTION, the heating element can be destroyed by an excessive value for the HEAT parameter. Thus, run the RAMP test in order to set a valid HEAT parameter. We recommend the program consists of two cycles as follows. First cycle: HEAT = RAMP test value, PULL = 0, VELOCITY = 30, TIME = 250, and second cycle: HEAT = RAMP test value, PULL = 50, VELOCITY = 120, TIME = 200. The parameters need no further modification once they have been established.
8. If sperm debris is visible, sonicate the sample until the debris is dissolved completely. ATTENTION: Do not remove both the supernatant and sperm debris after centrifuging.
9. When a concentrated sperm extract is required, the sample can be lysed in a lower volume of PBS after dialysis into water and subsequent lyphilization. Alternatively, the sample concentration can be improved by ultrafiltration.
10. Droplet of DMEM is necessary for washing and storage of the injection capillary, because one injection pipette can be repeatedly used for several ova.
11. In order to minimize the volume of PVP to introduce into the ovum, release the PVP solution until the sperm in the capillary reaches the capillary tip.
12. CAUTION: injection of the capillary into the deep part of the germinal disc is detrimental to ovum survival.
13. A rough estimate of the injection speed is 6 nL/min. Because the germinal disk of quail eggs is opaque, the completion of injection was confirmed visually by observing swelling at the injection site using a stereomicroscope.
14. The injection volume can be estimated by the calculation of the volume released into the PVP solution with certain manipulations.
15. Use a 70 mL plastic cup for chicken and 20 mL cup for quail.
16. Thick albumen can be used as a culture medium instead of DMEM. CAUTION: adjust to pH 7.4 by bubbling CO_2 using a CO_2 spray or an air gun attached to a pressure-regulated CO_2 cylinder and add 10,000 IU penicillin and 50 mg streptomycin/L of albumen.

17. Use a circular template of 34 mm diameter and a similar sized chicken eggshell (approx. 50–60 g whole egg weight) for chicken, and a circular template of 18 mm diameter and a similar sized quail eggshell (approx. 11–13 g whole egg weight) for quail.
18. Use a circular template of 40–45 mm and an egg shell from an approx. 30 g heavier egg compared with the expected egg for chicken. For quail, use a small-sized surrogate chicken shell (less than 45 g whole egg weight) cut off one-third of the distance from the blunt end of the shell.
19. This is the time when the embryo makes a hole in the chorio-allantoic membrane with its beak.

References

1. Birkhead TR, Sheldon BC, Fletcher F (1994) A comparative study of sperm-egg interaction in birds. J Reprod Fertil 101:353–361
2. Kuroki M, Mori M (1997) Binding of spermatozoa to the perivitelline layer in the presence of a protease inhibitor. Poult Sci 76:748–752
3. Robertson L, Brown HL, Staines HJ, Wishart GJ (1997) Characterization and application of an avian in vitro spermatozoa-egg interaction assay using the inner perivitelline layer from laid chicken eggs. J Reprod Fertil 110:205–211
4. Robertson L, Wilson YI, Lindsay C, Wishart GJ (1998) Evaluation of semen from individual male domestic fowl by assessment of sperm: perivitelline interaction in vitro and in vivo. Bri Pout Sci 39:278–281
5. Wishart GJ (1997) Quantitative aspects of sperm:egg interaction in chickens and turkeys. Anim Reprod Sci 48:81–92
6. Yanagimachi R (2005) Intracytoplasmic injection of spermatozoa and spermatogenic cells:its biology and applications in humans and animals. Reprod Biomed Online 10:247–288
7. Morozumi K, Shikano T, Miyazaki S, Yanagimachi R (2006) Simultaneous removal of sperm plasma membrane and acrosome before intracytoplasmic sperm injection improves oocyte activation/embryonic development. Proc Natl Acad Sci U S A 103:17661–17666
8. Iwao Y (2012) Egg activation in physiological polyspermy. Reproduction 144:11–22
9. Hrabia A, Takagi S, Ono T, Shimada K (2003) Fertilization and development of quail oocytes after intracytoplasmic sperm injection. Biol Reprod 69:1651–1657
10. Mizushima S, Takagi S, Ono T, Atsumi Y, Tsukada A, Saito N, Shimada K (2009) Phospholipase Cζ mRNA expression and its potency during spermatogenesis for activation of quail oocyte as a sperm factor. Mol Reprod Dev 76:1200–1207
11. Mizushima S, Takagi S, Ono T, Atsumi Y, Tsukada A, Saito N, Sasanami T, Okabe M, Shimada K (2010) Novel method of gene transfer in birds: intracytoplasmic sperm injection for green fluorescent protein expression in quail blastoderms. Biol Reprod 83:965–969
12. Mizushima S, Hiyama G, Shiba K, Inaba K, Dohra H, Ono T, Shimada K, Sasanami T (2014) The birth of quail chicks after intracytoplasmic sperm injection. Development 141:3799–3806
13. Ono T (2001) Ex ovo culture of quail embryos and its application for embryo manipulation. Anim Sci J 72:361–371
14. Kato A, Mizushima S, Shimada K, Kagami H, Ono T (2014) Simple culture system for bobwhite quail and Japanese quail embryos from the blastoderm stage to hatching using a single surrogate eggshell. J Poult Sci 51:203–206
15. Woodard AE, Mather FB (1964) The timing of ovulation, movement of the ovum through the oviduct, pigmentation and shell deposition in Japanese quail (*Coturnix Coturnix japonica*). Poult Sci 43:1427–1432
16. Sasanami T, Murata T, Ohtsuki M, Matsushima K, Hiyama G, Kansaku N, Mori M (2007) Induction of sperm acrosome reaction by perivilelline membrane glycoprotein ZP1 in Japanese quail (*Coturnix japonica*). Reproduction 133:41–49

Chapter 17

In Vitro and Ex Ovo Culture of Reptilian and Avian Neural Progenitor Cells

Wataru Yamashita, Toyo Shimizu, and Tadashi Nomura

Abstract

Reptiles and birds have been highlighted as excellent experimental models for the study of developmental biology; however, due to technical limitations in cellular analysis, dynamics of neural stem/progenitor cells of these animals remain unclear. In this chapter, we introduce the protocols for neurosphere culture and ex ovo embryonic culture of developing reptilian and avian embryos, which are modified from the method originally established for rodent embryos. Applications of these techniques provide powerful strategies for the study of comparative neural development of amniotes.

Key words Neural stem cells, Neurosphere assay, Ex ovo culture, Gecko, Turtle, Chick

1 Introduction

In the past few decades, biologists have unveiled extensive conservation of the developmental programs among animal taxa. Similar gene expression patterns in various species enable us to identify homologous structures that emerge during embryonic stages. However, it still remains unclear how such conserved body plans have contributed to the morphological diversification of animals. Heterotopic and heterochronic changes in the common developmental programs have been proposed as potential mechanisms to create novel anatomical architectures, although only a few pioneering studies have succeeded in obtaining experimental evidence to prove these hypotheses [1–3].

Reptiles and aves (birds) are included in the clade Amniota. "Reptiles" include several phylogenetic lineages, such as lepidosaurs (snakes and lizards), archosaurs (crocodiles and birds) and chelonians (turtles; an outgroup of archosaurs). Here we distinguish reptiles from birds according to the criteria of general public. Reptiles, birds, and mammals are thought to originate from common ancestor(s), and after diversification, several unique characteristics evolved in each of the amniote lineages [4–6].

Guojun Sheng (ed.), *Avian and Reptilian Developmental Biology: Methods and Protocols*, Methods in Molecular Biology, vol. 1650, DOI 10.1007/978-1-4939-7216-6_17,

Extensive morphological diversities are also observed in amniote brains. Reptiles and birds have unique brain characteristics as a result of evolutionary adaptation. For instance, the dorsal pallium, a homologue of the mammalian neocortex, is poorly laminated in reptilian and avian telencephali, whereas the ventral pallium, a homologue of the mammalian claustrum/amygrdala complex, protrudes toward the ventricle and forms the dorsal ventricular ridge (DVR) [6–8]. These anatomical differences in the telencephalon could be due to distinct behaviors of neural progenitor cells during embryogenesis, although precise comparative analyses have not been performed [9]. To answer this question, experimental techniques for the analyses of neural progenitor cells are required.

Here, we introduce the culture method for chicken neural progenitors and whole reptilian embryos. Neural progenitor cells can be maintained as floating aggregates, called neurospheres, in vitro. The neurosphere assay is a suitable method to evaluate self-renewal and differentiation potentials of neural progenitor cells [10]. In addition, the whole-embryo culture technique is an excellent model system that enables direct access and manipulation of developing mammalian and avian embryos. Recently, we have established ex ovo cultures for developing reptilian embryos by using rodent whole-embryo culture instruments [11]. We introduce the method for culturing neurospheres derived from the embryonic chick dorsal pallium and a convenient method for ex ovo culture of developing turtle embryos. These methods can be combined with gene introduction techniques such as electroporation, which broaden the experimental strategies for the analyses of comparative neural development.

2 Materials

2.1 Reagents and Instruments for Neurosphere Culture of Chicken Neural Progenitors

1. Forceps (Dumont, Montignez, Switzerland) and ophthalmologic scissors (ASONE, Osaka, Japan).
2. Dulbecco's modified Eagle medium (DMEM with high glucose, ThermoFisher Scientific, MA, USA) supplemented with 10% Fatal bovine serum (FBS, ThermoFisher Scientific) and antibiotics (Penicillin-Streptomycin, Nacalai Tesque, Kyoto, Japan), in the following referred to as dissection medium.
3. Sterile petri dishes (35 mm; BD Falcon, NJ, USA).
4. Sterile plastic tubes (1.5 mL and 15 mL, BD Falcon).
5. Cell strainer (40 μm mesh size, BD Falcon).
6. Neurobasal medium (Gibco) supplemented with GlutaMAX™ (ThermoFisher Scientific), B-27 supplement (ThermoFisher Scientific), human recombinant FGF-2 (final concentration: 20 ng/mL; Wako, Osaka, Japan), and antibiotics (Penicillin-

Streptomycin, Nacalai Tesque), in the following referred to as neurosphere medium.

7. TrypLE™ Express (ThermoFisher Scientific).
8. Noncoating 24-well culture plate (Watson, Tokyo, Japan).

2.2 Reagents and Instruments for Ex Ovo Culture of Madagascar ground geckoes and Chinese Softshell Turtles

1. Dissection instruments (same as Subheading 2.1).
2. Transfer pipettes (SARSTEDT, Tokyo, Japan; the tip is cut to widen the diameter to 10 mm).
3. Evaporating dish (50–60 mm in diameter; ASONE, Japan).
4. Hanks' balanced salt Solution (HBSS) with Mg^{2+} and Ca^{2+} (Wako), in the following referred to as HBSS.
5. HBSS supplemented with antibiotics (Gentamycin, Wako, and Penicillin-Streptomycin, Nacalai Tesque), in the following referred to as culture medium.
6. Whole-embryo culture system (Ikemoto Rika Inc., Tokyo, Japan).
7. Whole-embryo culture bottles (Ikemoto Rika Inc.) and rubber caps (Ikemoto Rika Inc.).

2.3 Reagents and Instruments for Electroporation

1. Electroporator (CUY21 EDIT II, BEX CO., LTD, Tokyo, Japan).
2. Glass capillary (G-1, Narishige, Tokyo, Japan).
3. Mouth-controlled pipette (Drummond Scientific, PA, USA).
4. Glass pipette puller (PC-10, Narishige).
5. Electrodes (a chamber-type: LF520P25, or a forceps-type: LF650P3, BEX).
6. DNA solution (1–5 μg/μL in phosphate-buffered saline) with 0.05% fast green (Wako).

3 Methods

3.1 Neurosphere Culture for Avian Neural Progenitor Cells

3.1.1 Establishment of Primary Neurosphere Cultures

1. Incubate fertilized chicken eggs (White Leghorn) at 37 °C in a humidified incubator for 5 days. Developmental stages are determined according to Hamburger-Hamilton (HH) stages [12] (*see* **Note 1**).
2. Transfer HH25 chicken embryos to a petri dish filled with ice-cold dissection medium. Under a dissection microscope, remove the chicken brain by using fine forceps. Transfer the brain to another petri dish containing fresh ice-cold dissection medium and carefully remove the meninges from the brain.
3. Cut the brain into small pieces by using fine ophthalmologic scissors.

4. Transfer the brain pieces into the 1.5 mL plastic tube containing fresh ice-cold dissection medium. Repeat **steps 2–4** to collect enough samples (a cell suspension from a single brain is enough for 2–3 wells of a 24-well culture plate).
5. Remove the dissection medium from the 1.5 mL tube and add 300 μL TrypLE™ Express into the tube. Incubate samples for 10 min at room temperature.
6. Add 1 mL of dissection medium into the tube to stop the enzymatic reaction. Transfer the medium to a 15 mL tube. Dissociate the tissue pieces using a 1 mL blue tip by pipetting for up to 10 times.
7. Centrifuge the cell suspension at 100 × *g* for 5 min at room temperature. Remove the supernatant and resuspend the cells in a 1 mL neurosphere medium.
8. Count cell numbers under the microscope. If cell aggregates and debris are still observed, filter the cell suspension through a cell strainer. Plate cells in uncoated 24-well culture plates at a density of 5×10^5 cells/mL in each well containing neurosphere medium. Culture in a humidified incubator at 37 °C with 5% CO_2. Change half of the medium twice in a week. Do not withdraw any cells. Passage the cells when the majority of the spheres become 200–300 μm in diameter (Fig. 1a, b; also *see* the next section).

3.1.2 Passage of Neurospheres

1. Transfer the medium containing the neurospheres to the 1.5 mL tube and centrifuge the tube at 100 × *g* for 5 min at room temperature.
2. Remove the supernatant and add 300 μL TrypLE™ Express. Resuspend the neurospheres and incubate the tube for 10 min

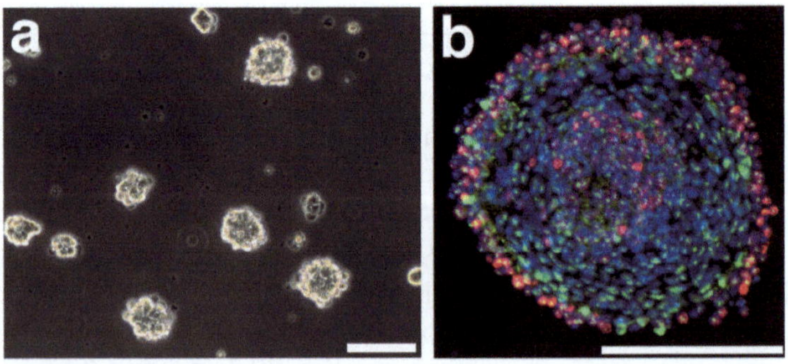

Fig. 1 Neurospheres derived from the chick dorsal pallium. (**a**) A phase-contrast image of chicken neurospheres that were cultured for 3 days in vitro. (**b**) Whole-mount immunostaining of a neurosphere with anti-Pax6 (rabbit polyclonal, MLB, 1:500; *green*) and anti-Tbr1 (chicken polyclonal, Millipore, 1:500; *red*) antibodies. Scale bars: 100 μm

at room temperature. To stop the enzymatic reaction, add 300 μL neurosphere medium and dissociate the spheres by pipetting.

3. Centrifuge the tube at 100 × *g* for 5 min at room temperature. Remove the supernatant and resuspend the cells with 1 mL of neurosphere medium.
4. Plate the cells in a noncoated 24-well culture dish (5 × 10^5 cells/mL in per one well). Incubate the cells at 37 °C with 5% CO_2.

3.2 Ex Ovo Culture of Reptilian Embryos

3.2.1 Isolation of Reptilian Embryos

1. Incubate fertilized eggs of the Madagascar ground gecko (*Paroedura pictus*) or Chinese softshell turtle (*Pelodiscus sinensis*) at 30 °C until 10–14 days post-oviposition (d.p.o.) (*see* **Note 2**).
2. Sterilize the surface of the eggs with 70% ethanol. Examine the position of the embryo in the egg by illuminating the egg with a fiber light.
3. Transfer the egg to an evaporating dish filled with HBSS. Crack the shell gently and remove the piece of shell close to the embryo with forceps. Avoid injuring the blood vessels of the yolk sac.
4. Cut the umbilical cord and remove the yolk sac from the embryo with fine ophthalmologic scissors. The bleeding from the umbilical cord will stop soon after the surgery. Transfer the embryo to fresh HBSS using a plastic pipette.

3.2.2 Setting Up the Whole-Embryo Culture System

1. Transfer the embryo to the culture bottle containing 2 mL of culture medium. Set the bottle on the rotator for the whole-embryo culture system.
2. Rotate the bottle (18 r.p.m.) and incubate the embryo with oxygen (95% O_2 and 5% CO_2, 75 cc/min) at 30 °C. The heartbeat and blood stream of the embryos can be maintained for 24 h (Fig. 2a, b).

3.3 Application: Gene Introduction by Electroporation

Ex ovo culture can be combined with the technique of gene transfection (*see* **Note 3**). Figure 2c shows the expression of a red fluorescent reporter protein (mRFP) derived from an expression vector (pCAGGS-mRFP) in the dorsal pallium of a developing turtle. Before electroporation, reptilian embryos are precultured in the whole-embryo culture system for 1 h, and the embryo is transferred to the evaporating dish filled with sterilized HBSS.

3.4 Electroporation of Exogenous Genes into Reptilian Brains

1. Prepare micropipettes by pulling glass capillary tubes using a glass pipette puller and pinch off the tips with forceps.
2. Inject the DNA solution (1–2 μL) into the lateral ventricle using mouth-controlled pipettes. Place the target region of

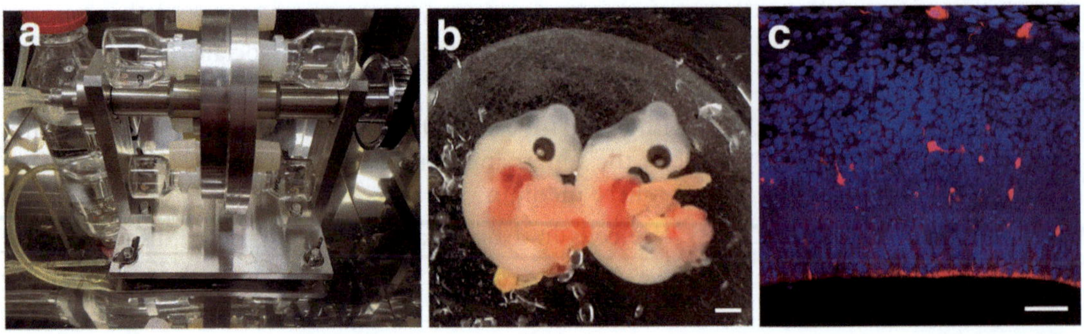

Fig. 2 Ex ovo culture of turtle embryos. (**a**) A rotator for the whole-embryo culture system (Ikemoto Rika Inc.) (**b**) Turtle embryos that were cultured for 24 h in the rotating culture system. (**c**) RFP expression in the developing turtle pallium. The pCAGGS-mRFP vector was electroporated into the turtle pallium. After electroporation, embryos were cultured for 24 h. Scale bars: 1 mm (**b**); 50 μm (**c**)

the brain between the electrodes and apply square-wave current pulses (*see* **Note 4**).

3. After applying electric pulses, transfer the embryo to the culture bottle to continue ex ovo culture (~24 h).

4 Notes

1. Fertilized chicken eggs can be stored at 14–15 °C for up to 1 week before starting incubation.
2. Fertilized gecko eggs are obtained by breeding adult males and females in our laboratory or from the animal facility of Riken Center for Developmental Biology (CDB). Fertilized turtle eggs are obtained from a local breeder (Daiwa Yoshoku, Saga, Japan). Reptilian eggs are placed on sand or expanded vermiculite. Turtle eggs must be incubated under moist condition. The embryonic stage of Madagascar gournd gecko and Chinese softshell turtles are determined by according to the reference [13, 14].
3. Ex ovo electroporation can be performed after ex ovo culture of reptilian embryos.
4. Parameters for electroporation are as follows: 80 V, 50 ms for pulse-on time, 950 ms pulse-off time, 4 sets of pulses (LF520P25); 28 V, 50 ms for pulse-on time, 950 ms pulse-off time, 3 sets of pulses (LF650P3).

Acknowledgments

We thank Ms. Misato Kawami for technical assistance, Dr. Masanori Uchikawa for pCAGGS-mRFP vector, and all members of Prof.

Ono lab for valuable discussion. This work is supported by the fellowship of Japan Society for the Promotion of Science (JSPS) to WY, KAKENHI (The Empathetic Systems, #26118510, Thermal biology, #16H01394) to TN.

References

1. Dyer MA, Martins R, da Silva Filho M, Muniz JA, Silveira LC, Cepko CL, Finlay BL (2009) Developmental sources of conservation and variation in the evolution of the primate eye. Proc Natl Acad Sci U S A 106(22):8963–8968. doi:10.1073/pnas.0901484106
2. Keyte AL, Smith KK (2010) Developmental origins of precocial forelimbs in marsupial neonates. Development 137(24):4283–4294. doi:10.1242/dev.049445
3. Nagashima H, Sugahara F, Takechi M, Ericsson R, Kawashima-Ohya Y, Narita Y, Kuratani S (2009) Evolution of the turtle body plan by the folding and creation of new muscle connections. Science 325(5937):193–196. doi:10.1126/science.1173826
4. Carroll RL (1988) Vertebrate paleontology and evolution. W.H.Freeman and Company, New York
5. Modesto SP, Scott DM, MacDougall MJ, Sues HD, Evans DC, Reisz RR (2015) The oldest parareptile and the early diversification of reptiles. Proc Biol Sci 282(1801):20141912. doi:10.1098/rspb.2014.1912
6. Nomura T, Kawaguchi M, Ono K, Murakami Y (2013) Reptiles: a new model for brain evo-devo research. J Exp Zool B Mol Dev Evol 320(2):57–73. doi:10.1002/jez.b.22484
7. Ulinski PS (1983) Dorsal ventricular ridge: a treatise on forebrain organization in reptiles and birds. John Wiley & Sons, New York
8. Ulinski PS (1990) The cerebral cortex of reptiles. In: Cerebral Cortex, vol 8A. Plenum, New York, pp 139–215
9. Nomura T, Gotoh H, Ono K (2013) Changes in the regulation of cortical neurogenesis contribute to encephalization during amniote brain evolution. Nat Commun 4:2206. doi:10.1038/ncomms3206
10. Pastrana E, Silva-Vargas V, Doetsch F (2011) Eyes wide open: a critical review of sphere-formation as an assay for stem cells. Cell Stem Cell 8(5):486–498. doi:10.1016/j.stem.2011.04.007
11. Nomura T, Yamashita W, Gotoh H, Ono K (2015) Genetic manipulation of reptilian embryos: toward an understanding of cortical development and evolution. Front Neurosci 9:45. doi:10.3389/fnins.2015.00045
12. Hamburger V, Hamilton HL (1992) A series of normal stages in the development of the chick embryo. 1951. Dev Dyn 195(4):231–272. doi:10.1002/aja.1001950404
13. Noro M, Uejima A, Manabe M, Tamura K (2009) Normal developmental stages of the Madagascar ground gecko Paroedura pictus with special reference to limb morphogenesis. Dev Dyn 238(1):100–109. doi:10.1002/dvdy.21828
14. Tokita M, Kuratani S (2001) Normal embryonic stages of the chinese softshelled turtle Pelodiscus sinensis (Trionychidae). Zool Sci 18(5):705–715. doi:10.2108/zsj.18.705

Part IV

New Model Systems

Chapter 18

Lifting the Veil on Reptile Embryology: The Veiled Chameleon (*Chamaeleo calyptratus*) as a Model System to Study Reptilian Development

Raul E. Diaz Jr., Federica Bertocchini, and Paul A. Trainor

Abstract

Living amniotes comprise three major phylogenetic lineages: mammals, birds, and non-avian reptiles. Mouse and avian embryos continue to be the primary species used in experimental settings to further our knowledge and understanding of the genetics and embryology of amniotes. In comparison, non-avian reptiles, which constitute up to 40% of all living amniotes, have played a comparatively minor role. Studies of non-avian reptiles are, however, paramount for providing insights into the evolutionary changes that occurred in the transition from reptilian-like amniote ancestors to derived mammalian and avian species. Here, we introduce the Veiled Chameleon, a squamate reptile, as a new experimental model for examining fundamental questions in development, evolution, and disease.

Key words Chameleon, Reptile, Embryo, Amniote, In situ hybridization, Immunohistochemistry, Organ culture, Limb

1 Introduction

Living amniotes comprise three major clades: mammals, birds, and non-avian reptiles [1–4]. Most of our understanding of amniote embryonic development has been derived from studies in mouse (a placental mammal) and chick (an avian reptile) model systems. Avian species constitute about half of all living reptiles (~10K species), with the remaining species being composed of the diverse yet understudied Crocodilia (crocodiles, alligators, and caimans), Rhynchocephalia (tuatara), Squamata (lizards, snakes, amphisbaenians), and Testudinata (turtles and tortoises).

Squamate reptiles are the most speciose (9900+ species) and exhibit a diverse array of phenotypes and reproductive life histories, while occupying a broad range of ecological niches. Reptiles are therefore an invaluable resource for furthering our understanding of the developmental pathways underpinning anatomical and physiological features shared by avians and mammals (including

Guojun Sheng (ed.), *Avian and Reptilian Developmental Biology: Methods and Protocols*, Methods in Molecular Biology, vol. 1650, DOI 10.1007/978-1-4939-7216-6_18, © Springer Science+Business Media LLC 2017

Fig. 1 Female adult veiled chameleon showing modified features for an arboreal lifestyle, such as cleft hands/feet, prehensile tail, and laterally flattened body

humans) as well as the mechanisms through which the development and evolution of derived morphologies and novelties have arisen. For example, lizards belonging to the family Chamaeleonidae present with traits that are shared by all amniotes, as well as traits typical of only reptiles, or traits that are unique to the arboreal family of lizards.

Chameleon evolution involved the development of several peculiar features such as independently moving eyes, a prehensile tail, and modified wrist and ankle bones that enabled them to adapt to a life of climbing shrubs and trees (Fig. 1; [5, 6]). In addition, chameleons developed complex features such as sexual dimorphism in cranial vault bones and coloration/patterning that are important for intraspecies communication. In the past few years, *Chamaeleo calyptratus,* more commonly known as the Veiled Chameleon, has become a model for studying several aspects of reptilian physiology, from feeding to communication and locomotion (reviewed in [5]). This particular species of reptile has also emerged as a potential model organism to study embryonic development in non-avian reptiles and in comparative studies involving avian and mouse model systems [6, 7].

An important question is why, until now, have non-avian reptiles not been widely used for studying early developmental processes and events? The principal reason for this is the fact that embryogenesis commences within the oviduct prior to oviposition. Furthermore, species-specific differences in developmental stage at oviposition are broad across squamate reptiles, ranging from early gastrula to live birth! In general, the embryos of most egg-laying squamate species have already developed to the limb bud stage at the time of oviposition [8]. In order to analyze earlier embryonic stages, it is necessary to sacrifice gravid females to retrieve fertilized eggs. In contrast, the embryos of a veiled chameleon have typically only developed to the pre-gastrula or early gastrula stage at the time of oviposition, thus overcoming the usual squamate limitation. Together with large clutch sizes (40–90 eggs) and multiple clutches (up to 4) per female per year [5], this species is quickly being recognized as an ideal model system for developmental, evolutionary, and ecological biologists. In this chapter, we describe the major developmental and experimental characteristics of veiled chameleon embryos. We include methods for harvesting and culturing whole embryos and organs, processing them for gene expression and protein activity, as well as dynamically assessing cell and developmental process such as proliferation and cell death, which our labs have successfully employed.

2 Materials

2.1 In Situ Hybridization

All reagents for in situ hybridization should be prepared using high-quality purified autoclaved deionized water. However, all the solutions especially those used for fixation, dehydration, and rehydration should be RNAse free and have been made with DEPC-treated water where applicable (*see* **Note 1**). Reagents (unless specified otherwise) should be stored at room temperature. Washes are conducted on a rocker at room temperature or under refrigeration.

2.2 Fixation and Dehydration Solutions

1. *1×PBS, pH 7.4*: Dissolve five (5) Phosphate-Buffered Saline (PBS) tablets in 1 L deionized water as recommended by the manufacturer. Store at room temperature (*see* **Note 2**).
2. *4% Paraformaldehyde*: 4% paraformaldehyde w/v in 1×PBS.
3. *Post fixative*: 4% formaldehyde +0.1% Glutaraldehyde in PBT (PBS Tween-20) (*see* **Note 3**).
4. *Absolute Methanol.*

2.3 In Situ Buffers

1. *0.1% PBT*: Add 1 mL Tween-20 in 1 L of 1×PBS, store at room temperature.
2. *TBST*: Dissolve 0.8 g NaCl, 0.02 g KCl, 2.5 mL Tris–HCl 1 M (pH 7.8) and 1 mL Tween-20 in deionized water up to 1 L.

2.4 In Situ Hybridization Solution

1. Add 500 mL formamide, 65 mL SSC 20 (pH 5.0), 10 mL EDTA 0.5 M, 2.5 mL tRNA 20 mg/mL (final concentration 50 μg/mL), 2.0 mL Tween-20, 5 mL SDS 20%, 2.0 mL Heparin 50 mg/mL (final concentration 100 μg/mL), in deionized water up to 1 L. Aliquot into 50 mL Falcon tubes and store at −20 °C.

2.5 In Situ Revelation Solution

1. *NTMT*: Add 1 mL NaCl 5 M, 2.5 mL Tris–HCl 2 M (pH 9.5), 0.5 mL Tween-20, 2.5 mL $MgCl_2$ 1 M, in deionized water up to 50 mL. Make it fresh every time.

2. *Revelation solution (NTMT-BCIP-NBT)*: Mix 135 μL NBT (50 mg/mL) and 70 μL of BCIP (50 mg/mL) in up to 20 mL of NTMT. Keep in the dark.

2.6 In Situ Proteins

1. *Proteinase K*: stock solution of 10 mg/mL in PBT to be diluted to a final concentration of 10 μg/ml.
2. *Anti-DIG antibody in blocking solution*: 1:5000.

2.7 In Situ Blocking Solution

1. 5% heat inactivated serum + Bovine Serum Albumen (BSA) 1 mg/mL in TBST.

2.8 Immuno histochemistry Buffers

1. *0.1% PBTr*: Mix 1 mL of Triton X-100 in Up to 1 L of 1 × PBS.

2.9 Immuno histochemistry Block Solution

1. *PBT*: *see* PBT for in situ.
2. *Block solution*: 2.5% donkey serum, 2.5% goat serum, 3% BSA in PBT.

2.10 Immuno histochemistry Antigens and Conjugates

1. *Phalloidin*: 488 Alexa Fluor: 1:100 in PBT.
2. *Vectashield*: antifade mounting medium with DAPI.

2.11 Immuno histochemistry for Post-Gastrula Stage Embryos (Modified from [11])

1. Hydrogen Peroxide (H_2O_2).
2. *Dimethylsulfoxide* (DMSO).
3. Dent's *Bleach*: Mix 100% Methanol, DMSO and 30% Hydrogen Peroxide in a 4:1:1 Ratio.
   (a) *1 M Tris–HCl pH 7.5 Stock Solution*: Add 60.57 g Tris-Base to 350 mL deionized water and mix at RT until dissolved. Adjust pH to 7.5 with HCl and add deionized water up to 500 mL. Store at 4 °C for 4–6 months.
4. *1 M Sodium Chloride* (NaCl).
5. *Bovine Serum Albumen.*
6. *TN-Blocking Buffer*: Add 50 mL of 1.0 M Tris–HCl, pH 7.5, 4.3875 g of NaCl (for final 0.15 M NaCl), 15 g of Bovine Serum Albumen (for final 3% BSA) in deionized water up to 500 mL final volume.
7. *DAPI, Dilactate stock solution*: dilute 2 mg in 1 mL deionized water, keep in 4 °C.
8. *DAPI, Dilactate working solution*: dilute 1 μL in 1 mL of 1 × PBS.
9. *DAB Substrate Kit.*
10. *Glycerol.*

2.12 In Vitro Whole Embryo and Organ Culture

1. 6-Well Tissue Culture Plates, Flat bottom.
2. Millipore® 6 Well MilliCell® hanging cell culture inserts 1 μm pore and PET transparent.

3. *Ascorbic Acid Stock*: Dissolve 50 mg in 1 mL of deionized water (can store at 4 °C for 1 week, keep in the dark).
4. *Culture Media*: Mix 0.5 mL Fetal Bovine Serum (heat denatured) with 9.3 mL DMEM/F-12 (1:1), L-Glutamine, 10 μL of Ascorbic Acid (final concentration of 50 μm/mL), and 200 μL Penicillin-Streptomycin (for final concentration of 2%).
5. *Incubator*: Incubate at 26 °C, filling any empty wells with water or placing an open dish filled with water in the incubator to create high ambient humidity (*see* **Note 4**).

2.13 Nile Blue Sulfate (NBS) Stain

1. 1.5% Stock Solution: Dissolve 1.5 g Nile Blue Sulfate in 100 mL deionized water.
2. Dilute stock solution 1:1000 in culture media or 1×PBS.

3 Methods

All the procedures are carried out at room temperature unless otherwise noted.

3.1 Embryo Harvesting

Incubation time for the hatching of Veiled chameleons is roughly 200 days (±1–2 weeks) at 26 °C. Hatching does not occur in synchrony across the clutch [5]. The construction of a staging system for the Veiled chameleon, and squamates in general, is complicated by the broad range of embryonic stages present within a single clutch of eggs at any point during the incubation period (Diaz et al., in preparation). This phenotypic disparity may be due to the order in which eggs were fertilized along the oviduct. The larger the clutch the longer the time gap between fertilization and onset of development between the first and last eggs in the clutch.

Veiled chameleon eggs are soft and leathery to the touch, unlike the heavily mineralized and hard-shelled eggs typical of avians. Squamate eggs differ by initially growing in size after oviposition (Fig. 2a). Eggs are maintained under high internal pressure

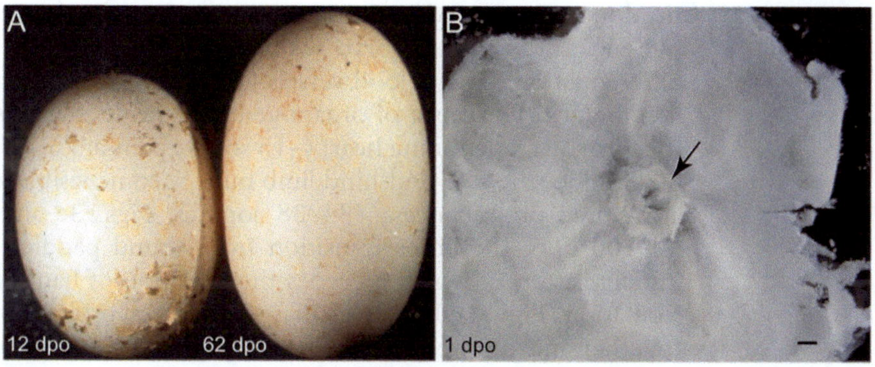

Fig. 2 (**a**) Two eggs (12 and 62 dpo) highlighting the increase in size after oviposition (with the smaller egg in A still larger than at 0 dpo [not shown]), (**b**) Day 1 gastrula with extraembryonic membranes, in ventral view (*arrow*)

and any puncture will result in expulsion of its contents with probable shearing and damage to the embryo. Candling (using any strong light) can be used to locate the site of the circular gastrula/embryo and membrane adhesion to the inner surface of the egg shell at early stages. A pinhole in the shell is made with forceps and fine scissors are then used to cut circumferentially around the eggshell at a safe distance from embryo (generally 90° away from the embryo with the cut made around the longest circumference of the oval egg). This procedure is performed with the egg submerged in 1×PBS, which helps to equilibrate the internal egg pressure. Limb bud stage and older, (>70 dpo) embryos, although easier to dissect, are still harvested while submerged in 1×PBS. The eggshell and underlying membranes have separated sufficiently by these stages to allow for a pinhole to be made through eggshell, and a circumferential cut made around the egg, leaving the embryo, yolk, and extraembryonic membranes intact. Figure 2b shows a Day 1 embryo retrieved from the egg with the extra embryonic structures still attached (ventral view shown, posterior end to the right).

At oviposition, veiled chameleon embryos have typically developed to a pregastrula or early gastrulation stage (Fig. 3a), at which time internalization of prospective mesendoderm has begun [7]. During the first 2 weeks of development, the gastrula is disc-shaped overall. Subsequent elongation along the antero-posterior axis renders the gastrula more oval-shaped (Figs. 3a and 4a, b), arrow; *see* [7]. During the second month of gestation, the posterior margin of the gastrula begins to flatten in a perpendicular direction to the anterior-posterior axis in concert with enlargement of the blastopore, the structure through which gastrulation occurs (Figs. 3b, arrow, and 4c [7]). Unlike other amniotes, the amniotic membrane is already present at oviposition, forming prior to and during gastrulation in chameleons [9]. Between 60 and 70 days post oviposition (dpo), veiled chameleon embryos undergo neurulation, cranial flexure, and segmentation of the first 2–4 pairs of somites lateral to the neural tube (Table 1, Fig. 3c in ventral view).

Figure 3d depicts an embryo in the process of turning. While the dorsum is covered by the amnion, complete rotation will lead to full enclosure of the embryo within the amnion with subsequent differentiation of key blood vessels linking to the extra embryonic membranes. The onset of organogenesis is evident at this stage in the form of a functioning heart (~15 somite pairs at 75–80 dpo). By 80–85 dpo, the fore-and-hind limb buds become visible as bilaterally paired prominences (~34–38 somites; Figs. 3e and 4d). By 90–100 dpo, the limbs transition from round paddles to being distally flattened, which precedes the initiation of hand/foot cleft formation (Fig. 3f). By 100–105 dpo, the facial prominences come together and fuse to give rise to the primitive face. Between 110 and 130 dpo the autopodia (hands/feet) complete morphogenesis.

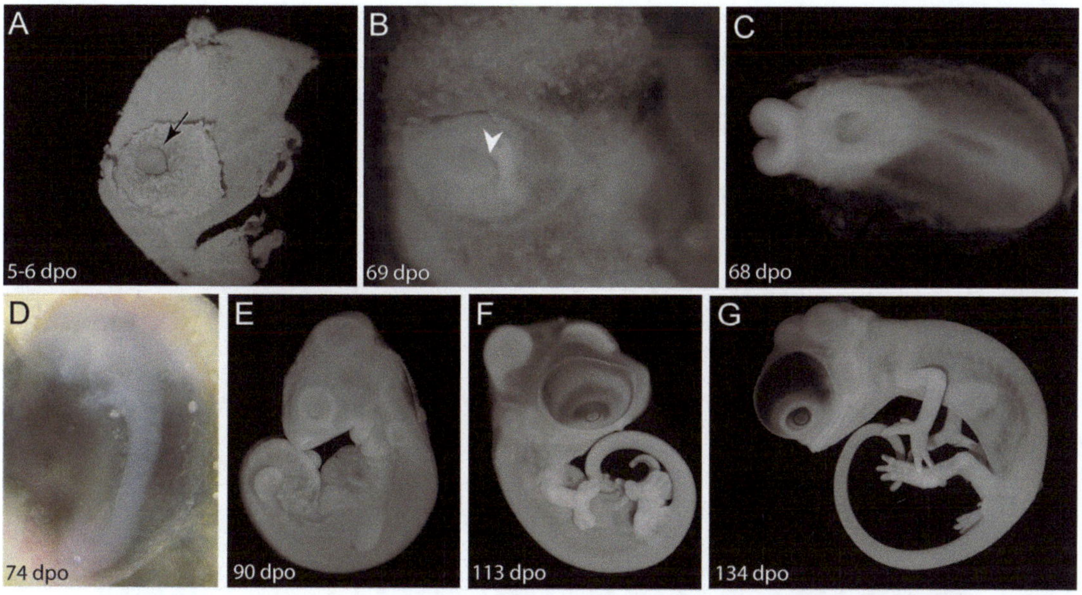

Fig. 3 (**a**) Ventral view of a 5–6 dpo gastrula (*arrow*), (**b**) Ventral view of a 69 dpo gastrula displaying an enlarged blastopore (*arrowhead*); (**c**) 68 dpo embryo in ventral view showing open anterior neuropore, 3–4 pairs of somites and developing heart field. Despite being a day younger in incubation relative to the embryo in (**b**), this embryo is morphogenetically more advanced, which highlights the variability in embryonic stage within a clutch; (**d**) 74 dpo turning embryo whose dorsal body is covered by the amniotic membrane; (**e**) 90 dpo embryo at the limb bud stage showing full complement of pharyngeal arches and development of the olfactory, optic, and auditory sensory systems; (**f**) 113 dpo embryo with very large eyes, clefting autopodia, somites having differentiated into distinct tissues and coiling tail; (**g**) 134 dpo older stage embryo with developing claws, eyelids, everted hemipenes, and cranial casque at early stages of growth

The interdigital tissue between non-cleft digit pairs becomes more compact, while the distal claws become well differentiated with keratin (Fig. 3g). At 130 dpo, skin and scale differentiation begins to reduce embryonic translucence and by 150 dpo, melanin becomes visible along the trunk. Morphogenesis and growth continue until hatching at 200 dpo (Table 1).

Chameleon embryos up to early limb stage can be easily processed for canonical staining techniques such as whole-mount in situ hybridization (Fig. 4a and b) and antibody staining (Fig. 4c and d). A series of basic staining techniques are summarized below, including whole-mount in situ hybridization, antibody staining, and phalloidin staining.

From gastrulation (Fig. 4a–c) through somitogenesis (Fig. 4d), and prior to the appearance of limb buds (<80 dpo), veiled chameleon embryos are relatively flat (both with and without the amnion) and can be cultured on filter supports similar to those used in basic organ or tissue explant culture approaches (Fig. 5a and b). To date, we have not cultured embryos beyond 4 or 5 days; however, longer term cultures are likely to be possible with current or modified and

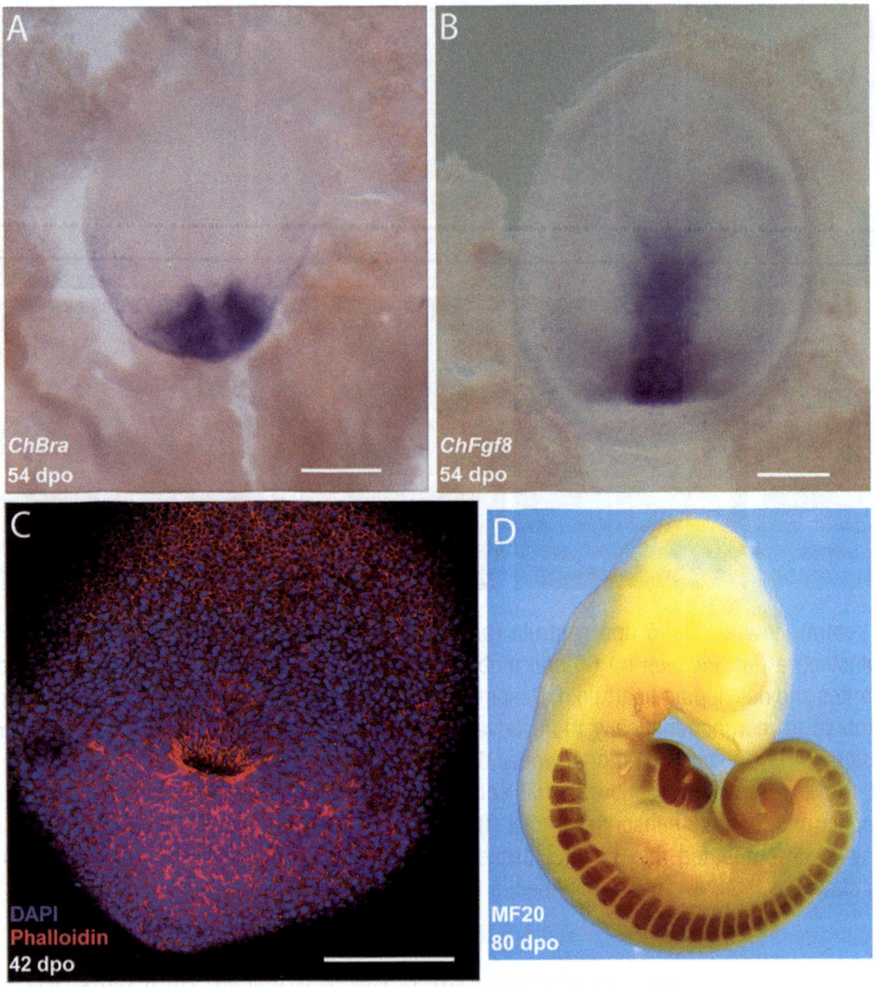

Fig. 4 54 dpo chameleon gastrulae showing mRNA distribution for Chameleon *Brachyury* (**a**) and *Fgf8* (**b**) (*dorsal views*). Confocal fluorescent image of a 42 dpo gastrula (*ventral view*) with nuclei stained by DAPI (*blue*) and F-Actin with Phalloidin (*red*) (**c**). (**d**) A limb bud stage embryo wholemount immunostained for MF20 (muscle Myosin Heavy Chain) using DAB revelation to show differentiation of the somite derived myotome tissue

optimized culture conditions. Figure 5a shows an excised and "filleted" pair of forelimbs at a round paddle stage, which have been grown on MilliCell tissue culture filters for 5 days. Adjacent is an ~11 somite stage embryo within its amnion, which was cultured on a filter (Fig. 5b). Once the limb bud stage is reached, body size and length increase leading to embryonic coiling, which no longer allows for an embryo flattening on a filter. However, at these later stages of development it becomes easier to separate the eggshell from the underlying embryo and its associated extra embryonic membranes such that the whole complex can be cultured semi-floating in growth media (Fig. 5c). For short-term experiments

Table 1
Key developmental and morphogenetic events during Veiled chameleon embryogenesis (at 26°C)

Pre-oviposition	
Mating	Egg clutch is laid in an underground nest 3–4 weeks after mating. Development begins and continues to the early gastrulation stage at the time of egg laying. Females are able to store sperm and will generally produce another egg clutch after approximately 2 months

Post-oviposition Days	**Somite pairs**	**Key events in morphogenesis**
Gastrulation		
0–70	–	Process of gastrulation takes ~2 months
Neurulation/beginning of organogenesis		
60–70	–	Neural fold formation
	–	Neurulation
	2–4	Anterior and posterior neuropores open
		Cranial flexure
65–75	~13–15	Cervical flexure
75–80	~15	Heart forms
80–84	19	Lens forms from ectoderm
80–84	24	Olfactory pit
80–85	~30	Pharyngeal arches/clefts complete
80–85	~34–38	Limb buds appear
90	–	Round limb paddle stage
95	–	Distally flattened paddle with digit rays distinct
100	–	Facial prominences make contact/begin fusion
		Splitting of syndactylous autopodial bundles
102–105	–	Clefting between digit bundles reach carpals/tarsals
113–115	–	Interdigital mesenchyme between digits compacts
		Claws begin to be distinct
130	–	Skin/scales begin differentiating and make embryo less translucent
150	–	Melanin begins to appear in trunk skin
200	–	Hatching

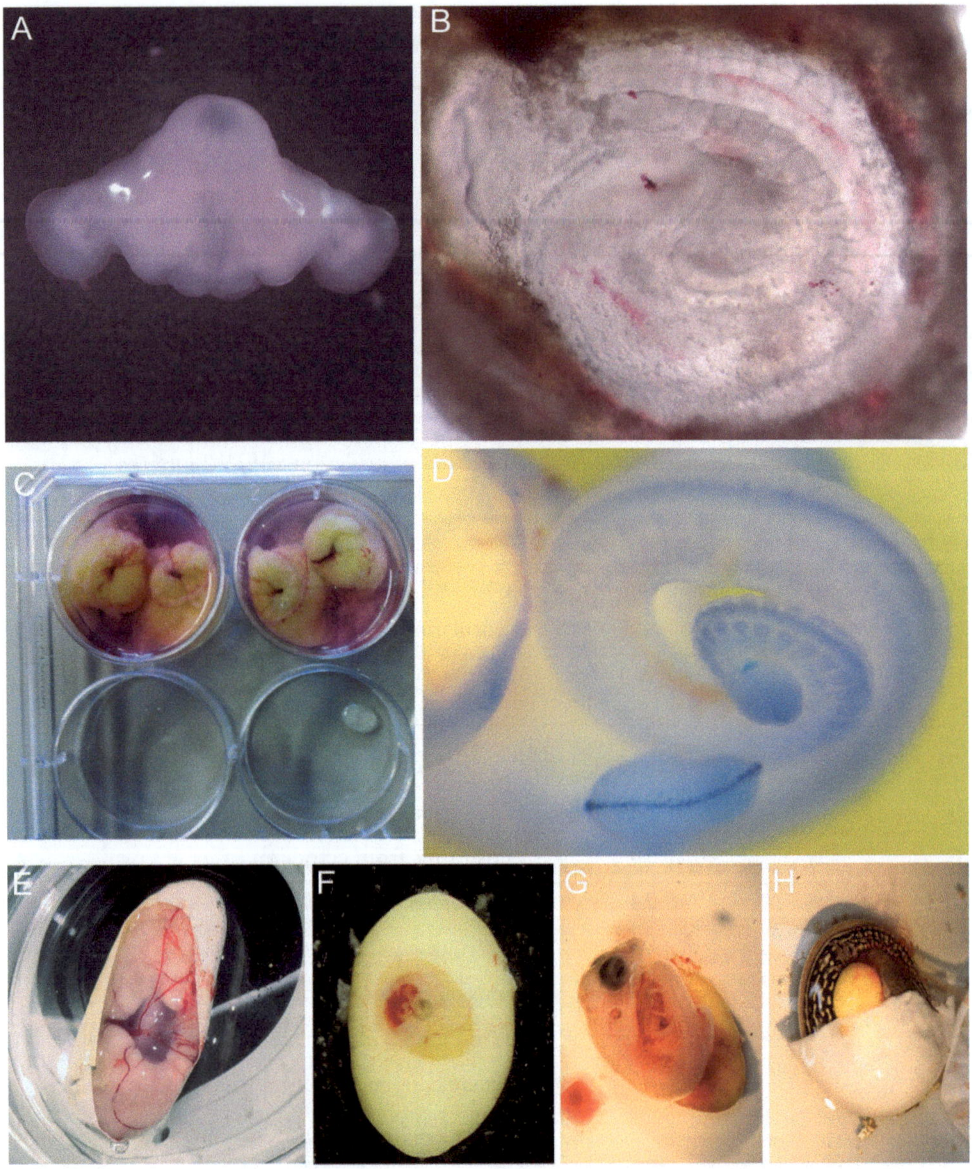

Fig. 5 (**a**) Filter-based organ culture of paddle stage chameleon forelimbs (3 days after incubation at 26 °C), (**b**) ~13 somite stage Chameleon embryo cultured within the amnion for 5 days using the filter organ culture method, (**c**) ~99 day chameleon embryos cultured in vitro suspended in growth media within extraembryonic membranes, (**d**) embryo stained with Nile Blue Sulfate, with blue staining highlighting the distribution of high macrophage activity, (**e**) Sinaloan Milk Snake embryo within extra embryonic membranes in 1×PBS, similar to (**c**) and (**f**), (**f**) Desert Grassland Whiptail lizard at 8–9 dpo in 1×PBS, and older stage embryos (**g**, **h**) showing that when yolk volume decreases, the embryo becomes more visible within the extraembryonic membranes (due to a reduction in both the yolk which had previously taken up much egg volume as well as the embryo now being larger). Eventually yolk sac and remaining yolk become internalized along the ventral midline

(such as for BrdU incorporation) embryos can be cultured in 1×PBS (or growth media) without membranes for a few hours at room temperature (Fig. 5d). Similarly, Nile Blue Sulfate staining of tissues to show cell death associated macrophage activity (by cells endocytosing the NBS granules), for veiled chameleon embryos was successful when embryos were cultured for 40–60 min in 1×PBS or growth media (*see* **Note 5**).

The feasibility of ex-ovo culturing depends on the particular stage of embryonic development of interest. Early stage snake and lizard embryos are completely surrounded by egg yolk and exhibit the same general changes in extraembryonic development (Fig. 5c for a Veiled chameleon, 5e for a Sinaloan Milksnake, and 5f for a Desert grassland whiptail lizard). During later stages of embryogenesis, the conversion of yolk energy into tissue gives way to a smaller yolk sac that exposes the older embryo within transparent and thin membranes (Fig. 5g and h). This renders these and older embryos amenable to many techniques such as bead implantation, electroporation, and/or viral transfection delivered through the membranes using a fine pipette or needle.

3.2 Fixation of Embryos

1. Upon harvesting embryos in 1×PBS, embryos are fixed overnight in 4% PFA at 4 °C with gentle shaking on a rocker.
2. Embryos are washed 2× 10 min in PBT.
3. Embryos are dehydrated in an increasing series of Methanol in PBT, 25%, 50%, 75%, and 100%, washes, each for 30 min on a rocker at 4 °C.
4. Embryos are then washed 3× 30 min in 100% Methanol (*see* **Note 6**).
5. Store embryos at −20 °C until needed for immunohistochemistry or in situ hybridization.

3.3 In Situ Hybridization

The protocol for chameleon embryos is based on the protocol developed for chicken embryos [10], with a few modifications, and has successfully worked on mouse, chick, snake, and lizard species.

1. In situ Day 1.
   (a) Rehydrate embryos through a decreasing concentration gradient of Methanol from 100% to 75%, 50%, and 25% in PBT (PBS + 0.1% Tween) for 10 min each (*see* **Note 7**).
   (b) Wash embryos in PBT 2× 20 min on a rocker.
   (c) Add Proteinase K to a final concentration of 10 μg/ml in PBT. Leave 1 h at room temperature (RT) (*see* **Note 8**).
   (d) Rinse embryos with PBT.
   (e) Add postfix solution to embryos and gently rock for 30 min at room temperature.

(f) Rinse embryos 2× in PBT.

(g) Place embryos in a hybridization solution and in a water bath (or hyb oven) at 68–70 °C for a minimum of 1 h (up to 6 h) to prehybridize tissues. Embryos in the hybridization solution can be stored at −20 °C.

(h) Replace the hybridization solution with hybridization solution combined with ribonucleic acid probe.

(i) Incubate overnight in the water bath at 68–70 °C.

2. In situ Day 2 (for **steps a–c** embryos are kept in the water bath at 68–70 °C, while **steps d–g** are carried out at RT).

   (a) Remove the riboprobe-Hybridization solution and rinse twice with the pre-warmed Hybridization solution.

   (b) Wash embryos 2× 30 min with the pre-warmed hybridization solution.

   (c) Wash embryos with the pre-warmed 1:1 hybridization solution: TBST for 10 min on a rocker.

   (d) Rinse embryos 3× in TBST.

   (e) Wash embryos 3× 30 min in TBST.

   (f) Block embryos in 5% heat inactivated serum + BSA 1 mg/mL in TBST via rocking for 3 h at RT.

   (g) Replace Blocking solution with Anti-DIG antibody diluted 1:5000 in blocking buffer and incubate overnight at 4 °C.

3. In situ Day 3.

   (a) Rinse embryos 3× in TBST.

   (b) Wash embryos 3× 1 h in TBST.

   (c) Wash embryos 2× 10 min in NTMT on rocker.

   (d) Incubate embryos in NTMT-BCIP-NBT solution until stained to desired degree (keep in the dark).

3.4 Wholemount Immuno histochemistry on Gastrula (Up to 70 Day of Development) Stage Embryos

This protocol has been described in Stower et al. [7]. All the washes are carried out at RT.

1. Fix the embryos in 4% PFA overnight at 4 °C.
2. Wash embryos 3× 10 min in PBTr.
3. Wash embryos in 0.25% PBTr for 30 min.
4. Wash embryos 3× 10 min in PBT.
5. Block the embryos in 2.5% donkey serum, 2.5% goat serum, 3% BSA in PBT for 1 h.
6. Incubate embryos with the desired primary antibody for 24 h at 4 °C.
7. Wash embryos 3× 10 min in PBT.

8. Wash embryos in PBT for 30 min.
9. Incubate embryos overnight at 4 °C with the desired secondary antibody.

Before mounting, samples can be incubated in *phalloidin*, according to the following protocol:

10. Incubate embryos with 1:100 phalloidin-488 (Invitrogen) in PBT overnight at 4 °C.
11. Wash embryos 4× 10 min in PBT.
12. Mount embryos in Vectashield mounting medium containing DAPI.

3.5 Wholemount Immuno histochemistry for Post-Gastrula Stage Embryos (After 70 Day of Development) (Modified from [11])

1. *Day 1 (~ 5 h).*
   (a) Embryos which have been fixed overnight and dehydrated into 100% Methanol (or taken from storage −20 °C; as described in *Fixation of embryos*, above) are taken from the Methanol and placed directly into Dent's Bleach solution.
   (b) Incubate embryos in Dent's Bleach for 2 h at room temperature (*see* **Note 9**).
   (c) Replace Dent's Bleach with absolute Methanol.
   (d) Process embryos through a series of decreasing Methanol concentrations (75%, 50%, 25% in 1×PBS) ×30 min at each stage (*see* **Note 10**).
   (e) Replace 1×PBS with TN-block (TNB) for 2 h (*see* **Note 11**).
   (f) Introduce primary antibodies into fresh TN-block at appropriate dilutions and incubate overnight at 4 °C with gentle rocking.
2. *Day 2 (~5 h).*
   (a) Wash embryos in 1×PBS for 5× 1 h.
   (b) Replace 1×PBS with TN-block and apply secondary antibody at desired dilution overnight at 4 °C.
3. *Day 3 (~6–7 h).*

   For DAB using HRP-conjugated secondary antibodies

   (a) Wash embryos for 5× 1 h using 1×PBS.
   (b) Add embryos to DAB Substrate as directed by the manufacturer at 4 °C in the dark (*see* **Note 12**).
   (c) Upon reaching desired degree of staining, wash embryos in 1×PBS 2× 15 min.
   (d) Embryos can be stored in 4% PFA (or formaldehyde) in 1×PBS at 4 °C (*see* **Note 13**).

   For Fluorescence-based secondary antibodies

   (a) Wash embryos in 1×PBS 3× 20 min.

(b) Wash in 1×PBS-DAPI for 30 min—3+h at 4 °C.

(c) Embryos can be cleared for better fluorescent imaging by equilibrating in an increasing series of Glycerol: 1×PBS at 4 °C in dark (on rocker, until embryos sink).

(d) Long-term storage without significant loss of fluorescence can be attained by transferring embryos into absolute methanol and keeping at 4 °C under refrigeration.

3.6 In Vitro Organ Culture

1. MilliCell filters are hung individually in wells of the 6-Well tissue culture plates.
2. 2 mL of culture/growth media is added to the well, which should reach but not submerge the filter, thus maintaining any tissue or organ placed on the filter at the liquid-air interface (Fig. 5a).
3. Incubate at 26–28 °C for desired time.
4. Change media every 2 days (*see* **Note 14**).

3.7 In Vitro Whole Embryo Culture

1. Embryos ranging from gastrula through pre limb bud stages that remain flat, can be cultured on a MilliCell filter as described above (in or out of their amnion) (Fig. 5b).
2. Older embryos, however, can be cultured in a semi-floating state in growth media within the well to support the embryo and its membranes (with bottom in contact with the well floor and the top in touch with the surface, not freely floating) (Fig. 5c).

3.8 Whole Embryo Nile Blue Sulfate (NBS) Staining

1. Place embryos in 20–30× their volume of 1×PBS or culture media containing NBS for approximately 40–60 min or until the desired degree of staining is achieved.
2. Wash in 1×PBS 2× 10 min to remove excess NBS.
3. Photograph (*see* **Note 15**).

4 Notes

1. DEPC-treated water may have adverse effects during in vitro organ or embryo culture, thus ultrapure autoclaved water should be substituted for such experiments.
2. Dissolving PBS tablets in water may take up to 30 min with agitation at RT.
3. Make post-fix fresh as needed.
4. Squamate reptile embryos are more tolerant to hypoxic conditions compared to mammals (with specific Oxygen and Carbon Dioxide levels). We have successfully cultured whole embryos and organs/tissues at 26–28 °C under normal atmospheric conditions.

5. Nile Blue Sulfate (NBS) concentration in solution decreases as embryonic tissues take up the NBS granules into regions of programmed cell death. A solution change (with preheated NBS if culturing in an incubator) may facilitate stronger and more complete staining in larger embryos.
6. Squamate embryos develop relatively dense ectoderm/skin and we have found that extending the duration of ethanol/methanol washes, as well as increasing the number of washes aids in fully dehydrating and rehydrating embryos (relative to mouse and chick), especially at later embryonic stages.
7. Preserving embryos in Ethanol is another option as it makes the transition to paraffin sectioning straightforward without a Methanol to Ethanol step. Rehydration of embryos at this stage makes them "sticky" and embryos begin to stick to sides of plastic pipettes and plastic tubes. It is advised to use glass vials and glass pipettes at these stages. If not, be cautious of embryos sticking to the inside of pipette and tube walls.
8. Generally, if the target tissue is within or close to the ectoderm/skin surface (such as the Apical Ectodermal Ridge or Placodes) shorter digestion times with Proteinase K treatment are advised (~5–10 min). However, we have performed digestion times as long as 40–60 min when examining gene expression in the somites of snake embryos.
9. *See* [12]. This is a crucial step to bleach pigments, quench autofluorescence, remove endogenous peroxidase activity, and promote antibody penetration. During this step, there will be gas production, which in a closed tube can lead to pressure buildup and release upon opening, such that the contents may be ejected and lost. We place embryos/tissues in Dent's Bleach in open Eppendorf tubes (2 mL) or loosely capped 20 mL glass vials in the dark (generally in the fume hood).
10. As above, embryos become sticky during rehydration.
11. Stickiness is lost instantly upon washing with TNB.
12. Revelation times range from 6 to 15 min (*see* Fig. 4d for MF20 revelation using DAB at the limb bud stage) using the DAB Substrate Kit.
13. Upon DAB revelation, the embryos may become very dark with excess DAB buildup in nonspecific locations (no primary or secondary-HRP antibody binding). Placing embryos in 4% PFA in 1×PBS facilitates long-term storage of embryos long term in a refrigerator, but after 2–3 days we see loss (diffusion) of excess DAB stain, which leaves the specific regions stained. Thus, in cases of overstaining, allow excess DAB to diffuse out of the embryos before photographing.

14. Add water to any empty wells to maintain high humidity during culturing.
15. NBS diffuses out within a few hours, so photographing must be done within the hour (with 1×PBS washes allowing for diffusion of less specific staining before imaging). Embryos may subsequently be fixed and stored in Methanol as described above to be used for downstream in situ hybridization or immunohistochemistry processing.

Acknowledgments

The Diaz Lab is supported by the College of Arts and Sciences and the Department of Biology at La Sierra University (LSU). RED would like to thank Ryan Carney for helpful discussions during manuscript preparation and the staff at the Reptile & Aquatics Facility at SIMR as well as Nayab Riasat and Antoine Hanna at the LSU Reptile Facility for help with animal husbandry. F.B. is a Ramon Y Cajal Fellow at the Institute of Biomedicine and Biotechnology of Cantabria (IBBTEC), Santander, Spain. P.A.T. is supported by the Stowers Institute for Medical Research.

References

1. Hedges SB (2012) Amniote phylogeny and the position of turtles. BMC Biol 10:64. doi:10.1186/1741-7007-10-64
2. Lee MS (2013) Turtle origins: insights from phylogenetic retrofitting and molecular scaffolds. J Evol Biol 26(12):2729–2738. doi:10.1111/jeb.12268
3. Modesto SP, Anderson JS (2004) The phylogenetic definition of reptilia. Syst Biol 53(5):815–821. doi:10.1080/10635150490503026
4. Gauthier J, Kluge AG, Rowe T (1988) Amniote phylogeny and the importance of fossils. Cladistics 4(2):105–209
5. Diaz RE Jr, Anderson CV, Baumann DP, Kupronis R, Jewell D, Piraquive C, Kupronis J, Winter K, Bertocchini F, Trainor PA (2015) The veiled chameleon (*Chamaeleo calyptratus* Dumeril and Dumeril 1851): a model for studying reptile body plan development and evolution. Cold Spring Harb Protoc 2015(10):889–894. doi:10.1101/pdb.emo087700
6. Diaz RE Jr, Trainor PA (2015) Hand/foot splitting and the 're-evolution' of mesopodial skeletal elements during the evolution and radiation of chameleons. BMC Evol Biol 15:184. doi:10.1186/s12862-015-0464-4
7. Stower MJ, Diaz RE, Fernandez LC, Crother MW, Crother B, Marco A, Trainor PA, Srinivas S, Bertocchini F (2015) Bi-modal strategy of gastrulation in reptiles. Dev Dyn. doi:10.1002/dvdy.24300
8. Andrews RM (2004) Patterns of embryonic development. In: Deeming DC (ed) Reptilian incubation: environment, evolution, and behaviour. Nottingham University Press, Nottingham, UK, pp 75–102
9. Pasteels JJ (1957) La formation de l'amnios chez les Cameleons. Ann Soc Zool Belg 87:243–246
10. Stern CD (1998) Detection of multiple gene products simultaneously by in situ hybridization and immunohistochemistry in whole mounts of avian embryos. Curr Top Dev Biol 36:223–243
11. Sandell LL, Butler Tjaden NE, Barlow AJ, Trainor PA (2014) Cochleovestibular nerve development is integrated with migratory neural crest cells. Dev Biol 385(2):200–210. doi:10.1016/j.ydbio.2013.11.009
12. Dent JA, Polson AG, Klymkowsky MW (1989) A whole-mount immunocytochemical analysis of the expression of the intermediate filament protein vimentin in Xenopus. Development 105(1):61–74

Chapter 19

Model Clades Versus Model Species: *Anolis* Lizards as an Integrative Model of Anatomical Evolution

Thomas J. Sanger and Bonnie K. Kircher

Abstract

Anolis lizards, known for their replicated patterns of morphological diversification, are widely studied in the fields of evolution and ecology. As a textbook example of adaptive radiation, this genus has supported decades of intense study in natural history, behavior, morphological evolution, and systematics. Following the publication of the *A. carolinensis* genome, research on *Anolis* lizards has expanded into new areas, toward obtaining an understanding the developmental and genetic bases of anole diversity. Here, we discuss recent progress in these areas and the burgeoning methodological toolkit that has been used to elucidate the genetic mechanisms underlying anatomical variation in this group. We also highlight the growing number of studies that have used *A. carolinensis* as the representative squamate in large-scale comparison of amniote evolution and development. Finally, we address one of the largest technical challenges biologists are facing in making *Anolis* a model for integrative studies of ecology, evolution, development, and genetics, the development of ex-ovo culturing techniques that have broad utility. Ultimately, with the power to ask questions across all biological scales in this diverse genus full, anoles are rapidly becoming a uniquely integrative and powerful biological system.

Key words Eco-evo-devo, Evolution, Macroevolution, Dimorphism

1 Introduction

Deep in El Yunque, the expansive Puerto Rican rainforest, a green lizard sits perched on a broad, green leaf around eye level to a human. The lizard's body proportions, such as its limb length, dimensions of its adhesive toe pads, and tail length, have become optimized over the course of its evolutionary history for living in this particular part of the forest canopy. Higher up, another lizard species four times the first lizard's size moves slowly through the branches searching for fruits and insects to consume, its body proportions optimized for this part of the arboreal canopy. Another species occupies the base of the tree, another the small twigs and vines, and another still is found within the grasses along the forest's edge. In each case the species' body proportions are distinct from

Guojun Sheng (ed.), *Avian and Reptilian Developmental Biology: Methods and Protocols*, Methods in Molecular Biology, vol. 1650, DOI 10.1007/978-1-4939-7216-6_19, © Springer Science+Business Media LLC 2017

those living in other parts of the forest, each well adapted to its unique place of the habitat (Fig. 1). Similar scenarios to this play out across the islands of the Greater Antilles: Jamaica, Hispaniola, Cuba in addition to Puerto Rico. Strikingly, all of the lizards in the scene described above are the members of the genus *Anolis*, or anoles as they are commonly referred. Examination of the evolutionary history and ecology of the lizards on each island and comparisons of the species between islands have fascinated biologists for decades [1]. But, more recently, a new community of researchers has also began investigating the developmental, genetic, and genomic bases of anole diversity. Rather than a single anole species becoming molded into another clichéd "model species," this community is using anoles as a model clade, capable of testing evolutionary hypotheses in a rigorous comparative phylogenetic framework. The community has developed the ability to move across evolutionary scales, from comparisons within species, to comparisons among sister groups on different islands, to comparisons among distantly related clades. Below we highlight the way that this phylogenetic approach has advanced the study of anatomical evolution and several of the technical challenges that our community needs to overcome to continue moving forward with the development of *Anolis* as an integrative model clade of morphological evolution.

Fig. 1 (**a**) A schematic of the Puerto Rican anole community. Note the differences in body size and the relative position of species across the forest canopy. Tree modified from Losos [1]. (**b**) Representatives of the Puerto Rican habitat specialists: (*1*) *A. cuvieri*, crown-giant; (*2*) *A. evermanni*, trunk-crown; (*3*) *A. occultus,* twig; (*4*) *A. gundlachi*, trunk-ground; (*5*) *A. krugi*, grass-bush. Habitat specialists are named for the microhabitat most often inhabitiated by the species

1.1 Anolis Lizard Diversity and Evolution

Anolis is a group of approximately 400 species distributed throughout the Caribbean, Central and South America, and the southeastern United States, but it is the anoles of the Greater Antilles that have received the most attention by biologists [1]. On each of the four islands a suite of distinct habitat specialists independently evolved by dividing the arboreal habitat based on perch height and perch width. The near perfect convergence in body proportion, color, body size, and behavior among species living in similar microhabitats on different islands is remarkable. Phylogenetic analysis of these species has repeatedly confirmed that this similarity is due to convergence and not common ancestry [2–4]. The evolutionary history of this group has, therefore, created a natural experiment with which to test the predictability of evolution at different biological levels.

Many fields in biology are driven forward by new technologies that can advance research areas or open up entirely new areas of investigation. The creation of these resources often requires great financial, intellectual, and personnel investment in a single species, such as *Danio rerio* "the" zebrafish, *Mus musculus* "the" mouse, or *Drosophilla melanogaster* "the" fly. Although other vertebrate lineages have well-established model organisms, squamates (i.e., lizards and snakes) have never had a representative species become a model for developmental or genetic studies (although there was a wealth of descriptive embryology performed during the mid-twentieth century [5–7]). Following the publication of the *A. carolinensis*, the green anole, genome [4], it can be argued that this species became the first squamate species with the potential of being developed for experimental embryology and functional genomics (i.e., knock-down or overexpression studies). The genome sequence opened up the possibility for investigators to more readily clone genes for expression analysis [8, 9], analyze coding and noncoding DNA sequences [8, 10, 11], and provide the needed out-group to polarize genomic comparisons across amniotes [4, 8, 12, 13]. A number of trait-specific descriptive developmental studies [14–16] of the green anole and its relative *A. sagrei*, the brown anole, have also followed since this the publication. With the advent of new tissue and ex-ovo embryo culturing techniques [8, 17, 18] (*see* below), the anole community is poised to make the next step toward establishing the ability to functionally test gene function, not just examine gene expression.

Following from the previous decades of research, one of the greatest intellectual strengths of *Anolis* lizard is providing a framework for testing evolutionary hypotheses at different phylogenetic scales: within species, among closely related anole species, and as a representative squamate for comparisons among distantly related amniotes. We have organized our discussions to emphasize this evolutionary hierarchy.

1.2 Variation Within Species: The Evolution of Sexual Dimorphism

One of the most striking features of anoles is the variable level of sexual dimorphism observed throughout the genus. In many species males are larger in body size, possess relatively longer limbs and larger head dimensions, and have an extensible, colorful throat fan called a dewlap that is used in communication [19–22]. In other species males and females are nearly indistinguishable without examining the genitalia. Thus, one of the discrete strengths of using the genus *Anolis* to examine the developmental bases of sexual dimorphism is the ability to drill deep into developmental bases of dimorphism in the green anole model and then compare these findings among closely related species with variable levels of dimorphism (a similar approach has also been applied to studies of neuromuscular reproductive physiology [23, 24]).

A general belief that transcends biological disciplines is that male-biased dimorphism—whether in size, shape, color, or behavior—is associated with sexual differences in testosterone. Within anoles there appears to be a correlation between male body size and levels of circulating testosterone [25]. Testosterone stimulates male growth and, in turn, increases the degree of dimorphism observed. Conversely, in squamate species with no dimorphism or female-biased dimorphism, testosterone may inhibit male growth allowing for the exaggeration of female size. The flexible role of testosterone on organismal growth has been termed the "bi-potential regulation hypothesis" [25].

The developmental basis of tissue-specific shape dimorphism in anoles does not readily fit within this paradigm. While males and females of many anoles species have subtle differences in facial length, two lineages have reached extreme levels of facial length dimorphism, surpassing two standard deviations from the average value [22]. One of these lineages reaches extreme levels of dimorphism through modification to a widespread, likely ancestral, mechanism whereby male and females diverge early in growth and maintain those differences throughout life (Fig. 2). The other lineage has evolved a novel mechanism not observed elsewhere in the genus. In this novel strategy, males and females do not diverge until sexual maturity and continue to diverge throughout their reproductive life. Rather than changes in androgen signaling, this novel growth mechanism appears to be associated with changes in estrogen signaling, particularly at the level of estrogen receptor beta [9]. There were no differences in the expression of androgen receptor or its complimentary molecules in the diverging facial tissues. Ultimately, use of the comparative developmental analyses of dimorphic characters in anoles has the potential to add greater resolution to the mechanisms by which mosaic patterns of dimorphism arise and evolve. Examination of males and females of a single species or of a single sex among many species will never offer the same explanatory power.

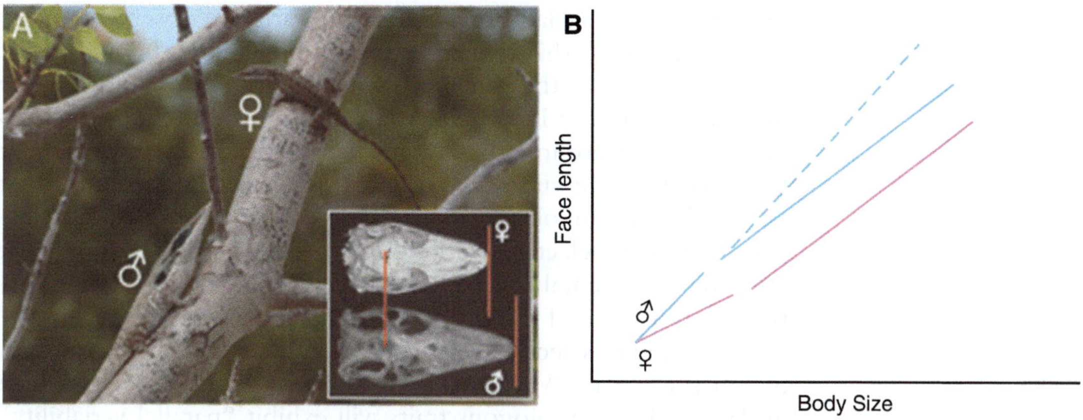

Fig. 2 (**a**) Male *A. brunneus* have much longer faces than their female conspecifics. (**b**) Differences in face length arise early in development in many anole species (*solid lines*). However, extreme levels of facial length dimorphism in the *carolinensis* clade are reached by a novel mechanism observed only in this lineage, late divergence of males and females (*dashed line*)

1.3 Variation Among Anolis Species: The Developmental Bases of Anole Diversity

One of the objectives of twenty-first century evolutionary biology is to understand the evolutionary processes that coordinately shape genomic, developmental, and phenotypic variation [26]. The phenotypic convergence of lizards living in similar habitats across the Greater Antilles provides a powerful model to assess the relative predictability and contingency of evolution at different levels of the biological hierarchy (tissues, cells, gene expression, nucleotide, etc.). Investigators have yet to uncover the genes responsible for the anatomical diversification of *Anolis* lizards, presumably because of the lack of molecular technologies and because distantly related species are not readily amenable to genetic mapping crosses. Despite those caveats, progress has been made toward understanding the developmental *processes* contributing to morphological divergence and convergence for several traits.

Relative limb length (limb length proportional to body size) is one of the most important morphological traits for conferring anole species their habitat-specific performance capabilities [1]. Despite the many ways limb length variation could be generated—through patterning, differential pre-hatching growth, or differential post-hatching growth—Sanger et al. [27] found that the divergence in limb length on each island consistently occurs through developmental modification to limb bud patterning, before the formation of the cartilaginous anlagen. This deep conservation of the processes generating evolutionarily relevant variation is a similar pattern as described for the widespread, potentially ancestral mechanism that underlies craniofacial dimorphism in *Anolis* [22].

Without formal phylogenetic analysis one might conclude that the same developmental processes have been independently recruited on each island to generate the variation observed among anoles. Although this remains a formal possibility, the alternative hypothesis is that the same generative mechanisms underlie the production of limb length variation across the radiation of anoles because they were inherited from their common ancestor. Under this scenario, the successive speciation events never erased that ancestral signature of *variation*, the actual fodder of natural selection. Repeated selection on limb length, therefore, recruited that same developmental mechanisms because of this ancestral signature, not because of those developmental processes offered something unique to selection relative to its other options. This pattern is consistent with Vavilov's Law of Homologous Series, which explained that homologous traits will exhibit "parallel variability" among closely related species [28]. In other words, closely related species tend to vary along similar dimensions not by chance, but because of their shared history. The challenge for the next round of biologists to investigate these traits will be to accurately relate changes at the genotypic level to the previously described patterns at the developmental and phenotypic levels. These researchers must sample deeply enough to tease apart the effects of lineage-specific mutations, molecular convergence across islands, and latent genetic polymorphisms inherited from their common ancestor.

1.4 Variation Among Amniotes: Anole as a Representative Squamate

Anolis lizards have provided an important benchmark for researchers interested in the evolution of amniotes. For later part of the twentieth century researchers drew conclusions about the developmental evolution of amniotes based on comparisons between the chick and mouse model systems, not allowing for objective determination of the amniote ancestral condition. As the utility of anoles has grown, they have been used to polarize comparisons among distantly related amniote lineages in studies of heart development [14, 29], tail regeneration [15, 30, 31], longitudinal body axis formation [12, 13], and external genital (i.e., phallus) development [8, 10, 32]. Reviewing the findings across this diversity in structures is beyond the scope of this chapter, but we briefly highlight the significance of anoles in our understanding of external genital evolution because of the collaboration of one author (TJS) with several studies in this area.

Among amniotes there is a tremendous degree of variation in the adult anatomy of external genitalia, which has confounded the accurate interpretation of its evolutionary history [32, 33]. Whether a phallus evolved once at the origin of amniotes or several times independently in distinct amniote lineages remained an open question. The answer to this lingering question came not through technical advances, but through detailed comparative embryology. By close examination of squamate (including *Anolis*), alligator, avian, and mammalian phallus embryology, it was found that in spite of the confounding variation in adult anatomy, all amniote phalluses have a common embryological origin as paired swellings flanking the cloaca [16, 34–38]. Even species that have secondarily lost their phallus maintain this embryological signature [33, 38], suggesting that the phallus evolved once and was later modified within each of the amniote radiations.

The developing limb buds and genital tubercle, the developmental precursor to the phallus, express many of the same genes [8, 39]. Recent developmental studies comparing gene expression and gene regulation across amniotes have raised questions about whether these shared expression profiles arise because of a shared embryological (i.e., cellular) origin of limbs and external genitalia or because the limbs and phallus share a common cis-regulatory landscape associated with appendage development [8, 10]. Use of the green anole and its genome have also been critical for polarizing the comparisons of transcriptomic and regulatory profiles between the limbs and external genitalia among evolutionarily distant lineages of birds, mammals, and reptiles.

Since the publication of the green anole genome several other squamate genomes have been completed, although they are not all annotated at this time [40–43]. These additional sequences will improve our power to dissect amniote genome evolution and to more precisely determine the conservation and flexibility of gene regulatory elements. Evolutionary-developmental biology has, however, moved beyond correlative studies at the sequence and expression levels. The field has emphasized the need to experimentally validate those differences. Reaching this level of experimental rigor is likely the largest hurdle facing the anole community in coming years.

1.5 Raising the Bar: Can Functional Genomics Be Performed on Anolis Lizards?

Gene function can evolve. Genes can also have multiple functions in one organ or distinct functions in different organs in the same species. Therefore, differential expression analysis is not adequate as the only source of information to define the underlying causes of anatomical evolution. Likewise, sequence analysis cannot alone determine when and where genes are expressed as spatiotemporal differences in gene expression are often due to changes in gene regulation, not changes in protein-coding regions [44]. One of the hallmarks of the most successful studies in evolutionary-developmental biology has been the functional validation of expression differences.

Functional validation can be performed using a range of experimental methods. In genetic systems such as *Mus*, *Danio*, and *Drosophila* functional analysis has been done with true knock-out and knock-in experiments where segments of the genome are removed, added, or edited. The opportunities for these experiments have significantly expanded with the recent advances in CRISPR-Cas9 genome-editing technology in both model and non-model species [45, 46]. In species that are not amendable to genetics, knock-down and up-regulation experiments are possible using pathway-specific small molecule inhibitors, electroporation of expression vectors [17, 47], or through the implantation of protein-soaked beads. In each of these cases, however, the critical hurdle to overcome is access to the developing embryo. In chicks, for example, the egg can be windowed and the embryo can be cultured in its natural shell until hatching. Squamates possess leathery eggs with high turgor pressure limiting our ability to window the egg without damaging the embryo. Although the utility of these experimental techniques could be extended to anole embryos in principle, the anole community's first hurdle will be to find reliable culturing techniques that allow long-term access to embryonic tissues.

Several published attempts have been made regarding culturing protocols for anole embryos and their cells (with varying degrees of experimental detail). Transfection of expression constructs into micromass cultures may prove to be a powerful tool to examine the conservation gene regulatory networks in anoles [17]. Park et al. [17] described micromass culturing for cells derived from early limb buds. For these experiments cells from the limb buds were disaggregated using trypsin, concentrated by centrifugation, and plated in a DMEM/F12 culture media. Cultured cells were incubated for up to 24 days in 5% atmospheric carbon dioxide conditions. After 2 weeks in culture differentiation of cartilage nodules were readily visible, which was slower than chicken cultures raised under the same conditions. These cultures were successfully electroporated with expression vectors driving GFP under the control of the CMV promoter and an experimental regulatory element of Pitx1. Together these results show great promise toward testing whether tissue-specific gene regulatory networks known in other more widely studied taxa are conserved in *Anolis* lizards.

Micromass cultures are ideal for testing regulatory and expression hypotheses, but cannot readily advance our understanding of anatomical morphogenesis where maintaining three-dimensional context is critical. Nomura et al. [47] described a protocol for windowing Madagascar ground gecko, *Paroedura pictus*, eggs for electroporation, but this has not been successful with *Anolis* eggs that are much smaller and under greater turgor pressure. Tschopp et al. [8] infected green anole embryos at the early stages of morphogenesis with a GFP cell-lineage tracing lentivirus using an ex-ovo culturing technique (Fig. 3a). In this experiment, partially shelled eggs were dissected from gravid females. The opaque layers of the shells were dissected away, leaving the internal membranes intact. The yolk mass and embryo were then transferred to an egg-shaped impression in dish of Nobel agar dissolved in media. The cultures were maintained in a humidified chamber at 28 °C for up to 12 days. The number of embryos that survived this incubation period was not reported, but our attempts to replicate this length of

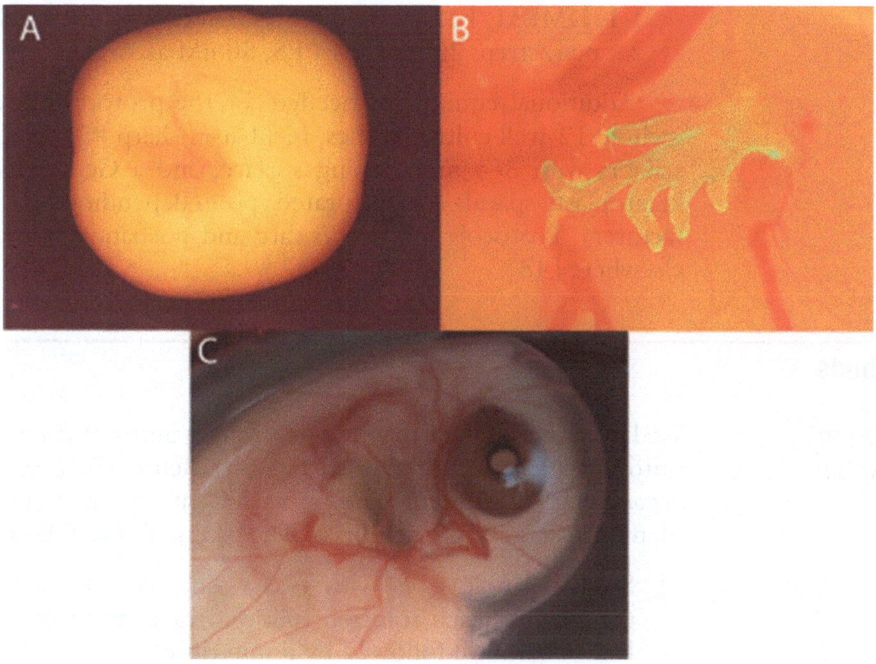

Fig. 3 (**a**) Early stage *A. sagrei* ex-ovo culture following the protocol of Tschopp et al. [8] alive after 24 h incubation. (**b**) Stage 12 *A. sagrei* limb stained with Vybrant cell labeling solution explanted in a chicken embryo after 24 h. (**c**) *A. sagrei* embryo at the air-liquid interface, incubated following the ex ovo protocol by Diaz and Trainer [18]. Embryo was incubated 12 days prior to shell removal and incubated for 24 h. Note the fully intact membranes surrounding the embryo

time have not been successful. We would also like to develop more sustainable protocols that will not require the euthanasia of gravid females. However, the use of this ubiquitously expressed GFP lentivirus is the first proof-of-concept that viral transgenesis may be a viable mechanism to manipulate the genome of anole embryos.

2 Materials

We have tested two additional ex-ovo culturing techniques that we feel have potential for advancing the experimental repertoire of anole biologists. The reagents required for the techniques described below are as follows.

- 5% Clorox bleach.
- 1× Phosphate-buffered saline (PBS).
- 2× Phosphate-buffered saline (PBS).
- Vybrant cell labeling solution (Life Technologies).
- Fertilized chicken eggs (UConn Poultry Farm, CT or Sunnyside Farms, WI).

- 1:1 DMEM/F12 (Life Technologies) with 10% calf serum, 2× pen/strep, 20 mM HEPES, 50 μM ascorbic acid.

Additional equipment needed for the protocols listed below include 12-well culture dishes, incubator, sharp #5 forceps, spring scissors with 3–4 mm cutting surface, and a Geuder perforated keratoplasty spatula (or perforated spoon depending on the manufacturer). Protocols for *Anolis* care and husbandry are described elsewhere [48].

3 Methods

3.1 Embryonic Tissue Explant

We have successfully explanted developing limbs of *Anolis* embryos onto the chicken embryo host (Fig. 3b). Before you begin with the *Anolis* egg dissection, window a chicken egg with Hamburger Hamilton stage 17–20 embryo. The protocol is as follows:

1. Sterilize anole eggs with two washes of 5% Clorox bleach (5 min in bleach solution followed by 5 min rinse, 5 min in bleach solution followed by 5 min rinse).
2. Dissect the embryo from its shell and extra-embryonic membranes while submerged in sterile PBS using #5 forceps and spring scissors. Transfer embryo to clean PBS using the spatula.
3. *Optional:* Because anole limbs are significantly smaller than chicken embryos and may be difficult to visualize, stain the anole tissue with Vybrant cell labeling solution. Dilute Vybrant following manufacturer's instructions. Incubate whole embryo for 30 min at 30 °C while rocking. Rinse twice with PBS.
4. Wound the blood vessels near the posterior of the chicken embryo using forceps or needle. Transfer the stained explant to the wounded area. The natural healing processes of the embryonic tissues lead to the establishment of a blood supply toward the anole tissue within 24 h when incubated at 30 °C (intermediate between the standard incubation temperatures of anole and chicken embryos).

3.2 Ex-Ovo Embryo Culture

We have also successfully cultured whole brown anole embryos with minor modification of Diaz and Trainor's [18] protocol used for chameleon embryos (Fig. 3c).

1. Sterilize in two washes of 5% Clorox bleach as above.
2. Dissect the opaque outer shell away from the yolk mass and embryo while submerged in sterile PBS using #5 forceps and spring scissors (*see* **Note 1**).
3. Transfer the embryo and yolk mass to a 12-well dish with DMEM/F12 growth media. Position the embryo at the air-liquid interface.

4. Incubate in normal air at 28 °C. Perform a full media change daily. We have observed a steady heartbeat for up to 10 days of incubation (*see* **Notes 2** and **3**).

4 Notes

1. From our experience, the largest hurdle for successful ex-ovo culturing protocols appears to be dissecting the opaque shell away from the yolk mass without damaging the underlying blood vessels. The success of this procedure can be very stage-specific. The inner and outer membranes appear to be most easily dissected before oviposition and after 5–7 days of incubation. At oviposition the membranes are partially adhered to one another making removal of the outer shell extremely difficult. Soaking the eggs for 2 min in 2× PBS may help to separate the inner and outer membranes, but is not always necessary.
2. The embryos readily move within their membranes. The embryos appear hardy when remaining enclosed within their membranes, but we have observed membranes rupture several times.
3. The embryos develop at a pace slightly slower than a normal.

Acknowledgments

We would like to thank P. Tschopp, R. Diaz, and M. Cohn for valuable discussion on the culturing protocols discussed herein. R. Dale supplied us with the bleaching protocol based on his work on zebrafish. This chapter is supported by laboratory start-up funds from Loyola University in Chicago to T.J.S. and an NSF Graduate Research Fellowship to B.K.K.

References

1. Losos J (2009) Lizards in an evolutionary tree. University of California Press, Berkeley, CA
2. Losos J, Jackman T, Larson A et al (1998) Contingency and determinism in replicated adaptive radiations of island lizards. Science 279:2115–2118
3. Mahler D, Revell L, Glor R, Losos J (2010) Ecological opportunity and the rate of morphological evolution in the diversification of Greater Antillean Anoles. Evolution 64:2731–2745
4. Alföldi J, Di Palma F, Grabherr M et al (2011) The genome of the green anole lizard and a comparative analysis with birds and mammals. Nature 477:587–591
5. Gans C, Billet F, Maderson P (1985) Biology of the reptilia, vol 14. John Wiley & Sons, New York
6. Gans C, Billet F (1985) Biology of the Reptilia, vol 15. John Wiley & Sons, New York
7. Sanger T, Losos J, Gibson-Brown J (2008) A developmental staging series for the lizard genus Anolis: a new system for the integration of evolution, development, and ecology. J Morphol 269:129–137
8. Tschopp P, Sherratt E, Sanger TJ et al (2014) A relative shift in cloacal location repositions

external genitalia in amniote evolution. Nature 516:391–394

9. Sanger T, Seav S, Tokita M et al (2014) The oestrogen pathway underlies the evolution of exaggerated male cranial shapes in Anolis lizards. Proc Biol Sci 281:20140329
10. Infante C, Mihala A, Park S et al (2015) Shared enhancer activity in the limbs and phallus and functional divergence of a limb-genital cis-regulatory element in snakes. Dev Cell 35:107–119
11. Gamble T, Geneva A, Glor R, Zarkower D (2014) Anolis sex chromosomes are derived from a single ancestral pair. Evolution 68:1027–1041
12. Eckalbar W, Lasku E, Infante C et al (2012) Somitogenesis in the anole lizard and alligator reveals evolutionary convergence and divergence in the amniote segmentation clock. Dev Biol 363:308–319
13. Kusumi K, May C, Eckalbar W (2013) A large-scale view of the evolution of amniote development: insights from somitogenesis in reptiles. Curr Opin Genet Dev 23:491–497
14. Koshiba-Takeuchi K, Mori A, Kaynak B et al (2009) Reptilian heart development and the molecular basis of cardiac chamber evolution. Nature 461:95–98
15. Ritzman T, Stroik L, Julik E et al (2012) The gross anatomy of the original and regenerated tail in the green anole (*Anolis carolinensis*). Anat Rec 295:1596–1608
16. Gredler M, Sanger T, Cohn M (2015) Development of the cloaca, hemipenes, and hemiclitores in the green anole, *Anolis carolinensis*. Sex Dev 9:21–33
17. Park S, Infante C, Rivera-Davila L, Menke D (2014) Conserved regulation of hoxc11 by pitx1 in Anolis lizards. J Exp Zool B Mol Dev Evol 322:156–165
18. Diaz R, Trainor P (2015) Hand/foot splitting and the "re-evolution" of mesopodial skeletal elements during the evolution and radiation of chameleons. BMC Evol Biol 15:184
19. Butler MA, Sawyer SA, Losos JB (2007) Sexual dimorphism and adaptive radiation in Anolis lizards. Nature 447:202–205
20. Butler M, Losos J (2002) Multivariate sexual dimorphism, sexual selection, and adaptation in Greater Antillean Anolis lizards. Ecol Monogr 72:541–559
21. Butler M, Schoener T, Losos J (2000) The relationship between sexual size dimorphism and habitat use in Greater Antillean Anolis lizards. Evolution 54:259–272
22. Sanger T, Sherratt E, McGlothlin J et al (2013) Convergent evolution of sexual dimorphism in skull shape using distinct developmental strategies. Evolution 67:2180–2193
23. Johnson M, Cohen R, Vandecar J, Wade J (2011) Relationships among reproductive morphology, behavior, and testosterone in a natural population of green anole lizards. Physiol Behav 104:437–445
24. Johnson M, Wade J (2010) Behavioural display systems across nine Anolis lizard species: sexual dimorphisms in structure and function. Proc Biol Sci 277:1711–1719
25. Cox R, Stenquist D, Calsbeek R (2009) Testosterone, growth and the evolution of sexual size dimorphism. J Evol Biol 22:1586–1598
26. Losos J, Arnold S, Bejerano G et al (2013) Evolutionary biology for the 21st century. PLoS Biol 11:e1001466
27. Sanger T, Revell L, Gibson-Brown J, Losos J (2012) Repeated modification of early limb morphogenesis programmes underlies the convergence of relative limb length in Anolis lizards. Proc R Soc B Biol Sci 279:739–748
28. Vavilov V (1922) The law of homologous series in variation. J Genet 12:47–89
29. Jensen B, van den Berg G, van den Doel R et al (2013) Development of the hearts of lizards and snakes and perspectives to cardiac evolution. PLoS One 8:e63651
30. Hutchins E, Markov G, Eckalbar W et al (2014) Transcriptomic analysis of tail regeneration in the lizard *Anolis carolinensis* reveals activation of conserved vertebrate developmental and repair mechanisms. PLoS One 9: e105004
31. Hutchins E, Eckalbar W, Wolter J et al (2016) Differential expression of conserved and novel microRNAs during tail regeneration in the lizard *Anolis carolinensis*. BMC Genomics 17:339
32. Gredler M, Larkins C, Leal F et al (2014) Evolution of external genitalia: insights from reptilian development. Sex Dev 8:311–326
33. Sanger T, Gredler M, Cohn M (2015) Resurrecting embryos of the tuatara, *Sphenodon punctatus*, to resolve vertebrate phallus evolution. Biol Lett 11:20150694
34. Perriton C, Powles N, Chiang C et al (2002) Sonic hedgehog signaling from the urethral epithelium controls external genital development. Dev Biol 247:26–46
35. Gredler M, Seifert A, Cohn M (2015) Morphogenesis and patterning of the phallus and cloaca in the american alligator, *Alligator mississippiensis*. Sex Dev 9:53–67
36. Leal F, Cohn M (2015) Development of hemipenes in the ball python snake *Python regius*. Sex Dev 9:6–20

37. Larkins C, Cohn M (2015) Phallus development in the turtle *Trachemys scripta*. Sex Dev 9:34–42
38. Herrera A, Shuster S, Perriton C, Cohn M (2013) Developmental basis of phallus reduction during bird evolution. Curr Biol 23:1065–1074
39. Cohn M (2011) Development of the external genitalia: conserved and divergent mechanisms of appendage patterning. Dev Dyn 240:1108–1115
40. Vonk F, Casewell N, Henkel C et al (2013) The king cobra genome reveals dynamic gene evolution and adaptation in the snake venom system. Proc Natl Acad Sci U S A 110:20651–20656
41. Tollis M, Hutchins E, Kusumi K (2014) Reptile genomes open the frontier for comparative analysis of amniote development and regeneration. Int J Dev Biol 58:863–871
42. Liu Y, Zhou Q, Wang Y et al (2015) *Gekko japonicus* genome reveals evolution of adhesive toe pads and tail regeneration. Nat Commun 6:10033
43. Georges A, Li Q, Lian J et al (2015) High-coverage sequencing and annotated assembly of the genome of the Australian dragon lizard Pogona vitticeps. Gigascience 4:45
44. Carroll S (2008) Evo-devo and an expanding evolutionary synthesis: a genetic theory of morphological evolution. Cell 134:25–36
45. Gilles A, Averof M (2014) Functional genetics for all: engineered nucleases, CRISPR and the gene editing revolution. EvoDevo 5:43
46. Bassett A, Tibbit C, Ponting C, Liu J-L (2013) Highly efficient targeted mutagenesis of Drosophila with the CRISPR/Cas9 system. Cell Rep 4:220–228
47. Nomura T, Yamashita W, Gotoh H, Ono K (2015) Genetic manipulation of reptilian embryos: toward an understanding of cortical development and evolution. Front Neurosci 9:45
48. Sanger T, Hime P, Johnson M et al (2008) Laboratory protocols for husbandry and embryo collection of Anolis lizards. Herp Rev 39:58–63

Chapter 20

The Feather Model for Chemo- and Radiation Therapy-Induced Tissue Damage

Zhicao Yue and Benhua Xu

Abstract

Chemo- and radiation therapy are the main modalities for cancer treatment. A major limiting factor is their toxicity to normal tissue, thus reducing the dose and duration of the therapy. The hair follicle, gastrointestinal tract, and hematopoietic system are among the target organs that often show side effects in cancer therapy. Although these organs are highly mitotic in common, the molecular mechanism of the damage remains unclear. The feather follicle is a fast-growing mini-organ, which allows observation and manipulation on each follicle individually. As a model system, the feather follicle is advantageous because of the following reasons: (1) its complex structure is regulated by a set of evolutionarily conserved molecular pathways, thus facilitating the effort to dissect the specific signaling events involved; (2) its morphology allows the continuity of normal–perturbed–normal structure in a single feather, thus "recording" the damaging effect of chemo- and radiation therapy; (3) further histological and molecular analysis of the damage response can be performed on each plucked feather; thus, it is not necessary to sacrifice the experimental animal. Here, we describe methods of applying the feather model to study the molecular mechanism of chemo- and radiation therapy-induced tissue damage.

Key words Feather, Morphology, Cancer therapy, Side effect, Chemotherapy, Radiation therapy, 5-Fluouracil, Cyclophosphamide, Ionizing radiation

1 Introduction

Common side effects in cancer chemo- and/or radiation therapy include hair loss (alopecia), nausea/lack of appetite, compromised immune function/reduced blood cell count, etc. These side effects are caused by therapy-induced damage to the respective actively growing tissue/organ. The molecular mechanism of the damage may involve activation of p53 signaling, cytokine production, increased reactive oxygen species level, and eventually cell apoptosis [1–5]. Cell senescence and specific damage to the stem cell compartment have also been documented [6–8]. Controversial opinions often arise from different investigations. For instance, it was once believed that by reducing the cell proliferation rate in normal tissue such as the hair follicle, the side effects might be alleviated [9].

Guojun Sheng (ed.), *Avian and Reptilian Developmental Biology: Methods and Protocols*, Methods in Molecular Biology, vol. 1650, DOI 10.1007/978-1-4939-7216-6_20, © Springer Science+Business Media LLC 2017

Perturbation of a specific pathway, the sonic hedgehog (Shh) signaling, in chemotherapy-induced tissue damage has also been proposed [10]. These studies will likely improve our understanding of the underlying pathological mechanism of the damage and help design novel methods for future clinical treatment of the side effects in cancer therapy.

To build a reasonable model for chemotherapy-induced side effects, several aspects have to be considered including metabolic differences between species, drugs of choice, dose, route and duration of the delivery, etc. For chemotherapy-induced hair loss (alopecia, CIA), the actively cycling adult hair follicle was considered more advantageous than the new born rat model, where the hair follicle was still in morphogenesis [2, 11]. A single intraperitoneal injection of a chemotherapeutic agent, cyclophosphamide (CYP), at a dose of 150 mg/kg body weight was considered optimal, which conveniently achieved complete hair loss in the mouse model [2]. Using this model, it was further shown that the damage may be regulated by genetic factors, and novel methods for pharmacological manipulation of this condition have been proposed (reviewed in 1). Similar considerations may also help build an optimal model for radiation damage. For instance, the mouse is more resistance to ionizing radiation (IR)-induced skin damage than the rat [12].

The adult feather follicle is an actively growing and naturally cycling mini-organ [13–15]. Compare to mammalian hair, the feather structure is far more complex. Recent advancement in technology has significantly improved our toolbox to analyze and manipulate the adult feather follicle. Whole-genome expression profiling based on next-generation sequencing allowed a better view of the molecular landscapes in various aspects of feather development [16, 17]. RCAS and *Lentivirus*-mediated gene overexpression and knockdown enabled molecular manipulation in the adult feather follicle [16, 18]. Based on these technological advancements, our understanding of the growth, regeneration, and pattern formation of this complex mini-organ has also been significantly improved. For instance, BMP/Noggin signaling regulates feather branching and rachis formation [18], and FGF/Sprouty signaling regulates the proximal-distal patterning of the feather follicle [19]. An anterior-posterior Wnt signaling gradient controls the bilateral symmetry of the feather follicle [20], while a Wnt/Dkk feedback loop in the dermal papilla regulates feather regeneration [16].

Recently, we developed the adult feather follicle as a model to study how chemo- and radiation therapy damage normal tissue [21]. This leads to several interesting discoveries, including the critical role of Shh signaling in chemotherapy-induced tissue damage, and the involvement of cytokine signaling in IR-induced tissue damage [10, 22]. The complex feather structure offers the opportunity to "record" the damaging effect which, when combined with developmental biology study, helps dissect the molecular details of the pathobiology. Here we describe in detail how to apply the adult feather follicle as a model for chemo-drug testing and IR-damage recording.

2 Materials

1. Adult chickens were bought from a local farm, where they were raised in a natural environment. They were usually 3–6 month old and sized 1.0–1.5 kg body weight. Chickens were housed under a 12 h diurnal condition with free access to water and food.
2. Common chemotherapeutic agents including cyclophosphamide (CYP), 5-fluorouracil (5-FU), and doxorubicin (Dox) were used. These were dissolved in PBS and intraperitoneally injected into the chicken using a 10 mL syringe. Fresh solutions were prepared every time. The concentrations used were: for CYP, 150 mg were weighted and dissolved in 10 mL PBS (15 mg/mL); for 5-FU, 100 mg were weighted and dissolved in 10 mL PBS (10 mg/mL); for Dox, 10 mg were weighted and dissolved in 5 mL PBS (2 mg/mL).
3. Nine megaelectronvolts high energy electron beam was obtained from a Varian Clinac 23 ex linear accelerator. A dose rate of 500 cGy/min was used. A series of homemade lead cover with various opening sizes (5 cm × 5 cm, 10 cm × 10 cm, 20 cm × 20 cm) were used to protect the rest of the body, depending on the size of the object to be irradiated. Chickens were anesthetized by pentobarbital (50 mg/kg) during the operation.
4. A commercial photo scanner such as UniScan B5300 from Tsinghua Ziguang was used to document the feather morphology.
5. A stereo dissection microscope such as ZSA0850 from Chongqing Optical Instrument Factory Co. was used to dissect the developing feather follicle.
6. Standard protocols were followed for histological or immunohistological analysis. Whole-mount feather branching was documented using a Nikon inverted fluorescence microscope, and histology/immunohistology images were taken under a Leica fluorescence microscope.

3 Methods

3.1 Inducing Actively Growing Feather Follicles in the Chicken

1. Fully grown feathers covering the animal body were usually in resting phase and not suitable for manipulation. Mechanical plucking was applied to induce regeneration and actively

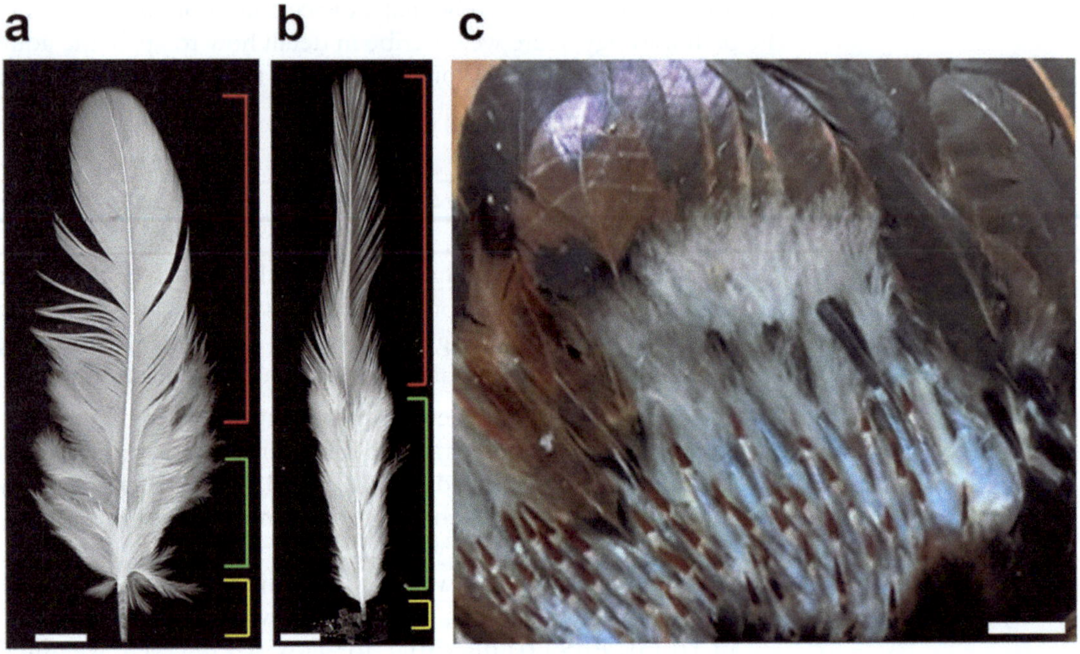

Fig. 1 The morphology of fully grown feathers and plucking-induced feather regeneration. (**a**) A fully grown wing contour feather and (**b**) a fully grown saddle feather. *Red bracket*: pennaceous vane; *green bracket*: plumulaceous vane; *yellow bracket*: calamus. (**c**) Plucking-induced feather regeneration in the wing contour region. Photographed 2 weeks after plucking. These actively growing feather follicles were ready for chemo- or radiation perturbation (adopted from Ref. 22). Bar = 1 cm

growing feather follicles. This was usually performed in the wing contour region (for irradiation) and/or saddle region (for chemo-drug testing; *see* **Note 1**).

2. After 1–2 weeks, feathers entered actively growing phase as judged from the gross morphology (Fig. 1). Twenty to forty follicles from a local region were plucked, which allows simultaneous time-serial analysis of histological and molecular events inside individual follicles, as well as document the morphological diversity of feather phenotype after perturbation (*see* **Notes 2** and **3**).

3.2 Delivering Chemotherapeutic Agents or Ionizing Radiation

1. For chemotherapeutic agents, an initial test of different doses was performed (CYP 50, 100, 150, 200 mg/kg; 5-FU 50, 100, 150 mg/kg; Dox 5, 10, 20 mg/kg). These doses are usually about five to ten times higher than those used in clinic. Drugs were dissolved in PBS, and a single intraperitoneal injection was performed. Upon analyzing the dose response, the choice of final dose for each drug was CYP 150 mg/kg, 5-FU 100 mg/kg, and Dox 10 mg/kg.
2. For ionizing radiation, the actively growing feather follicles were irradiated with the rest of the body protected by a lead

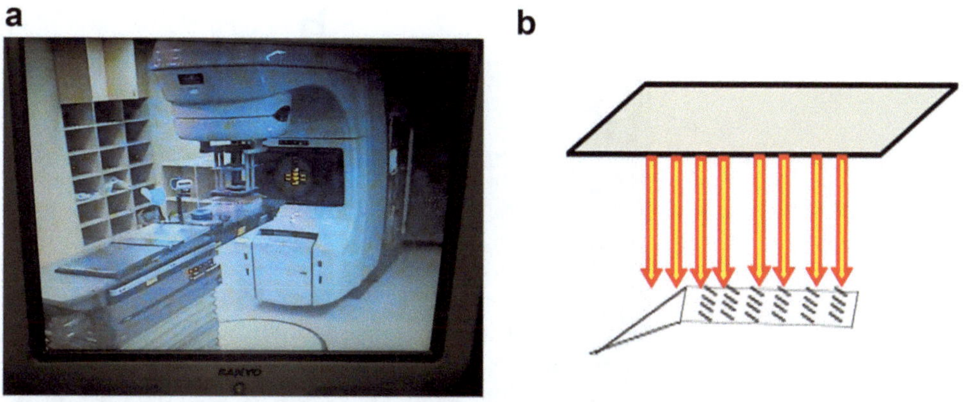

Fig. 2 Irradiate the feather follicle using a linear accelerator. (**a**) The apparatus in operation viewed from a remote monitor screen (adopted from Ref. 12). (**b**) A diagram showing the lead cover protected the rest of the body, while the wing contour region was irradiated through the opening

cover (Fig. 2). A dose rate of 500 cGy/min was used. The irradiation dose was tested in pilot experiments (5, 10, 20 Gy in a single delivery). Upon analyzing the response, the choice of final doses was 10 and 20 Gy.

3.3 Recording the Feather Phenotype

1. Unlike the mammalian hair which fall out after chemotherapy and the skin/scalp appeared bald, the feather did not fall out and finished the grow cycle normally. Thus after about 1 month, the feather finished the growth cycle and entered resting phase. The feathers were then plucked and their morphology was documented. This was achieved by using a photo scanner at 600 or 1200 dpi (Fig. 3). Alternatively, the feather morphology can also be documented using a stereo microscope or scanning electron microscope.
2. For irradiation, the feathers may fall out after high-dose exposure (20 Gy), which may occur 4 days to 1 week postirradiation. Thus, the irradiated feather follicles were followed more carefully. The feathers may shed off (which were then collected), or remain in the follicle and finished the growth cycle normally (after which they were plucked and photographed; Fig. 4, *see* **Note 4**).

3.4 Histological and Molecular Analysis of the Perturbation

1. A major advantage of the feather model is that each follicle allows morphological, histological, and molecular analysis individually. This can be achieved simply by plucking the feather and proceed for whole-mount open-prep, histology, or RNA extraction. The plucked feather contains the branching epithelium, the rachis, and the mesenchymal pulp that are the key components for feather development and morphogenesis (Fig. 5). Thus the plucked feather allows detailed investigation for the damage responses, without the necessity to sacrifice the animal.

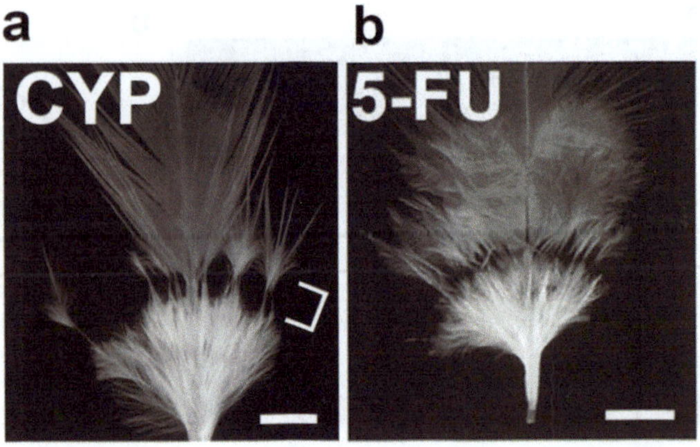

Fig. 3 Examples of chemotherapy-damaged feather. After treating the bird with chemotherapeutic agents (**a**) CYP; (**b**) 5-FU, the feathers finished the growth cycle normally (adopted from Ref. 10). An "isochrone," meaning cells along this line were born at the same time, was produced in the feather vane. The feather branches (barbs) were disrupted or broken, but the rachis appeared normal which held the feather in place and did not fall out. Bar = 1 cm

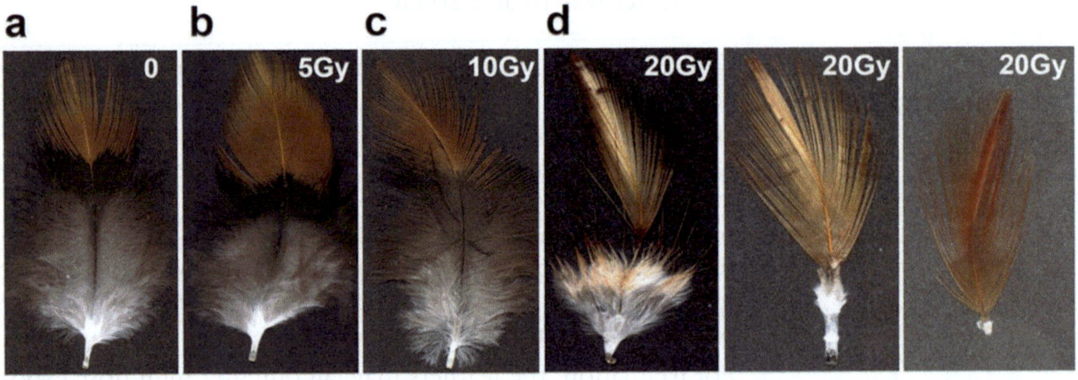

Fig. 4 Examples of radiotherapy-damaged feather. Dose-dependent damage response was documented in the feather model for IR-induced tissue damage (adopted from Ref. 22)

2. Open-prep of the feather branching epithelium: after plucking, the feather was dissected to expose the mesenchymal pulp cord under a stereo dissection microscope. Remove the pulp with care and avoid damaging the branching epithelium. After dissection, the feather epithelium was fixed in 4% PFA for 1 h at 4 °C, counterstained with 0.5 mg/mL DAPI in PBS for 1 h, mounted on a cover slip, and photographed under a Nikon inverted fluorescence microscope (Fig. 5).
3. The plucked feather was fixed in 4% PFA at 4 °C overnight for histology or homogenized in TRIzol reagent for RNA extraction. Standard protocols were followed for these experiments (Fig. 6).

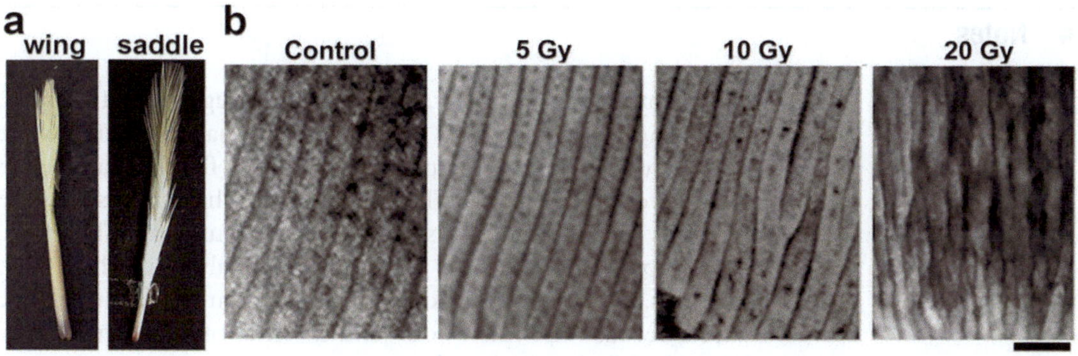

Fig. 5 The plucked feather and open-prep of the branching feather epithelium. (**a**) The feather could be plucked and analyzed individually anytime during the experiment. (**b**) The plucked feather was dissected to expose the branching epithelium (open-prep). The branching morphogenesis was perturbed by high-dose IR exposure (adopted from Ref. 22). Bar = 100 μm

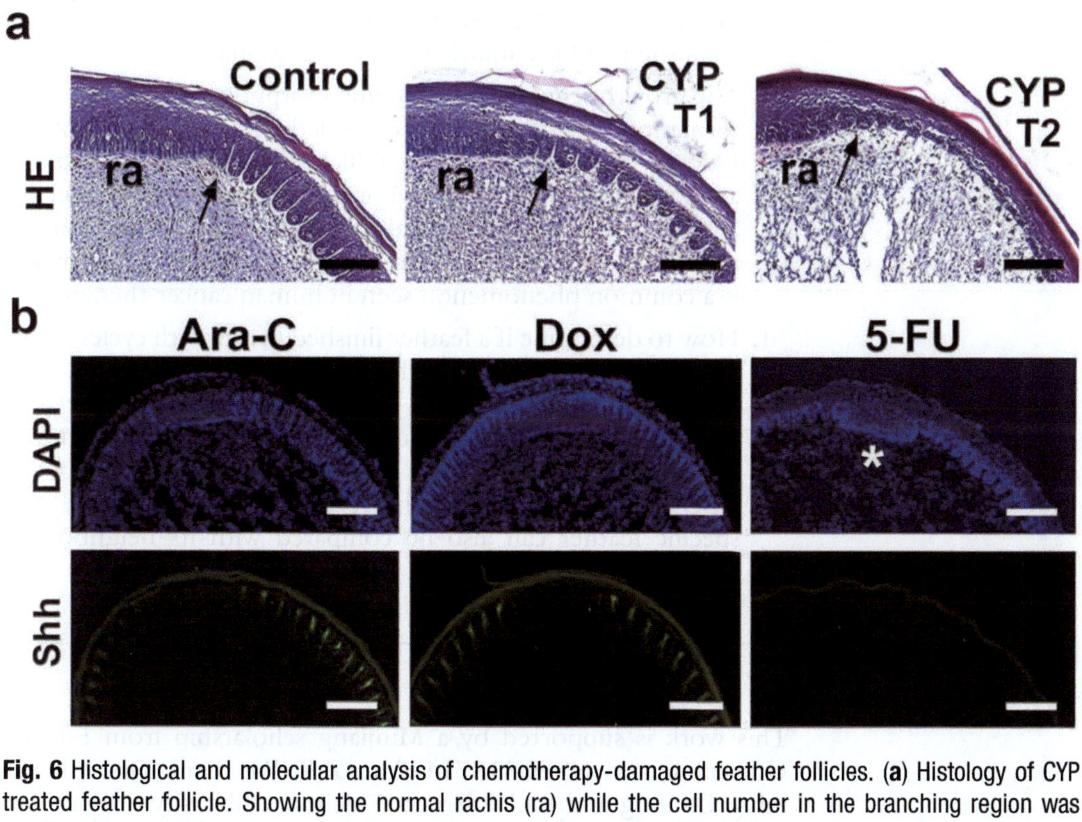

Fig. 6 Histological and molecular analysis of chemotherapy-damaged feather follicles. (**a**) Histology of CYP treated feather follicle. Showing the normal rachis (ra) while the cell number in the branching region was significantly reduced. (**b**) 5-FU treatment kept the normal rachis (marked by *asterisk*) while reduced the branching epithelium and *Shh* gene expression in the branching region. Immunofluorescence staining using a Shh antibody and counterstained by DAPI (adopted from Ref. 10). Bar = 100 μm

4 Notes

1. The feather morphology shows dramatic regional differences: the wing contour feather has a typical pennaceous vane, while the saddle feather contains plumulaceous (fluffy) branches in the lower part (Fig. 1). The molecular mechanism responsible for this morphological distinction remains unclear. Therefore, to simplify the analysis, we limit our analysis in these two regions. For irradiation, only the wing contour feathers were used because the main body and internal organs can be protected while irradiating the wing region.
2. Feather regeneration is a fast process: after plucking, it takes about 4 days to form a feather anlage, after which the feather will elongate at a speed of about 1–2 mm per day. From 1–3 weeks post wounding, the feather follicle remains actively growing. Thus, the perturbation can be performed any time in this window. After about 1 month, the feather growth cycle is finished and the follicle enters resting phase.
3. The growth regulation, branching morphogenesis, and pattern formation are individually controlled in each feather follicle. Due to its size, the feather follicle can be manipulated and analyzed individually. These include histological/molecular analysis, and morphological documentation. Variations in response to the perturbation can also be documented, which is a common phenomenon seen in human cancer therapy.
4. How to determine if a feather finished the growth cycle? Along the proximal-distal axis of the feather, there are distinct morphological features (*see* Fig. 1). Using the saddle feather as an example, it starts as pennaceous vane, gradually switches to plumulaceous vane, and to calamus without branches before enters complete resting. The relative length and shape of a specific feather can also be compared with its neighbors to help define its cycling stage.

Acknowledgment

This work is supported by a Minjiang scholarship from Fuzhou University, and NSFC31371472 to Z.Y. We thank Drs. Cheng-ming Chuong and Radall Bruce Widelitz (University of Southern California, Los Angeles, USA) for helpful input.

References

1. Paus R, Haslam IS, Sharov AA, Botchkarev VA (2013) Pathobiology of chemotherapy-induced hair loss. Lancet Oncol 14:e50–e59
2. Paus R, Handjiski B, Eichmuller S, Czarnetzki BM (1994) Chemotherapy-induced alopecia in mice. Am J Pathol 144:719–734

3. Ryan JL (2012) Ionizing radiation: the good, the bad, and the ugly. J Invest Dermatol 132:985–993
4. Hymes SR, Strom EA, Fife C (2006) Radiation dermatitis: clinical presentation, pathophysiology, and treatment 2006. J Am Acad Dermatol 54:28–46
5. Schaue D, Kachikwu EL, McBride WH (2012) Cytokines in radiobiological responses: a review. Radiat Res 178:505–523
6. Inomata K, Aoto T, Binh NT, Okamoto N, Tanimura S, Wakayama T et al (2009) Genotoxic stress abrogates renewal of melanocyte stem cells by triggering their differentiation. Cell 137:1088–1099
7. Cao X, Wu X, Frassica D, Yu B, Pang L, Xian L et al (2011) Irradiation induces bone injury by damaging bone marrow microenvironment for stem cells. Proc Natl Acad Sci U S A 108:1609–1614
8. Pineda JR, Daynac M, Chicheportiche A, Cebrian-Silla A, Sii Felice K, Garcia-Verdugo JM et al (2013) Vascular-derived TGF-beta increases in the stem cell niche and perturbs neurogenesis during aging and following irradiation in the adult mouse brain. EMBO Mol Med 5:548–562
9. Davis ST, Benson BG, Bramson HN, Chapman DE, Dickerson SH, Dold KM et al (2001) Prevention of chemotherapy-induced alopecia in rats by CDK inhibitors. Science 291:134–137
10. Xie G, Wang H, Yan Z, Cai L, Zhou G, He W et al (2015) Testing chemotherapeutic agents in the feather follicle identifies a selective blockade of cell proliferation and a key role for sonic hedgehog signaling in chemotherapy-induced tissue damage. J Invest Dermatol 135:690–700
11. Hussein AM, Jimenez JJ, McCall CA, Yunis AA (1990) Protection from chemotherapy-induced alopecia in a rat model. Science 249:1564–1566
12. Liao C, Xie G, Zhu L, Chen X, Li X, Lu H et al (2016) p53 is a direct transcriptional repressor of keratin 17: lessons from a rat model of radiation dermatitis. J Invest Dermatol 136:680–689
13. Lucas AM, Stettenheim PR (eds) (1972) Avian anatomy—integument. Agriculture handbook. U.S. Department of Agriculture, Washington, DC
14. Yu M, Yue Z, Wu P, Wu DY, Mayer JA, Medina M et al (2004) The developmental biology of feather follicles. Int J Dev Biol 48:181–192
15. Lin SJ, Wideliz RB, Yue Z, Li A, Wu X, Jiang TX et al (2013) Feather regeneration as a model for organogenesis. Develop Growth Differ 55:139–148
16. Chu Q, Cai L, Fu Y, Chen X, Yan Z, Lin X et al (2014) Dkk2/Frzb in the dermal papillae regulates feather regeneration. Dev Biol 387:167–178
17. Ng CS, Chen CK, Fan WL, Wu P, SM W, Chen JJ et al (2015) Transcriptomic analyses of regenerating adult feathers in chicken. BMC Genomics 16:756
18. Yu M, Wu P, Widelitz RB, Chuong CM (2002) The morphogenesis of feathers. Nature 420:308–312
19. Yue Z, Jiang TX, Wu P, Widelitz RB, Chuong CM (2012) Sprouty/FGF signaling regulates the proximal-distal feather morphology and the size of dermal papillae. Dev Biol 372:45–54
20. Yue Z, Jiang TX, Widelitz RB, Chuong CM (2006) Wnt3a gradient converts radial to bilateral feather symmetry via topological arrangement of epithelia. Proc Natl Acad Sci U S A 103:951–955
21. Chiu CT, Chuong CM (2015) Feather on the cap of medicine. J Invest Dermatol 135:1719–1721
22. Chen X, Liao C, Chu Q, Zhou G, Lin X, Li X et al (2014) Dissecting the molecular mechanism of ionizing radiation-induced tissue damage in the feather follicle. PLoS One 9:e89234

Chapter 21

An Early Chick Embryo Culture Device for Extended Continuous Observation

Hans-Georg Sydow, Tobias Pieper, Christoph Viebahn, and Nikoloz Tsikolia

Abstract

Appropriate mechanical tension of the vitelline membrane as the culture substrate for the early chick embryo is frequently reported to be required for successful in vitro development. Here we describe a modified device, made of anodized aluminum, for in vitro culture which we used for studies of left-right symmetry breaking with emphasis on morphology and gene expression as readouts. The technique allows for easy, high-throughput tissue handling and provides a suitable tension in a stable and easily reproducible manner proven to be suitable for correct molecular left-right patterning and heart looping after long-term culture.

Key words Chick embryo, Gastrulation, Left-right symmetry breaking, In vitro culture

1 Introduction

Chick embryo development in the first 36 h provides an excellent model to study molecular and cellular mechanisms of vertebrate axis formation [1], epithelial-mesenchymal transition during gastrulation [2], and somitogenesis [3, 4]. Furthermore—including the complete first three developmental days—morphogenetic processes leading to neurulation [5], to development of the heart tube and intestinal portal [6] as well as the head folding can be studied [7].

Various methods of explantation and ex ovo cultivation of early chicken embryos provide easy accessibility to the embryo for many kinds of manipulation as well as for fate mapping analysis including live imaging. However, successful development of early chick embryo ex ovo requires appropriate tension of the vitelline membrane and the embryonic tissues [8], while it does not require—at least until stage 12—external delivery of nutrients. The need of appropriate tension for correct development is in line with data which show the role of mechanical stresses in tissue morphogenesis

Guojun Sheng (ed.), *Avian and Reptilian Developmental Biology: Methods and Protocols*, Methods in Molecular Biology, vol. 1650, DOI 10.1007/978-1-4939-7216-6_21, © Springer Science+Business Media LLC 2017

and differentiation [for review see 9, 10]. Two of the most common and reproducible methods are the culture developed by Dennis New [8, 11, 12] and the so-called early chick culture ("EC culture," [13]). New culture [11] is the best tested so far and is particularly reliable for culturing early embryos [8]. In this culture a glass ring is used and the embryo is positioned above the pool of albumen. Appropriate tension is achieved by wrapping the margins of the vitelline membrane over the edges of the glass ring and adding the correct amount of thin albumen to the culture dish prior to pressing the glass ring onto the bottom of the dish. The culture allows for a relatively high throughput and has been especially successful to investigate initial axis formation [14], primitive streak formation [1], and epithelial-mesenchymal transition [2]. The EC culture as the other widely used method was originally introduced by Sorokin (referred in 15) and modified by Chapman and co-workers. Here, the vitelline membrane is stretched over a filter paper ring [13]. One important advantage of this technique is the high speed by which the culture can be transferred to culture conditions after explantation. However embryos frequently develop head defects (12, 16, own observations), and the success of this technique decreases inversely with the age of the embryo. Moreover, once the vitelline membrane is stretched and fixed onto the paper, its tension cannot be modified any longer. A modified version of New culture, the so-called Gallera method [17], may enable better morphological development, but the setup appears to be technically difficult [16]. An interesting alternative is the modification of Cornish pasty culture which allows various treatments including electroporation and robust development of embryos as a vesicle without the vitelline membrane as a substrate [18]. Interestingly the embryos in this culture are able to self-organize the tension which is provided by vesicle extension through inward fluid transport by epiblast cells.

Left-right symmetry breaking and patterning cannot be properly studied in ovo since making the window in the egg shell at early stages leads to severe malformation including omphalocoele with randomization of heart looping (19, personal communication by B. Christ and J. Männer). Interestingly, these malformations do not occur if the embryo is cultured ex ovo. Left-right symmetry breaking and early patterning also take place during first 24 h of development, i.e., at Hamburger and Hamilton (HH) stage 4 [20], which does not display any known molecular and morphological left-right asymmetries, through stage 5, when the *nodal* asymmetry can be detected [21], until stage 7, when *nodal*-dependent transcription factor *pitx2* is expressed in the left lateral plate mesoderm [22, 23]. This asymmetric expression of *pitx2* has been shown to be causally involved in asymmetric morphogenesis of the heart tube, gut, and other visceral organs [24–26]. Interestingly, the left paraxial expression of *nodal* follows asymmetric morphogenesis of the node [21]. The node asymmetry is the first morphological sign of LR asymmetry and is highly robust [27, 28] which leads to the suggestion that the node asymmetry is related to the asymmetrical formation of the chick notochord extending from the (thicker) right side (also called "shoulder") of the node [29].

To study dynamics of left-right symmetry breaking in the chick, we decided to use a culture technique (cf. Figs. 1, 2, 3, and 4) modified from the device described by Seidl [30], which was originally used to minimize the difficulty of explantation and to enable long-term video recordings [30, 31]. The device has a number of important advantages especially in connection with experimental investigations of left-right symmetry breaking.

The main advantage of the technique proposed here is its simple tissue handling which allows rapid explantation and uniform, easily adjustable mechanical tension of the vitelline membrane. This well-controlled tension appears to be important because heart looping is particularly sensitive to variations in tension and to the position of the embryo in the New culture device [28, 32], i.e., more than 15% of embryos develop left-sided (abnormal) heart looping. To address this particular sensitivity of the embryo, the blastoderm has to be centered carefully within the ring, and embryos which have grown to touch the side of the ring should be eliminated from consideration [32]. The device described in this report enables an alternative solution of this

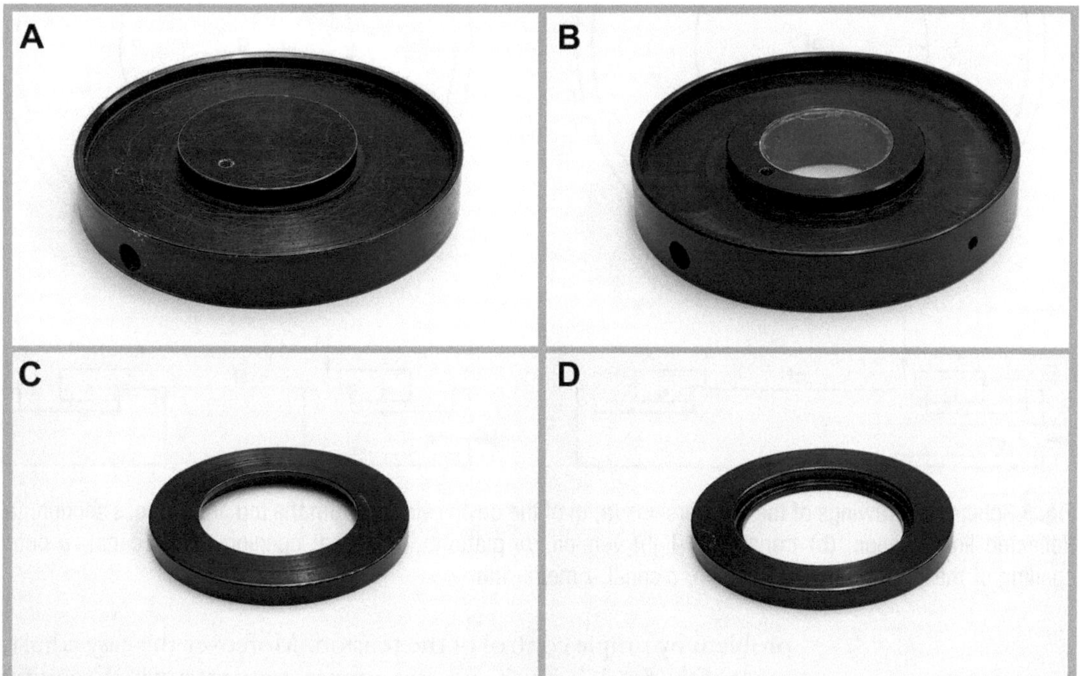

Fig. 1 Extended chick culture device consisting of a base (**a**, **b**) for either reflected-light illumination (**a**) or transmitted-light illumination (**b**) and a metal ring (**c**, **d**) shown here in an upright (**c**) and an inverted (**d**) position to reveal the groove in its inner side

Fig. 2 Oblique view of the device with the blastoderm (*b*) of a stage 9 chick embryo fully mounted and the vitelline membrane (*v*) under optimal tension. *i* internal opening of the canal filled with thin albumen, *p* plug, *r* metal ring

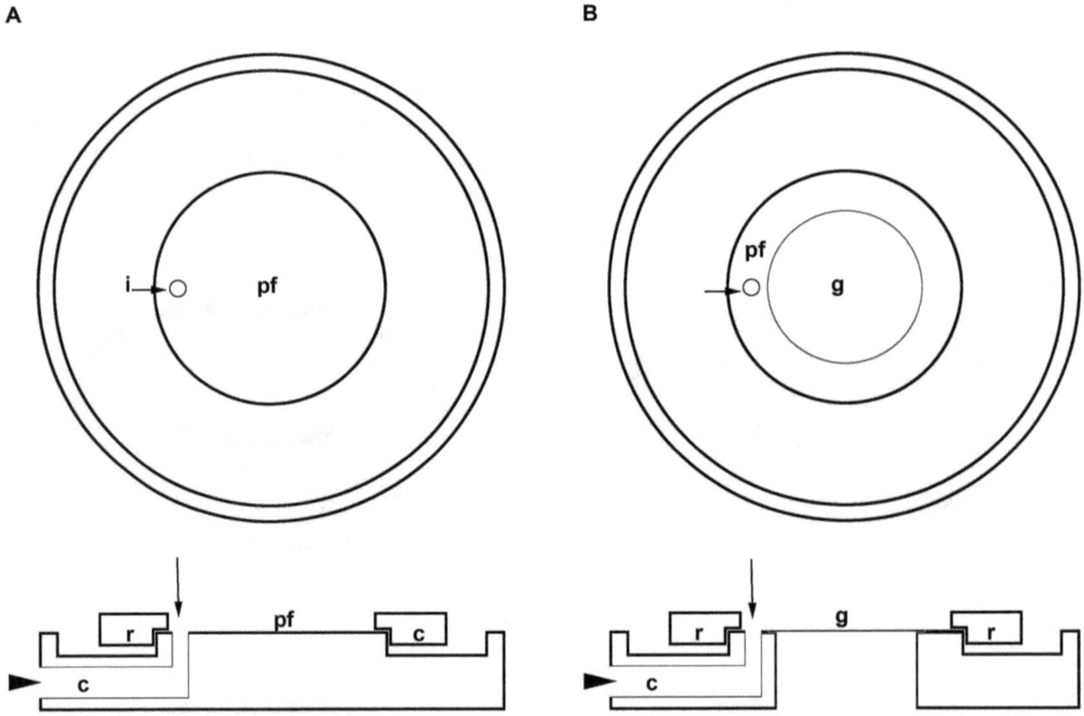

Fig. 3 Schematic drawings of the two versions (**a**, **b**) of the device viewed from the top and in cross section. (**a**) Reflected-light version; (**b**) transmitted-light version. *pf* platform, *i* internal opening of the canal, *o* outer opening of the canal, *g* glass coverslip, *c* canal, *r* metal ring

problem by simple control of the tension. Moreover the easy adjustment of the tension may be necessary for microsurgery which requires a transient decrease of tension during manipulation [33].

The technique described in this report was used in different experimental setups [21, 29, 34–36] and allowed robust

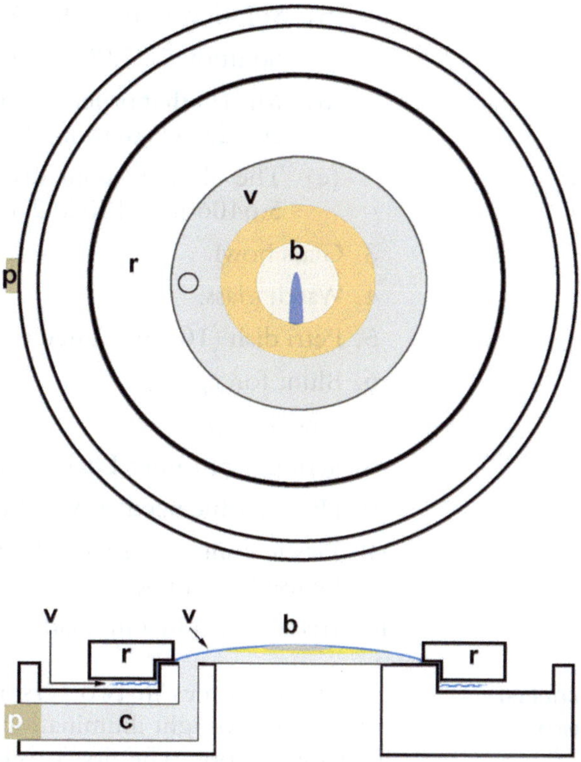

Fig. 4 Schematic drawing of the device with a HH stage 4—blastoderm (*b*) and vitelline membrane (*v*) in position (cf. Fig. 2). (**a**) View from above, (**b**) in cross section. *c* Canal filled with thin albumen, *p* plug in the outer opening of the canal, *angled arrow* cut edge of the vitelline membrane

development of embryos up to HH stage 16–17 (cf. **Note 1**). This robustness was especially important for left-right studies: we did not observe any case of right-sided or bilateral expression of *nodal* or *shh* mRNAs after hundreds of experiments (Pieper et al. unpublished). Similarly, node asymmetry and sidedness of the heart looping displayed a very high reproducibility (29, unpublished) which we attribute to the idea that a controlled and reproducible tissue tension is required for correct left-right morphogenesis and patterning. Finally, it is worth mentioning that the inverted topology of the tension (cf. Figs. 3 and 4) is comparable to the Gallera method [16].

2 Materials

2.1 Instruments and Consumables

1. Incubated chicken eggs.
2. Locke's saline.
   (a) Solution A: 9.427 g NaCl, H_2O to 1 L.

(b) Solution B: 6.0129 g KCl, H_2O to 500 ml.

(c) Solution C: 7.9043 g $CaCl_2.2H_2O$, H_2O to 500 ml.

(d) Mix (in that order): 1000 ml solution A, 37 ml solution B, and 21 ml solution C.

(e) The buffer contains thus 152.4483 mmol NaCl; 5.6406 mmol KCl; 2.1357 mmol CaCl.

3. Glass bowl.
4. Watch glass.
5. Petri dish (10 cm diameter).
6. Blunt forceps.
7. Fine forceps.
8. Syringe with Luer Lock (5 ml).
9. Hypodermic needles with Luer Lock.
10. Plastic connecting part (Luer Lock) of hypodermic needle to be used as a plug.
11. Iris scissors (11 mm blades).

2.2 Description of the Device

The device comes in two versions: one designed for microscopy using reflected-light illumination and another one for transmitted-light microscopy. The first version of the device (Figs. 1a and 3a) consists of (a) an aluminum base (7 mm high, 59 mm diameter) with an elevated rim (3 mm high) and a central platform (3 mm high, 30 mm diameter) to carry the vitelline membrane with the incubated blastoderm (cf. Figs. 2 and 4) and (b) an aluminum ring with a groove snugly fitting the edge of the platform. Two boreholes (4 and 2 mm diameter) at the side of the base and off-center in the surface of the platform, respectively, meet underneath the platform to provide a canal for albumen application to the surface of the platform, while the embryo is in place (cf. Fig. 4).

The second version of the device, which is made to enable transmitted-light microscopy (Figs. 1b and 3b), has a modified platform with an additional central borehole (18 mm in diameter), which spares the opening of the albumen canal in the periphery of the platform and has a small groove to receive a glass coverslip fixed with superglue.

Both versions of the device are made of anodized aluminum suitable for sterilization in an autoclave at 120 °C.

3 Methods

3.1 The Use of the Device for In Vitro Cultivation of the Chicken Embryo

1. Incubate the eggs at 38 °C under humidified conditions to the required stage according to HH [15] (cf. **Note 2**).
2. Puncture the egg shell with the hypodermic needle at its blunt end, and aspirate thin part of the albumen using 5 ml syringe.

3. Break the egg shell at its blunt end and discard the rest of the albumen.
4. Transfer the yolk gently into a glass bowl filled with Locke's saline (or any other buffer such as Pannett-Compton saline).
5. Remove carefully the rest of the albumen from the surface of the yolk by suctioning it into a syringe.
6. To remove the vitelline membrane carrying the blastoderm, place the yolk with the blastoderm facing upward, and make an incision in the membrane just below the equator using iris scissors. Cut the membrane along the equator while carefully rotating the yolk with forceps.
7. Using two pairs of fine forceps, slowly lift the membrane from the yolk, reflect it onto itself toward the blastoderm, and continue to pull the vitelline membrane across the area of the blastoderm; this latter step should be carried out as slowly as possible to prevent the detachment of the blastoderm from the membrane.
8. After removing the vitelline membrane from the yolk, flip it over gently so that the hypoblast is facing upward; transfer the membrane in a small volume of fluid to a watch glass.
9. Inject a small volume of the thin albumen into the opening of the device (cf. Fig. 4) until the albumen covers the major part of the platform surface, and leave the syringe connected to the device for later adjustment of the volume of albumen.
10. Spread the albumen evenly across the platform surface using a hypodermic needle, and moisten the well surrounding of the platform with Locke's solution.
11. To transfer the membrane to the platform, hold the watch glass with its edge reaching just beyond the edge of the albumen covering the platform and hold in position the periphery of the membrane with a pointed pair of forceps while gently removing the watch glass between membrane and albumen. When carried out at the correct (slow) speed, this movement will result in a gliding motion with the blastoderm keeping its shape during the procedure.
12. Spread the periphery of the membrane evenly over the platform edge, and carefully place the ring with its groove facing down onto the platform to secure the position of the membrane between the ring and the platform.
13. Inject further albumen (ca. 2–3 ml) onto the platform until the blastoderm is free of folds and under homogenous tension (cf. Figs. 2 and 4). To achieve this, the tension of the blastoderm can be modified by adding or removing albumen, while the syringe is still connected to the device.

14. When the tension is considered optimal, remove the syringe and quickly close the canal using a suitable plug (e.g., the one made from a connecting part of the hypodermic needle; *see* Subheading 2.1, **item 10**) (cf. **Note 3**).
15. Using a Pasteur pipette, rinse the entire surface of area pellucida carefully with a small amount of Locke's solution; superfluous solution will run across the ring and collects in the well surrounding the platform.
16. To start the culture, the device can be placed into suitable Petri dish and incubated at 38 °C and high humidity (cf. **Note 4**).

4 Notes

1. Culturing older embryos (between HH stages 12 and 17) can be carried out in a modified device with a larger platform diameter and by adding yolk at the margins of the area opaca.
2. The incubation time to the required stage may slightly vary depending on the season and the storage temperature of the eggs prior to culture.
3. Incisions or other microsurgeries should be carried out under decreased tension of the vitelline membrane (cf. 33).
4. If the embryo is to be treated with molecular compounds, place 5–10 μl drops of the diluted substance onto the blastoderm; make sure that the blastoderm never dries out during this procedure.

Acknowledgment

We thank Jörg Männer and Wolfgang Seidl for fruitful discussions.

References

1. Voiculescu O, Bertocchini F, Wolpert L, Keller RE, Stern CD (2007) The amniote primitive streak is defined by epithelial cell intercalation before gastrulation. Nature 449:1049–1052
2. Nakaya Y, Sukowati EW, Wu Y, Sheng G (2008) RhoA and microtubule dynamics control cell-basement membrane interaction in EMT during gastrulation. Nat Cell Biol 10:765–775
3. Dias AS, de Almeida I, Belmonte JM, Glazier JA, Stern CD (2014) Somites without a clock. Science 343:791–795
4. Dubrulle J, McGrew MJ, Pourquie O (2001) FGF signaling controls somite boundary position and regulates segmentation clock control of spatiotemporal Hox gene activation. Cell 106:219–232
5. Colas JF, Schoenwolf GC (2001) Towards a cellular and molecular understanding of neurulation. Dev Dyn 221:117–145
6. Manner J (2000) Cardiac looping in the chick embryo: a morphological review with special reference to terminological and biomechanical aspects of the looping process. Anat Rec 259:248–262
7. Varner VD, Voronov DA, Taber LA (2010) Mechanics of head fold formation: investigating tissue-level forces during early development. Development 137:3801–3811
8. Stern CD, Bachvarova R (1997) Early chick embryos in vitro. Int J Dev Biol 41:379–387
9. Beloussov LV (2015) Morphomechanics of development. Springer, Heidelberg

10. Beloussov LV (2008) Mechanically based generative laws of morphogenesis. Phys Biol 5:015009
11. New DAT (1955) A new technique for the cultivation of the chick embryo in vitro. J Embryol Exp Morph 3:326–331
12. Voiculescu O, Papanayotou C, Stern CD (2008) Spatially and temporally controlled electroporation of early chick embryos. Nat Protoc 3:419–426
13. Chapman SC, Collignon J, Schoenwolf GC, Lumsden A (2001) Improved method for chick whole-embryo culture using a filter paper carrier. Dev Dyn 220:284–289
14. Torlopp A, Khan MA, Oliveira NM, Lekk I, Soto-Jimenez LM, Sosinsky A, Stern CD (2014) The transcription factor Pitx2 positions the embryonic axis and regulates twinning. Elife 3:e03743
15. Menkes B, Micela C, Elias S, Deleanu M (1961) Researches on the formation of axial organs: I. Studies on the differentiation of the semites. Acad Repub Pop Rom Baza Cercet Timisoara Stud Cercet Stiint Med 8:7–34
16. Ghatpande SK (2008) Gallera method of chick embryo culture in vitro supports better growth compared with original new method. Develop Growth Differ 50:437–442
17. Nicolet G, Gallera J (1963) Under what conditions will the amnion of the chick embryo develop in culture in vitro? Experientia 19:165–166
18. Nagai H, Sezaki M, Nakamura H, Sheng G (2014) Extending the limits of avian embryo culture with the modified Cornish pasty and whole-embryo transplantation methods. Methods 66:441–446
19. Fischer M, Schoenwolf GC (1983) The use of early chick embryos in experimental embryology and teratology: improvements in standard procedures. Teratology 27:65–72
20. Hamburger V, Hamilton HL (1951) A series of normal stages in the development of the chick embryo. J Morphol 88:49–92
21. Tsikolia N, Schroder S, Schwartz P, Viebahn C (2012) Paraxial left-sided nodal expression and the start of left-right patterning in the early chick embryo. Differentiation 84:380–391
22. Ryan AK, Blumberg B, Rodriguez-Esteban C, Yonei-Tamura S, Tamura K, Tsukui T, de la Pena J, Sabbagh W, Greenwald J, Choe S et al (1998) Pitx2 determines left-right asymmetry of internal organs in vertebrates. Nature 394:545–551
23. St Amand TR, Ra J, Zhang Y, Hu Y, Baber SI, Qiu M, Chen Y (1998) Cloning and expression pattern of chicken Pitx2: a new component in the SHH signaling pathway controlling embryonic heart looping. Biochem Biophys Res Commun 247:100–105
24. Patel K, Isaac A, Cooke J (1999) Nodal signalling and the roles of the transcription factors SnR and Pitx2 in vertebrate left-right asymmetry. Curr Biol 9:609–612
25. Logan M, Pagan-Westphal SM, Smith DM, Paganessi L, Tabin CJ (1998) The transcription factor Pitx2 mediates situs-specific morphogenesis in response to left-right asymmetric signals. Cell 94:307–317
26. Piedra ME, Icardo JM, Albajar M, Rodriguez-Rey JC, Ros MA (1998) Pitx2 participates in the late phase of the pathway controlling left-right asymmetry. Cell 94:319–324
27. Dathe V, Gamel A, Manner J, Brand-Saberi B, Christ B (2002) Morphological left-right asymmetry of Hensen's node precedes the asymmetric expression of Shh and Fgf8 in the chick embryo. Anat Embryol (Berl) 205:343–354
28. Cooke J (1995) Vertebrate embryo handedness. Nature 374:681
29. Otto A, Pieper T, Viebahn C, Tsikolia N (2014) Early left-right asymmetries during axial morphogenesis in the chick embryo. Genesis 52:614–625
30. Seidl W (1977) Description of a device facilitating the in vitro culture of the chick embryo according to the new method. Folia Morphol (Praha) 25:43–45
31. Seidl W, Steding G (1978) Frühentwicklung der Herzanlage. Gallus domesticus – 1. Und 2. Bebrütungstag. Film E2507, Encyclopaedia Cinematographica. Institut für den wissenschaftlichen Film, Göttingen
32. Levin M, Johnson RL, Stern CD, Kuehn M, Tabin C (1995) A molecular pathway determining left-right asymmetry in chick embryogenesis. Cell 82:803–814
33. Psychoyos D, Stern CD (1996) Restoration of the organizer after radical ablation of Hensen's node and the anterior primitive streak in the chick embryo. Development 122:3263–3273
34. Steding G, Seidl W, Sydow H, Bahners W, Borowski D (1981) Experimental manipulations leading to cardiac malformations in chick embryo. In: Neubert D, Merker H-J (eds) Culture techniques. Applicability for studies on prenatal differentiation and toxicity. De Grueter, Berlin, New York, pp 539–551
35. Kremer JE (1988) Wachstumsverhalten des Entoderm beim Descensus der Darmpforte. Inaugural dissertation, Georg-August Universität, Göttingen
36. Meyer E (1989) Die embryonale Kopfdrehung in Abhängigkeit von der Lage des Herzens. Inaugural dissertation, Georg-August Universität, Göttingen

Chapter 22

A Sensitive and Versatile In Situ Hybridization Protocol for Gene Expression Analysis in Developing Amniote Brains

Pei-Shan Hou, Takuma Kumamoto, and Carina Hanashima

Abstract

The detection of specific RNA molecules in embryonic tissues has wide research applications including studying gene expression dynamics in brain development and evolution. Recent advances in sequencing technologies have introduced new animal models to explore the molecular principles underlying the assembly and diversification of brain circuits between different amniote species. Here, we provide a step-by-step protocol for a versatile in situ hybridization method that is immediately applicable to a range of amniote embryos including zebra finch and Madagascar ground gecko, two new model organisms that have rapidly emerged for comparative brain studies over recent years. The sensitive detection of transcripts from low to high abundance expression range using the same platform enables direct comparison of gene of interest among different amniotes, providing high-resolution spatiotemporal information of gene expression to dissect the molecular principles underlying brain evolution.

Key words Brain development, Amniote embryos, Gecko, Zebra finch, Pallium, Dual color in situ hybridization, Evolution

1 Introduction

An important aspect of developmental and evolutionary biology is to investigate the ontogeny of an organism at both microscopic (molecular and cellular) and macroscopic (tissue and species) levels. Among tissues, the brain, which is the processing center of external sensory information to generate coordinated motor outputs, has been an area of wide interest for many developmental neuroscientists. Until recently, comparative brain studies have relied on limited developmental animal models such as mouse (*Mus musculus*) and chicken (*Gallus gallus*). However, acceleration in sequencing technologies and bioinformatics has led to the emergence of new animal models and advanced comparative neuroscience to a new era [1–5]. The rapid increase in number of experimental animals has not only revealed species-specific brain cytoarchitecture

Guojun Sheng (ed.), *Avian and Reptilian Developmental Biology: Methods and Protocols*, Methods in Molecular Biology, vol. 1650, DOI 10.1007/978-1-4939-7216-6_22, © Springer Science+Business Media LLC 2017

(i.e., cellular patterning) but it has also uncovered common principles underlying brain assembly and acquisition of higher brain function such as vocal learning [6–9].

During early development, amniote brains undergo identical patterning events in which the dorsal and ventral subdivisions of the telencephalon are first specified. However, in contrast to the ventral telencephalon that gives rise to the basal ganglia and shows similar architecture, the dorsal telencephalon gives rise to the pallium, which shows divergence in cellular patterning between the amniote clades. First, extant amniotes that are divided into two lineages synapsids (which includes mammals) and diapsids (which includes reptiles and birds) have distinctive anatomical features, where diapsids contain a unique protrusion in the lateral pallium called the dorsal ventricular ridge (DVR), whose homolog in the mammalian brain has not been elucidated [10–12]. Second, the cellular organization ranges from a three-layered structure in reptiles, nuclear organization in avians, to a six-layered cortex in mammals [13], and these structural differences may contribute to species-unique behavior [14, 15]. However, the molecular mechanisms that direct lineage segregation leading to these anatomical and function specializations remain largely elusive.

Recently, new diapsid organisms have been introduced for comparative brain studies, including the Madagascar ground gecko (*Paroedura picta*) and zebra finch (*Taeniopygia guttata*) [16, 17]. Both gecko and zebra finch meet key criteria for developmental animal models, (1) easy breeding and colony maintenance in a conventional laboratory setup, (2) perennial reproductive capacity, (3) constant number of eggs laid during the mating period, and (4) feasibility to obtain and manipulate embryos under protection of hard egg shells [18–21]. Furthermore, the basis of cellular homology with mouse embryonic brains has begun to be characterized [16, 22, 23], facilitating direct comparison of genes of interest on an anatomically annotated platform.

The detection of RNA molecules in developing brain tissues by in situ hybridization has important applications in studying spatial and temporal gene expression regulation. With reduced cost and labor compared to generating antibodies for immunohistochemical protein expression analysis, probes can be designed based on primary nucleotide sequences to detect specific RNA molecules. This is particularly useful when researchers need to compare gene family among different amniote species, as it is difficult to find commercial antibodies that have unique reactivity in gene orthologs and paralogs. Instead, RNA probes can be easily constructed by standard molecular cloning using species-specific sequence information.

Here, we provide a sensitive and versatile in situ hybridization protocol to detect transcripts from low to high expression range in a wide range of aminote embryos. This platform provides high-resolution spatial information of expression of gene of interest, thereby facilitating direct comparison of cellular organization among different species. The protocol is also applicable to simultaneous detection of different RNA molecules, and in combination with thymidine analog detection and immunohistochemistry. The same probe can also be used in whole embryos for whole mount in situ hybridization to examine early patterning in three-dimensional brain [24, 25], in which antibody penetration is typically challenging.

While the protocol here provides tissue processing and mRNA detection procedure for gecko and zebra finch embryos, the same in situ hybridization method is applicable to mouse (embryonic day (E9–E18), chicken (E3–E14), and marmoset (gestational week: GW 12) embryonic brains without further modifications. The versatility of the protocol without requiring additional reagents is advantageous for high-throughput and parallel processing of different amniote tissues for comparative gene expression analysis.

2 Materials

All reagents up to the hybridization step should be prepared using nuclease-free equipment and solutions (baked glasswares, disposable plastic tubes, nuclease-free water, and DEPC-treated PBS).

2.1 Collecting Brains

1. Dissection tools: scissors, forceps (No. 3, No. 5), spatula, 1 ml syringes, 27G needles, Styrofoam plate and pins.
2. Sterile PBS.
3. 10× Tyrode's buffer (autoclaved, for 1 l):

   80 g sodium chloride.

   2 g potassium chloride.

   0.36 g sodium phosphate monobasic dihydrate.

   10 g glucose.

   2.71 g calcium chloride dihydrate.

   2 g magnesium chloride hexahydrate.

   in water.
4. 1× Tyrode's buffer: dilute 10× Tyrode's buffer with sterile water.
5. 4% paraformaldehyde in PBS stored at −20 °C.

2.2 Cryosection

1. Sterile 30% sucrose in DEPC-PBS.
2. Sterile PBS.
3. Tissue-Tek O.C.T. Compound (Sakura, 4583).
4. Tissue-Tek Cryomold (Sakura, 4557).
5. Dryice.

2.3 cDNA Preparation

1. Dissection tools: scissors, forceps (No. 3).
2. DEPC-PBS.
3. TRIzol LS reagent (Ambion, 10296010).
4. QIAgen RNeasy mini kit (74104).
5. Chloroform (Sigma).
6. PrimeScript™ II 1st Strand cDNA Synthesis Kit (TaKaRa, 6110A).
7. Thermal cycler.

2.4 Template for Probe Preparation

1. Digestion enzyme and buffer.
2. QIAquick PCR purification kit or QIAquick Gel Extraction kit.
3. Nuclease-free water.

2.5 In Vitro Transcription

1. RNase inhibitor (Roche, 03335399001).
2. T7 or Sp6 RNA polymerase (Roche, 10881767001 or 10810274001).
3. 10× transcription buffer (Roche).
4. 100 mM DTT.
5. 10× DIG or Fluorescence mix (Roche, 11685619910 or 11277073910).
6. Nuclease-free water.
7. 4 M LiCl (Sigma) in nuclease-free water.
8. 100% and 70% ethanol in nuclease-free water.
9. Isopropanol.
10. RNA-binding columns (Roche, *optional*).

2.6 In Situ Hybridization

1. Dry-heat sterilized tanks, HybriSlip (Sigma), Pap Pen.
2. 4% paraformaldehyde.
3. H_2O_2-MeOH: 1.5% hydrogen peroxide in methanol (prepare before use).
4. 0.2 M hydrochloric acid in nuclease-free water (prepare before use).
5. DEPC-PBS.
6. 10 μg/ml proteinase K (Thermo, EO0491) in DEPC-PBS (prepare before use).
7. Acidic solution (prepare before use) (60 ml): 57.75 ml nuclease-free water, 0.81 ml triethanolamine (Sigma, 90279), 0.105 ml 10 N hydrochloric acid (Sigma, 320331), 0.15 ml anhydrous acetic acid (Sigma, 320099).
8. Hybridization buffer: 50% formamide (Gibco), 2x SSC, 1× Denharts, 20% dextran sulfate, 0.5 mg/ml baker's yeast RNA (Sigma R6750), 0.5 mg/ml ssDNA.

9. 20x SSC (1 l): 175.3 g sodium chloride, 88.2 g sodium citrate, pH 7, autoclaved.
10. 5x, 2x and 0.1x SSC (prepare before use): dilute from 20x SSC.
11. 50% formamide in 2x SSC (prepare before use).
12. RNase A buffer: 0.5 M sodium chloride, 10 mM Tris–HCl pH 7.5, 5 mM EDTA.
13. 20 μg/ml RNase A (Roche) in RNase A buffer (prepare before use).
14. Wash buffer: 100 mM maleic acid, 150 mM sodium chloride, 0.05% Tween 20, pH 7 with Sodium Hydroxide.
15. 10% BB1 (store at 4 °C): dissolve Blocking Reagent (Roche 11096176001) in wash buffer at 60 °C for 1 h.
16. 1% BB1 (prepare before use): dilute 10% BB1 with wash buffer.
17. Anti-DIG-POD antibody (1:500, Roche) in 1% BB1 (prepare before use).
18. Anti-Fluorescein-POD antibody (1:500, Perkin Elmer NEF710) in 1% BB1 (prepare before use).
19. BB2 (store at −20 °C): 0.5% Casein, 150 mM sodium chloride, 100 mM Tris–HCl pH 7.5.
20. 1.8 mM Biotinyl-Tyramide (store at −20 °C).
21. Diluent buffer: 0.1 mM Imidazole HCl pH 7.6 in PBS.
22. Biotinyl-Tyramide solution (prepare before use): 18 μM Biotinyl-Tyramide, 0.001% hydroperoxide in diluent buffer.
23. Streotavidin-AP (1:750, Roche) in 1% BB1 (prepare before use).
24. Texas-red streptavidin (1:500, Vector) in 1% BB1 (prepare before use).
25. Fluorescein Streptavidin (1:500, Vector) in 1% BB1 (prepare before use).
26. 3% hydrogen peroxide in nuclease-free water.
27. Avidin blocker and biotin blocker in avidin-biotin blocking kit (Vector).
28. B3 solution (prepare before use): 0.1 M Tris pH 9.5, 0.1 M sodium chloride, 50 mM magnesium chloride.
29. B4 solution (prepare before use): 340 mg/ml NBT(Roche), 175 mg/ml BCIP (Roche), 1 mM levamisole in B3 solution.
30. CC/Mount™ tissue mounting medium (Sigma).
31. SlowFade® Gold Antifade Mountant with DAPI (Thermo, S36938).

3 Methods

Before starting the experiments, make sure to wear gloves to protect samples from RNase.

3.1 Collecting Brains

Here, we describe the dissection procedure for removing brains from zebra finch (*Taeniopygia guttata*) and gecko (*Paroedura pictus*) embryos. For breeding of colonies and staging of embryos, refer to Mak et al. [20] and Noro et al. [18]. The in situ hybridization protocol can be applied to different aminote embryonic brains including mouse, chicken, and marmosets.

3.1.1 Gecko Embryos

1. For breeding and staging of embryos, refer to Noro et al. [18].
2. Break eggshell carefully using No. 3 forceps. Open egg in 1× HBSS buffer.
3. To fix brains with intact morphology, dissect embryos depending on the stages:
   (a) Early stage embryos (until 30 d.p.o.): Brain tissues of these stages are small and fragile. Collect whole brains and punch a hole in the mesencephalon using No. 5 forceps. (*optional*) After making a hole, prefixing by 4% PFA for 1–2 h at 4 °C makes the following steps easier.
   (b) Late stage embryos (30 d.p.o. and after): Brain tissue has hardened enough. Isolate the cortical tissue from brain carefully using No. 3 forceps and spatula. From this stage onward, olfactory nerve has developed. Removing brains from tissues and skulls with intact olfactory nerve is very difficult because olfactory nerve is thin and fragile. Hold the skin of part of cerebral cortex using No. 3 forceps, then gently remove the brain tissue along olfactory nerve (from the brain side) using No. 5 forceps with the tips closed (*see* **Note 1**).

3.1.2 Zebra Finch Embryos

1. For breeding and staging of embryos, refer to Mak et al. [20] and Murray et al. [26].
2. Break eggshell carefully using No. 3 forceps (Fig. 1a, b). Open egg in 1× Tyrode's buffer.
3. To take out brains with tissue intact, the dissection procedure is divided into three stages depending on the embryonic day (before ED5, ED6–8, ED9 and after):
   (a) Before ED5: Brain tissues at this stage are small and fragile. Collect whole brain and make a hole in mesencephalon with No. 5 forceps (Fig. 1c). The hole provides proper infiltration of reagents during the following steps.

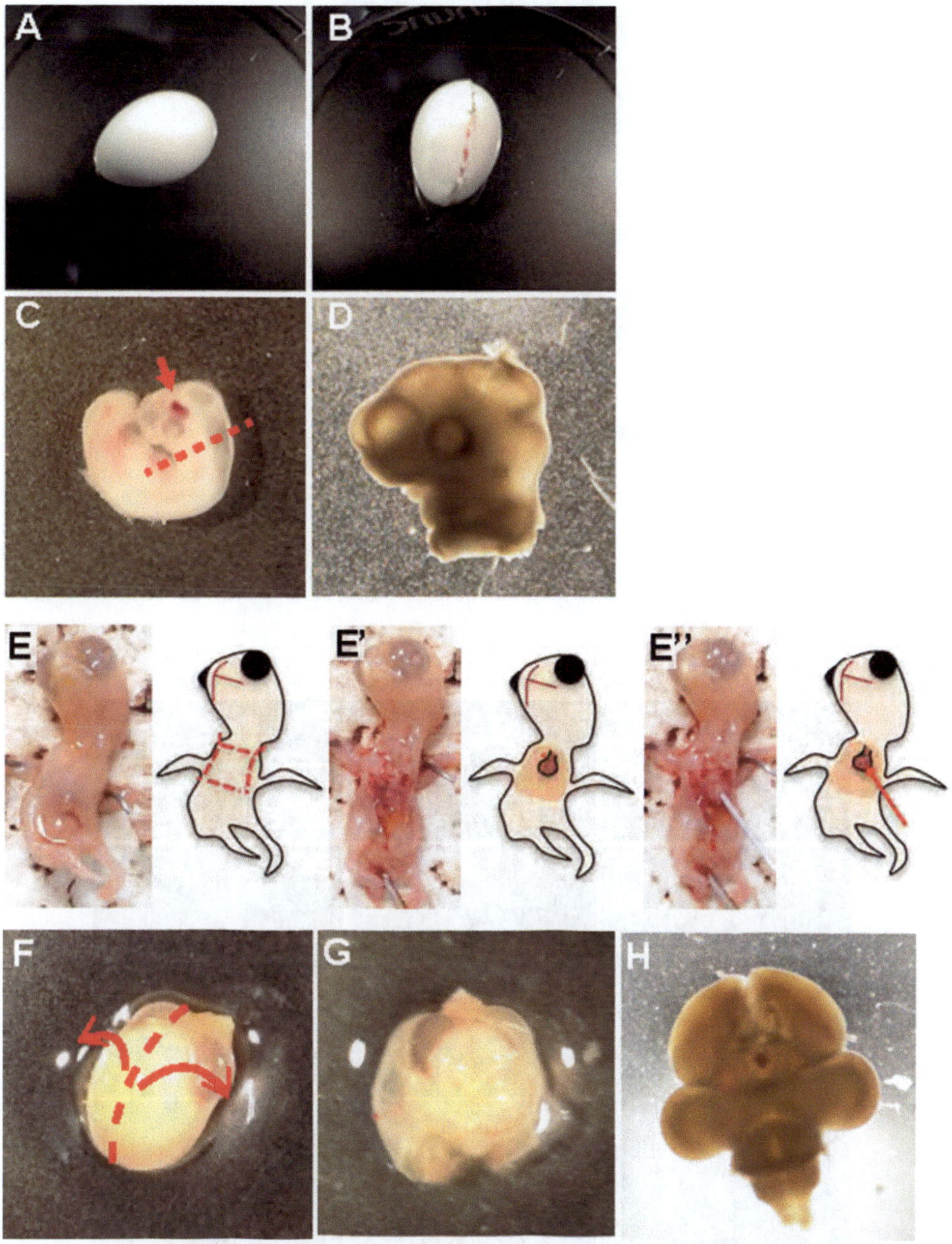

Fig. 1 Brain collection. (**a**) Zebra finch egg. (**b**) Break shells along the long-axis. (**c**) Embryonic day 4 zebra finch embryo. To collect brains, cut head from the trunk along the *dashed line* and make an incision as indicated by the *arrow*. (**d**) Dissected embryonic day 4 zebra finch brain. (**e**) Perfusion procedure. (**e**) Open chest. (**e'**) Identify the heart and make a small incision in the right atrium. (**e"**) Inject PBS and then 4% paraformaldehyde at indicated position. (**f**) Dissection of embryonic day 13 embryo. Cut along the *dashed line* and flip open the skin and skull. (**g**) Top view of the brains exposed. (**h**) Whole view of embryoniic day 13 zebra finch brain

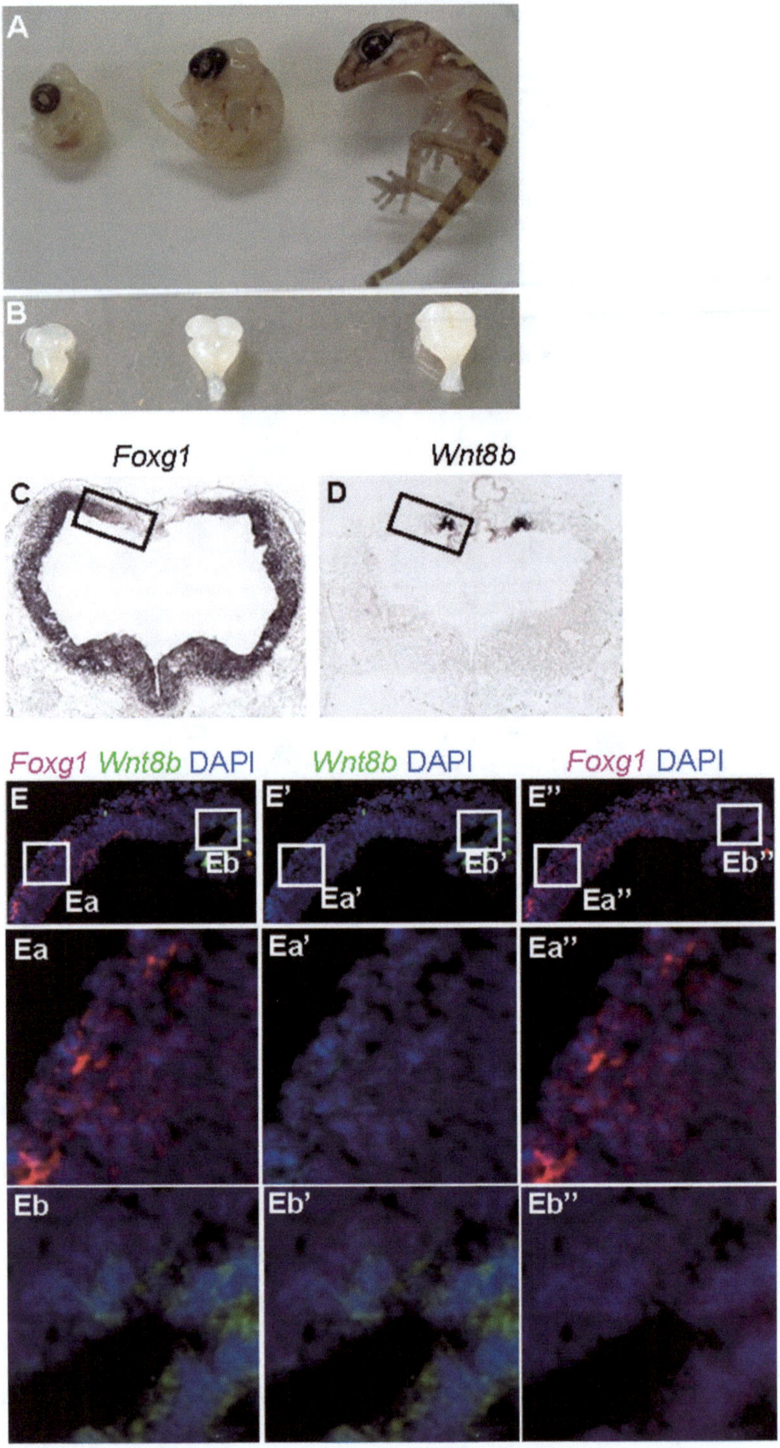

Fig. 2 Chromogenic and fluorescence dual in situ hybridization of Madagascar ground gecko embryonic brain. (**a, b**) Whole view (**a**) and dissected brain (**b**) of distinct embryonic stage Madagascar ground gecko. (**c, d**) Chromogenic in situ hybridization example of *Foxg1* (**c**) and *Wnt8b* (**d**) mRNA detection in E14 Madagascar ground gecko coronal brain sections. Boxes indicate regions that show complementary expression pattern that is shown in fluorescence dual in situ hybridization. (**e**) Dual color in situ hybridization using Texas *red* (*Foxg1*, in *red*) and Fluorescein (*Wnt8b*, in *green*) using the same cDNA templates. **Ea-Ea"** and **Eb-Eb"** indicate higher magnification views of the *boxed regions* shown in **e-e"**

(b) ED6-8: Brain tissues have hardened. Carefully isolate brain tissue from the head using No. 3 forceps and spatula.

(c) After ED9: As blood vessels in brain tissues have developed, perfusion is recommended to facilitate penetration and reduce background signals. To perform perfusion, disconnect yolk from embryos and fix embryos on Styrofoam plate with pins (Fig. 1e). Use small scissors to open the chest (Fig. 1e'). Expose the heart and make a small incision in the right atrium carefully as indicated in Fig. 1e". Carefully isolate the cortical tissue from brain with No. 3 forceps and spatula.

- Post-fix brain tissues in 4% paraformaldehyde overnight at 4 °C.

Step 3.2 and onward are common to gecko and zebra finch brain tissues.

3.2 Cryosection

1. Rinse brain tissues with sterile PBS at room temperature, 5 min, three times.
2. Immerse tissue in 30% sucrose-PBS for replacement at 4 °C until tissue sink (*see* **Note 2**). (*optional*) for early stage embryos, stepwise replacement in 10% → 20% → 30% sucrose in DEPC-PBS solution will facilitate penetration.
3. Apply Tissue-tek O.C.T. compound to cryomolds.
4. Briefly rinse the brain tissue with sterile PBS and gently remove residual PBS using Kimwipes (*see* **Note 3**). Transfer tissue into O.C.T. compound in cryomold.
5. Use No.3 forceps to swirl O.C.T. compound around the tissue to better adhere tissue to compound during freezing (*see* **Note 4**).
6. Orient the tissue at correct angles using forceps and place the cryomold on a block of dry ice for gradual freezing. After freezing, samples can be stored at −20 °C for short-term (1 week) or −80 °C for long-term storage (over 1 year).
7. Section tissue blocks into 12 μm thicknesses using a cryostat.

3.3 Preparing cDNA Templates for PCR Reaction (see Note 5)

Collecting tissue samples

1. Collect brain tissues from embryos at desired stages (embryonic stage and tissue should be selected depending on the target gene of interest).
2. Break eggshell carefully using No. 3 forceps and open the egg in DEPC-PBS (gecko embyro) or 1x Tyrode's buffer (zebra finch embryo).
3. Dissect embryos and collect brain tissue into 2 ml or 5 ml tube.

RNA extraction: TRIzol + QIAgen RNeasy mini kit

4. Add 1ml TRIzol LS reagent and vortex vigorously for homogenization. *(optional: Keep at −80 °C for long-term storage).*
5. Add 200 μl chloroform and shake vigorously for 15 s.
6. Sit 3 min at RT.
7. Extract RNA according to RNeasy kit protocol.
8. Elute RNA with 50 μl RNase-free H_2O and check concentration.

   cDNA synthesis: PrimeScript™ II 1st strand cDNA Synthesis Kit
9. Prepare reaction as below:

RNA template (<5 μg)	×	
Oligo dT primer	1	
10 mM dNTP mixture	1	
RNase-free water	y	
	10	μl

10. 65 °C, 5 min, incubate, then sit on ice.
11. Prepare reaction as below:

Reaction mix	10	
5 × PrimeScript II Buffer	4	
RNase Inhibitor (40 U/μL)	0.5	
PrimeScript II RTase (200 U/μL)	1	
RNase-free water	4.5	
	20	μl

12. Reaction by thermal cycler as below

42 °C	60 min
95 °C	5 min

13. Keep on ice.
14. Store at −20 °C.
15. cDNA templates are directly used for subsequent PCR and TA-cloning.

3.4 Template for Probe Preparation (See Note 5)

1. Prepare template for probe preparation from plasmids that contain target sequence and T7 or Sp6 promoter using (A) linear plasmid or (B) PCR product.

   (A)-1. Linearize 20 μg/probe plasmid with appropriate enzyme in 100 μl reaction at appropriate temperature for hours.

(A)-2. Perform electrophoresis with 5 μl reaction to ensure complete digestion.

(B)-1. Amplify DNA fragment using appropriate primers (*see* **Note 6**) in 20 μl reaction/probe.

(B)-2. Perform electrophoresis with 2 μl reaction to confirm a single DNA product.

6. Purify fragments/PCR product: use PCR purification kit in case of single DNA fragment, or use a Gel extraction kit if multiple fragments are detected. Elute with 20 μl Nuclease-free water.

3.5 In Vitro Transcription

1. Prepare reaction as below (*see* **Note 7**):

Template (1 μg) + nuclease-free water	12	
10× transcription buffer	2	
10× DIG or Fluorescence (FL) Mix	2	
RNase inhibitor	1	
10 mM DTT	1	
RNA polymerase	2	
	20	μl

Incubate at 37 °C for 2 h.

2. Add 2 μl DNase (RNase-free) to reaction, 37 °C, 2 h.
3. For precipitation, add:

Nuclease-free water	100	
4 M LiCl	10	
(DIG) 100% Ethanol	300	
(FL) Isopropanol	100	
(DIG) 432; (FL) 282		μl

Let sit at −20 °C for 1 h.

4. Spin 16,000 x *g* for 15 min at 4 °C (*see* **Note 8**).
5. Wash with 500 μl 70% ethanol once, 16,000 x *g* for 10 min at 4 °C.
6. Air-dry at room temperature for 10 min (*see* **Note 9**).
7. Dissolve the pellet in 100 μl or 75 μl (if using RNA-binding spin column) nuclease-free water.
8. (*Optional*) centrifuge RNA-binding spin column (Roche), 1,000 x *g*, 1 min at room temperature.
9. Apply probe solution to RNA-binding spin column.

10. 1,000 x *g*, 4 min at room temperature.
11. After measuring the concentration, preserve at −20 °C.

3.6 In Situ Hybridization (See Note 10)

3.6.1 Day 1

1. Thaw 4% paraformaldehyde in 37 °C water bath and keep on ice before using.
2. Dry slides at room temperature for 1 h.
3. Post-fix with 4% paraformaldehyde on ice for 10 min.
4. Wash with DEPC-PBS at room temperature for 5 min, twice.
5. 1.5% H_2O_2-MeOH at room temperature for 15 min.
6. Wash with DEPC-PBS at room temperature for 5 min, twice.
7. 0.2 M hydrochloric acid room temperature for 8 min.
8. Wash with DEPC-PBS at room temperature for 5 min, twice.
9. 10 μg/ml proteinase K at room temperature for 5 min.
10. Wash with DEPC-PBS at room temperature for 5 min.
11. Post-fix with 4% paraformaldehyde on ice for 10 min.
12. Wash with DEPC-PBS at room temperature for 5 min.
13. Acidic solution at room temperature for 10 min.
14. Wash with DEPC-PBS at room temperature for 5 min, three times.
15. Dilute RNA probe in hybridization buffer (50–100 ng/ml) and pre-warm at 80 °C for 2 min (*see* **Note 10**). Preheat the hybridization oven (or temperature adjustable incubator) at 55 °C (*see* **Note 11**). Prepare a humidity chamber using wet tissue paper with 50% formamide in nuclease-free water.
16. Transfer slides to a chamber and add hot probe (300 μl/slide). Apply HybriSlip on slides to cover the section. Incubate at 55 °C overnight in the hybridization oven (*see* **Note 12**).

3.6.2 Day 2 (See Note 13)

Pre-warm 37 °C and 65 °C water bath and 37 °C oven.

1. Remove hybrid-slip by holding slides vertically using slide forceps (*see* **Note 14**).
2. Wash in 5x SSC at 65 °C for 10 min.
3. Rinse in 50% formamide in 2x SSC at 65 °C for 30 min.
4. Wash with RNase buffer at 37 °C for 10 min, three times.
5. Treat with 20 μg/ml RNase A in RNase buffer at 37 °C for 30 min.
6. Wash with RNase buffer at 37 °C for 15 min.
7. 50% formamide in 2x SSC at 65 °C for 20 min, twice.
8. 2x SSC at 37 °C for 15 min.
9. 0.1x SSC at 37 °C for 15 min.

10. Wash with wash buffer at room temperature for 5 min.
11. Draw edge of sections using Pap Pen.
12. Transfer slides to chamber and add pre-warmed 1% BB1 solution (300 μl/slide) for blocking. Incubate at 37 °C for 30 min (*see* **Note 15**).
13. Apply Anti-DIG-antibody in 1% BB1 to slides (300 μl/slide) at 4 °C overnight.

3.6.3 Day 3 (for Chromogenic Detection)

Dissolve BB2 at 37 °C water bath. BB2 can be stored at 4 °C for 1 month.

1. Wash with wash buffer at room temperature, 30 s once and 5 min for three times.
2. Apply BB2 to slides (300 μl/slide) at 37 °C for 30 min.
3. Apply Biotinyl-Tyramide solution to slides (300 μl/slide) at 37 °C for 15 min. Prevent from light.
4. Wash with wash buffer at room temperature, 30 s once and 5 min for three times.
5. Apply Streptavidin-AP solution (300 μl/slide) at room temperature for 1 h. Prevent from light.
6. Wash with wash buffer at room temperature, 30 s once and 5 min for three times.
7. Apply B3 solution at room temperature for 5 min.
8. Apply B4 solution (300 μl/slide) at room temperature for signal development (*see* **Note 16**).
9. Stop reaction with PBS.
10. Mount with CC/Mount™ tissue mounting medium.

3.6.4 Day 3 (for Dual Color Fluorescence In Situ Hybridization)

Pre-warm BB2 solution in 37 °C water bath

1. Wash with wash buffer at room temperature, 30 s once and 5 min for three times.
2. Apply BB2 to slides (300 μl/slide) at 37 °C for 30 min.
3. Apply Biotinyl-Tyramide solution to slides (300 μl/slide) at 37 °C for 15 min. Prevent from light.
4. Wash with wash buffer at room temperature, 30 s once and 5 min for three times.
5. Apply Texas Red Streptavidin solution (300 μl/slide) at room temperature for 2 h. Prevent from light and check signal occasionally (*see* **Note 17**).
6. Wash with wash buffer at room temperature, 30 s once and 5 min for three times.

7. Inactivate with 3% hydrogen peroxide at room temperature for 30 min.
8. Wash with PBS at room temperature for 5 min, three times.
9. Apply Avidin Blocker (3 drops/slide) at room temperature for 15 min.
10. Wash with PBS at room temperature for 5 min.
11. Apply Biotin Blocker (3 drops/slide) at room temperature for 15 min.
12. Wash with PBS at room temperature for 5 min.
13. Add pre-warmed 1% BB1 solution (300 μl/slide) for blocking. Incubate at 37 °C for 30 min.
14. Apply Anti-Fluorescein-POD antibody in 1% BB1 to slides (300 μl/slide) at 4 °C overnight.

3.6.5 Day 4 (for Dual Color Fluorescence In Situ Hybridization)

Pre-warm BB2 at 37 °C water bath

1. Wash with wash buffer at room temperature, 30 s once and 5 min for three times.
2. Apply BB2 to slides (300 μl/slide) at 37 °C for 30 min.
3. Apply Biotinyl-Tyramide solution to slides (300 μl/slide) at 37 °C for 15 min. Prevent from light.
4. Wash with wash buffer at room temperature, 30 s once and 5 min for three times.
5. Apply FITC Streptavidin solution (300 μl/slide) at room temperature for 2 h. Prevent from light and check signal occasionally (*see* **Note 17**).
6. Wash with wash buffer at room temperature, 30 s once and 5 min for three times.
7. Stop reaction with PBS.
8. Mount with SlowFade® Gold Antifade Mountant with DAPI or appropriate aqueous mounting medium for fluorescence staining.

4 Notes

1. In geckos, from 28 d.p.o. and onward the olfactory nerve cords develop. If the olfactory bulbs need to be preserved, hold the brain surface using No. 3 forceps and gently remove the brain toward the olfactory bulb.
2. It is important to completely replace PBS to 30% sucrose-PBS. If solution exchange is not complete, tissues get fragile during subsequent cryosectioning and ISH procedures. The time required for replacement ranges from 6 h (early stage embryos)

to overnight (late stage embryos), and is judged by sinking of tissues. If tissues do not sink after 24 h, gentle tapping or shaking tubes may help.

3. While removing PBS, do not disturb the surface of brain tissue especially the dorsal surface.
4. Although the tips of No. 3 forceps are not too sharp, direct contact with brain tissues should be avoided. Ensure that O.C.T. solution has covered the entire tissue. If the O.C.T. solution is not homogeneous and residual PBS remains, the tissue can detach from the compound during subsequent cryo-sectioning procedure.
5. In order to prevent RNA and probe degradation, use Nuclease-free reagents/equipment and wear gloves during the entire procedure.
6. In the case of using pGEM-T plasmid backbones, M13-F and M13-R primers can be used for amplifying sequence and promoters.
7. Make sure to use appropriate RNA polymerase for anti-sense (signal detection) and sense (negative controls) transcription.
8. The centrifuge should be precooled at 4 °C to protect RNA probes from degradation.
9. Residual ethanol will affect the quality of RNA probes. Make sure to completely air-dry.
10. Two detection methods are provided here: chromogenic and fluorescence. For chromogenic detection, only one probe can be used in one slide; for fluorescence detection, two transcripts can be detected simultaneously using DIG- and Fluorescein-labeled probes.
11. If the signal and background are both high, higher hybridization temperature can be used (60–70 °C).
12. Do not let the slides from different probes contact each other and make sure slides are horizontally leveled.
13. Solutions used on day 2 should be pre-warmed in water bath at 37 °C or 65 °C.
14. If HybriSlips does not fall off, proceed to next step. Do not force to remove HybriSlips since this causes damage to sections. After incubation with 5x SSC, the HybriSlip should fall off naturally when taking out of the jars vertically.
15. If background signal is high, prolong 1% BBS incubation period.
16. For sensitive probes, the signal can be detected around 10 min. Prolonging developing time will increase background signal.
17. Signal can be checked after 1 h incubation.

References

1. Wang Z et al (2013) The draft genomes of soft-shell turtle and green sea turtle yield insights into the development and evolution of the turtle-specific body plan. Nat Genet 45 (6):701–706
2. Warren WC et al (2010) The genome of a songbird. Nature 464(7289):757–762
3. Green RE et al (2014) Three crocodilian genomes reveal ancestral patterns of evolution among archosaurs. Science 346 (6215):1254449
4. Hara Y et al (2015) Optimizing and benchmarking de novo transcriptome sequencing: from library preparation to assembly evaluation. BMC Genom 16:977
5. Zhang G et al (2014) Comparative genomic data of the avian Phylogenomics project. Gigascience 3(1):26
6. Whitney O et al (2014) Core and region-enriched networks of behaviorally regulated genes and the singing genome. Science 346 (6215):1256780
7. Wirthlin M et al (2014) Comparative genomics reveals molecular features unique to the songbird lineage. BMC Genom 15:1082
8. Pfenning AR et al (2014) Convergent transcriptional specializations in the brains of humans and song-learning birds. Science 346 (6215):1256846
9. Benichov JI et al (2016) The forebrain song system mediates predictive call timing in female and male zebra finches. Curr Biol 26 (3):309–318
10. Shein-Idelson M et al (2016) Slow waves, sharp waves, ripples, and REM in sleeping dragons. Science 352(6285):590–595
11. Karten HJ (2015) Vertebrate brains and evolutionary connectomics: on the origins of the mammalian 'neocortex'. Philos Trans R Soc Lond Ser B Biol Sci 370(1684)
12. Dugas-Ford J, Ragsdale CW (2015) Levels of homology and the problem of neocortex. Annu Rev Neurosci 38:351–368
13. Kumamoto T, Hanashima C (2014) Neuronal subtype specification in establishing mammalian neocortical circuits. Neurosci Res 86:37–49
14. Naumann RK et al (2015) The reptilian brain. Curr Biol 25(8):R317–R321
15. Montiel J et al (2016) From sauropsids to mammals and back: New approaches to comparative cortical development. J Comp Neurol 524(3):630–645
16. Nomura T, Gotoh H, Ono K (2013) Changes in the regulation of cortical neurogenesis contribute to encephalization during amniote brain evolution. Nat Commun 4:2206
17. Yoshida M et al (2016) Conserved and divergent expression patterns of markers of axial development in reptilian embryos: Chinese soft-shell turtle and Madagascar ground gecko. Dev Biol 415(1):122–142
18. Noro M et al (2009) Normal developmental stages of the Madagascar ground gecko Paroedura Pictus with special reference to limb morphogenesis. Dev Dyn 238(1):100–109
19. Nomura T et al (2015) Genetic manipulation of reptilian embryos: toward an understanding of cortical development and evolution. Front Neurosci 9:45
20. Mak SS et al (2015) Zebra finch as a developmental model. Genesis 53(11):669–677
21. Agate RJ et al (2009) Transgenic songbirds offer an opportunity to develop a genetic model for vocal learning. Proc Natl Acad Sci U S A 106(42):17963–17967
22. Chen CC et al (2013) Molecular profiling of the developing avian telencephalon: regional timing and brain subdivision continuities. J Comp Neurol 521(16):3666–3701
23. Jarvis ED et al (2013) Global view of the functional molecular organization of the avian cerebrum: mirror images and functional columns. J Comp Neurol 521(16):3614–3665
24. Pani AM et al (2012) Ancient deuterostome origins of vertebrate brain signalling centres. Nature 483(7389):289–294
25. Hébert JM, Fishell G (2008) The genetics of early telencephalon patterning: some assembly required. Nat Rev Neurosci 9(9):678–685
26. Murray JR et al (2013) Embryological staging of the Zebra Finch, Taeniopygia guttata. J Morphol 274(10):1090–1110

Chapter 23

Somitogenesis and Axial Development in Reptiles

Cindy Xu, Mariana B. Grizante, and Kenro Kusumi

Abstract

Among amniote vertebrates, reptiles display the greatest variation in axial skeleton morphology. Only recently have they been used in gene expression studies of somitogenesis, challenging previous assumptions about the segmentation clock and axial patterning. An increasing number of reptile genomes and transcriptomes are becoming available as next-generation sequencing becomes more affordable. Information regarding gene sequence and structure can be used to design and synthesize labeled riboprobes by in vitro transcription for gene expression analysis by in situ hybridization, thus, enabling the characterization of spatial and temporal expression patterns of genes involved in somitogenesis, a topic of great interest within evolutionary developmental studies of vertebrates.

Key words In situ hybridization, Reptile, Somitogenesis, Axial skeleton, Vertebra, Ribs, Development, In vitro transcription

1 Introduction

Reptiles have the greatest diversity in axial skeleton among amniotes, and there are many questions about the developmental mechanisms generating this variation. Whereas all known extant species of mammals have a modal number of 25 presacral vertebrae, including 7 cervical, 13 thoracic, and 5 lumbar segments [1], extant reptiles vary widely both in terms of number and vertebral shape and patterning [2]. Important differences in the axial skeleton can be identified both between and within reptile orders, specifically Crocodilia, Testudines, and Squamata. Species within Crocodilia, which includes alligators, crocodiles, and gharials, typically have 24 presacral vertebrae, including 9 cervical, 10 thoracic, and 5 lumbar segments [2, 3]. Species within Testudines, which includes turtles and tortoises, typically have 18 presacral vertebrae, including 8 cervical and 10 modified thoracic segments [2, 4]. A distinctive morphological characteristic of the Testudines is their bony shell, in which the fusion of dermal bones

Cindy Xu and Mariana B. Grizante contributed equally to this work.

Guojun Sheng (ed.), *Avian and Reptilian Developmental Biology: Methods and Protocols*, Methods in Molecular Biology, vol. 1650, DOI 10.1007/978-1-4939-7216-6_23, © Springer Science+Business Media LLC 2017

with 10 thoracic vertebrae and associated ribs forms the dorsal carapace and the fusion of the pectoral girdle and sternum forms the ventral plastron (reviewed in [5]). Notably, the greatest variation in axial skeleton is observed in Squamata, an order that includes lizards, snakes, and amphisbaenians. Along the evolution of squamates, independent events of axial elongation by an increase in the number of trunk vertebrae have occurred associated to limb reduction [6]. Trunk vertebral counts can range from 78 to 138 in amphisbaenian species, 114 to over 286 in snake species, and 16 to 116 in lizard species [7, 8]. To understand the evolution of diversity in the axial skeleton in reptiles, the underlying developmental mechanisms need to be examined. The mechanism generating the axial segments is somitogenesis, and the mechanism specifying axial identity is regulated by *Hox* genes.

The segmented vertebrae, axial musculature, and tendons are formed from segments of mesodermal tissue flanking the notochord, called somites [9]. During development, somites compartmentalize and differentiate into axial and limb skeletal muscle, the vertebral bones, cartilage, and tendons of the adult [10, 11]. Somite pairs are formed iteratively from head to tail, budding off from the rostral border of the presomitic mesoderm (PSM). Additional axial mesoderm is required to complete formation of the entire spine, and mesenchymal cells continue to be formed during somitogenesis from a structure called the tail bud [12, 13]. While there is underlying conservation of genetic networks regulating somitogenesis in vertebrates, the processes appear to vary between taxonomic orders, and the clock rate and final number of segments produced also differs considerably between species [14].

The developmental mechanism that drives the repeated formation of somites is termed the "segmentation clock" and is based on the "clock and wavefront model." There are two major components of this mechanism: (1) the "clock" that involves cyclical expression within the PSM of genes in the Notch, Wnt, and FGF signaling pathways, which interact with (2) a determination front or "wavefront" where segmental boundaries are predetermined. In the reptile segmentation clock, cycling expression of the Notch ligand delta-like 1 (*dll1*) and the *hairy* and *Enhancer of split* (HES) factor (*hes7*) is observed in the anole lizard [9]. However, the lunatic fringe (*Lfng*) gene shown to display cycling expression in the PSM of the mouse and chicken does not appear to be expressed in the PSM of either the green anole or alligator [15]. Differences have also been identified between the reptilian and mammalian determination front, which is marked by expression of the *Mesp2* gene [16]. While gradients of *fgf8* and *wnt3a* are present in both green anole and mouse, *hes6* is also expressed in a gradient in the lizard, as it is in anamniotes such as the *Xenopus* frog [15]. Opposing gradients of retinoic acid within the PSM produced by the somites have been described in other model systems (reviewed by [16]) but have not been examined in reptile embryos.

As somites are produced, they are simultaneously specified as to their identity along the rostral-caudal axis by differential expression of *Hox* genes. The specific combination of *Hox* genes determines vertebral identity (cervical, thoracic, lumbar, sacral, caudal) as well as the location of limb buds [17]. *Hox* gene patterns corresponding to axial transitions are generally conserved among vertebrates, with changes associated with divergence in the body plan as in crocodilians [3], the shell development in turtles [18], and the multiple instances of axial elongation and limblessness in squamates [19–23].

With the rapidly falling costs of next-generation sequencing, reptilian genomes and transcriptomes are becoming available at a fast pace. This will allow the exploration of the axial diversity present in reptiles in a manner that has never previously been possible. The access to gene sequences permits primers to be designed with high specificity to amplify templates for synthesizing riboprobes. The temporal and spatial patterns of gene expression within the embryo can then be examined by in situ hybridization in a species-specific manner. While in situ hybridization protocols have been published for developmental models such as the mouse and chicken, reptiles present a particular challenge due to the variability in eggs and embryos. Here, we present a streamlined protocol of riboprobe synthesis by in vitro transcription and its use for evolutionary analysis of reptilian development of the axial skeleton by whole mount in situ hybridization (Fig. 1). The protocol contains instructions for (1) collecting, dissecting, and fixing reptile embryos, (2) designing and testing primers to amplify fragments of genes of interest, (3) synthesizing riboprobes by in vitro transcription, and (4) performing whole mount in situ hybridization experiments.

2 Materials

RNA is highly sensitive to degradation from RNAses, which are found on the skin and secretions. Thus, contact with biological materials should be minimized. Gloves should be worn at all times. Lab benches and tools should be cleaned prior to use with solutions that inactivate RNAses. Since glassware can be difficult to make completely RNAse-free, plasticware that is RNAse-free or individually packaged should be used to make solutions. Investigators should separate reagents used for RNA studies. All aqueous solutions for RNA riboprobe steps should be prepared using commercial RNAse-free water. For other steps, purified, double-distilled water (ddH_2O) typically suffices. RNAse-free analytical grade reagents should be used whenever possible. Solutions can be prepared and stored at room temperature unless otherwise specified. Proper hazardous waste management and disposal should be strictly followed based on institution guidelines and policies.

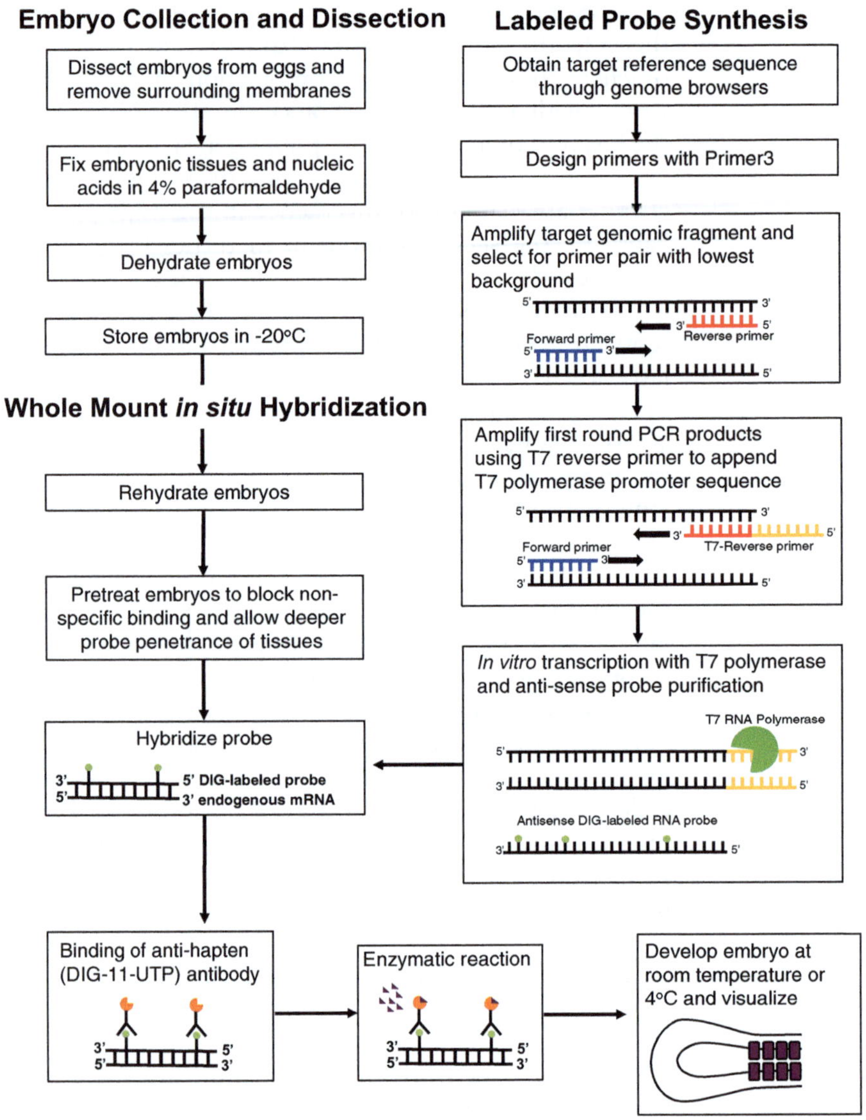

Fig. 1 An overview of the protocol for whole mount in situ hybridization of reptilian embryos

2.1 Embryo Collection, Dissection, and Fixation

Tools:

1. Surgical scissors (small).
2. Microdissecting forceps (Dumont #5 or finer).
3. Petri dishes.
4. Wide-bore plastic Pasteur transfer pipets (VWR cat #: 414004-032).
5. Microcentrifuge tubes (2 mL round bottom).

Solutions:

1. Phosphate-buffered saline (PBS, 0.01 M).
2. 4% paraformaldehyde (4% PFA) in PBS, pH 7.2–7.4, filtered (*see* **Note 1**).

3. 100% methanol.

Equipment:

1. Stereodissecting microscope.
2. Rocking platform or shaker (e.g., Nutator).

2.2 Primer Design and Selection

1. Geneious (http://www.geneious.com) or comparable sequence alignment tool.
2. Primer3 (http://bioinfo.ut.ee/primer3-0.4.0/).
3. Genomic or transcript sequence of interest.

2.3 Primer Testing and Sequence Amplification

1. For genomic DNA extraction, use Qiagen DNeasy Blood and Tissue Kit (cat #: 69504). For total RNA extraction and cDNA synthesis, use Ambion mirVana miRNA Isolation Kit, with phenol (cat #: AM 1560) and Invitrogen SuperScript III First-Strand Synthesis System (cat #: 18080051).
2. Roche GC-RICH PCR System Kit (cat #: 12314030600l).
3. dNTP mix (1.25 mM).
4. Forward primer oligonucleotide (20 μM).
5. Reverse primer oligonucleotide (20 μM).

2.4 Probe Synthesis and Purification

1. PCR amplification kit of choice.
2. Forward primer oligonucleotide (20 μM).
3. Reverse-T7 primer oligonucleotide (20 μM) (*see* **Note 2**).
4. Qiagen QIAquick PCR Purification Kit (cat #: 28104).
5. Ambion MAXIscript T7 In Vitro Transcription Kit (cat #: AM1314).
6. Roche digoxigenin-11-UTP (cat #: 11209256910).
7. Qiagen RNeasy MinElute Cleanup Kit (cat #: 74204).

2.5 Whole Mount In Situ Hybridization

Equipment:

1. Flat-bottom cell culture plates (12 well).
2. Netwell inserts (15 mm diameter, Corning, cat #: 3477).
3. Rocking platform or shaker.
4. Hybridization oven with shaker or rotating platform.
5. Heating block or shaking Thermomixer (Eppendorf).

Stock solutions (can be prepared and stored):

1. PBS containing 0.1% Tween-20.
2. 75%, 50%, and 25% methanol (MeOH) in PBT.
3. 4% paraformaldehyde in PBS, pH 7.2–7.4, filtered.
4. Pre-hybridization buffer: formamide, 20× SSC pH 4.5, Tween-20, heparin (100 mg/mL). Mix 250 mL formamide,

125 mL 20× SSC pH 4.5, 500 μL Tween-20, and 250 μL heparin. Add water to adjust volume to 500 mL. Store at 4 ° C.

5. Solution I: formamide, 20× SSC pH 4.5, 20% SDS.

   Mix 100 mL formamide, 40 mL 20× SSC pH 4.5, 10 mL 20% SDS. Add water to adjust final volume to 200 mL (*see* **Note 3**).
6. Solution II: 5 M NaCl, 1 M Tris pH 7.5, Tween-20. Mix 20 mL NaCl, 2 mL Tris, and 200 μL Tween-20. Add water to adjust final volume to 200 mL.
7. Solution III: formamide, 20× SSC pH 4.5. Mix 100 mL formamide and 20 mL 20× SSC pH 4.5. Add water to adjust final volume to 200 mL.
8. TBS containing 0.1% Tween-20 (TBST, 1×).
9. Alkaline phosphatase developing buffer (NTMT, 1×): 5 M NaCl, 1 M Tris pH 9.5, 1 M $MgCl_2$, Tween-20 (*see* **Note 4**).
10. PBS (pH 5.5): PBS (0.01 M), HCl (12 N).
11. 50% glycerol in PBS.

Fresh solutions (must be prepared on the day of the experiment):

1. 6% hydrogen peroxide in PBT.
2. 4.5 μg proteinase K per mL PBT.
3. 2 mg glycine per mL PBT.
4. 0.2% glutaraldehyde in 4% paraformaldehyde solution.
5. Hybridization buffer: Add torula yeast RNA (10 μL/mL) and probe to pre-hybridization buffer. Heat to 70 °C.
6. 100 μg RNAse A per mL solution II.
7. Pre-blocking solution: 10% fetal bovine serum (FBS, heat inactivated) in TBST.
8. Blocking solution: 1% fetal bovine serum (FBS, heat inactivated) in TBST.
9. Antibody: Roche anti-DIG-AP Fab fragments antibody (cat #: 11093274910; dilute 1:2000).
10. Developer solution: Add 4.5 μL nitro blue tetrazolium chloride (NBT) and 3.5 μL 5-bromo-4-chloro-3′-indolyl phosphate p-toluidine salt (BCIP) per mL NTMT.

3 Methods

3.1 Embryo Collection, Dissection, and Storage

Reptiles vary considerably in the type and size of eggs, seasonal timing, and clutch size. In addition, reptiles differ in the embryonic stage at egg-laying and optimal temperature for development, which influences incubation conditions. Therefore, it is important to carry out a literature search for each species used in developmental studies during experimental design, as shown in Table 1.

Table 1
Relevant information for planning an in situ hybridization experiment using oviparous reptile species

Species	Common name	Egg-laying season	Eggs per clutch	Temperature of incubation	Incubation period (days)	Developmental process at egg laying	Somite pairs	Limb bud	Neural tube	Ecol. & Dev. Citations	G	T	NGS citation
Squamata													
Anolis carolinensis	Green anole	May–Aug [24]	1	28 °C [25]	34 [9]	Neurulation	25–30	Initial FL buds	Open	[9, 24, 25]	X	X	[26, 27]
Gekko japonicus	Schlegel's Japanese gecko	Apr–Jul [28]	2 [29]	24–32 °C [30]	62 ± 24	_	_	_	_	[28–30]	X	X	[31]
Pogona vitticeps	Bearded dragon	Oct [32]	11–47 [32, 33]	22–36 °C [34]	93 ± 4 [32]	_	_	_	_	[32–34]	X	X	[35]
Ophisaurus gracilis	Chinese glass lizard	Sep [36]	4–20 ($\bar{x}$=11) [37]	_	63 ± 7	_	_	_	_	[36, 37]	X	_	[38]
Python molurus	Burmese python	Dec–Mar	35 [39]	32 °C	64 ± 6	Neurulation	~260	Absent		[39, Pers. Obs.]	X	X	[40]
Ophiophagus hannah	King cobra	Mar–Jul [41]	27	30–32 °C	_	_	_	Absent	_	[41]	X	X	[42]
Pantherophis guttatus	Corn snake	Apr–Jul [43]	9–19	28 °C [44]	73 ± 5 [43]	_	110–215 [44]	Absent	_	[43, 44]	X	X	[45]
Testudines													
Chrysemys picta bellii	Western painted turtle	Jun–Aug [46]	5–22 [47]	26–30 °C	53.5 [48]	Gastrulation	Absent	Absent	Absent	[46–48]	X	X	[49]

(continued)

Table 1
(continued)

Species	Common name	Egg-laying season	Eggs per clutch	Temperature of incubation	Incubation period (days)	Developmental process at egg laying	Somite pairs	Limb bud	Neural tube	Ecol. & Dev. Citations	G	T	NGS citation
Trachemys scripta	Red-eared slider	Apr–Jul [50]	12.5–15.1 [51]	26–32 °C [52]	65 [53]	Gastrulation [51]	Absent	Absent	Absent	[50–53]	_	X	[54]
Pelodiscus sinensis	Chinese soft-shelled turtle	Jun–Aug [55]	9–15	28–32 °C [56]	47 ± 2	Gastrulation	Absent (4 on 2nd day)	Absent	Absent	[55, 56]	X	_	[57]
Chelonia mydas	Green sea turtle	Dec–Jul [58]	100 [37]	24–33 °C [59]	71 ± 23	Gastrulation	Absent	Absent	Absent	[58, 59]	X	_	[57]
Crocodilia													
Gavialis gangeticus	Indian gharial	Apr [60]	16–61 ($\bar{x}$=40)	31 °C (25–37 °C)§	82 ± 11	_	_	_	_	[60]	X	_	[61]
Crocodylus porosus	Saltwater crocodile	Jul–Aug [60]* Nov–Mar**	25–72 ($\bar{x}$=42)* 16–71 ($\bar{x}$=50)**	32 °C (25–37 °C)* 30.1 °C (25–37 °C)** §	80 ± 10* 89 ± 9**	Neurulation	16–18	Absent	Open	[60]	X	_	[61]
Alligator mississippiensis	American alligator	Jun [60]	2–68 ($\bar{x}$=39.8)	30 °C (23.3–35.7 °C)§	64 ± 1	Neurulation	16–18	Absent	Open	[60]	X	X	[61]
Alligator sinensis	Chinese alligator	Jun–Aug [60]	10–40 ($\bar{x}$=26)	29 °C [62]	75 ± 5 [60]	Neurulation	16–18 [63]	Absent	Open	[60, 62]	X	_	[63]

Somite pairs, limb bud, and neural tube descriptions correspond to egg-laying stage for each species. Abbreviations: *G* genome, *T* transcriptome, *NGS* next-generation sequencing, *FL* forelimb. Symbols: *Ceylon/**Australia; §range of egg cavity temperature (from [37])

Reptile eggs can be collected in the field, from laboratory-housed specimens, or obtained from breeders or commercial vendors. The investigator should work with their institutional animal regulatory officials to ensure that the required collection, housing, and/or transportation permits are obtained. For some species, additional regulations governing the transport or shipment of live eggs across state or national boundaries will be applicable, which may not apply to the transport of fixed embryonic tissues.

Reptile eggs are susceptible to desiccation and must be kept moist until the moment of dissection, but they must also have sufficient contact with air to permit gas exchange. A good substrate that meets both these requirements is vermiculite mixed with water in equal parts. Below we describe the dissection of a squamate egg, which typically has leathery shell. The procedure can be slightly different for geckos, crocodilians, tortoises, and some turtle eggs that have hard shells.

1. Thaw 4% PFA in a 37 °C water bath, if frozen, and keep it on ice. Place a Petri dish filled with PBS at stereodissecting microscope.
2. With surgical scissors, cut an opening at the top of the egg (*see* **Note 5**). Gently wash part of the yolk out of the egg through the opening using the Pasteur pipet. Open the rest of the eggshell with a scissor until you see the "embryonic disk." With caution, isolate it using fine forceps. Using a clean Pasteur pipet, transfer the "embryonic disk" to a new Petri dish filled with cold PBS.
3. Using forceps, dissect the embryo from the allantois and amniotic membranes. For embryos with newly closed neural tubes, puncture the rhombic roof of the hindbrain, as this cavity often traps riboprobe leading to background signal.
4. Using a clean Pasteur pipet, transfer the dissected embryo to a container (tube or plate well with Costar Netwells) with 4% PFA (*see* **Note 6**). Place container on rocking platform at 4 °C for 2 h to overnight depending on the size of the embryo.
5. Wash embryos 2× in PBS for 5 min to remove PFA. Dehydrate embryos in a series of 5 min washes with PBS: methanol mixtures (MeOH 25%, 50%, 75%, and finally 2× 100% MeOH). Store fixed embryos in round-bottom 2 mL tubes at −20 °C.

3.2 Primer Design and Selection

1. Sequence retrieval: Download nucleotide sequence of interest from public genome browsers. Common database portals include the National Center for Biotechnology Information (NCBI) (http://www.ncbi.nlm.nih.gov/), ENSEMBL (http://www.ensembl.org), or the University of California Santa Cruz (UCSC) Genome Browser (http://www.genome.ucsc.edu).

Obtain nucleotide sequence by using species and gene name as search words. This will be used as the reference sequence for primer selection (*see* **Note 7**). Depending on the sequence length and quality of the genome or transcriptome of interest, different methods can be used to obtain the target reference sequence.

(a) Annotated genome and/or transcriptome: A well-annotated genome or transcriptome includes information about location of exons and intron boundaries, protein coding sequences, and untranslated regions (UTRs). The sequence of interest can be downloaded directly from genome browsers in FASTA format (*see* **Note 8**).

(b) Unannotated transcriptome: If an annotation is not available for the transcriptome, the cDNA sequence of interest can be extracted from assembled contigs using nucleotide Basic Local Alignment Search Tool (http://blast.ncbi.nlm.nih.gov/Blast.cgi). Using the sequence of an orthologous gene of a closely related species as a query, BLAST identifies regions of sequence similarity and calculates the statistical significance between the query and assembled contigs. Regions of strong sequence similarity identified in the assembled contigs can be copied and saved as a text file to be used for primer design. It can be assumed a sequence downstream of the alignment is the 3′ UTR.

(c) Unannotated genomes are particularly challenging for designing primers. If some cDNA sequence is available for a related reptilian species, it may be possible to identify putative protein coding exons.

2. Download and label sequence files according to gene and species of origin, making careful note of whether the sequences represented are sense or antisense, genomic or cDNA, and of features such as intron-exon boundaries, coding sequence, and alternate splice forms.
3. Copy target sequence into Primer3 (http://bioinfo.ut.ee/primer3-0.4.0/). PCR-generated riboprobes typically range between 200 and 1000 nucleotides, with 600–700 base pairs being a recommended starting point (*see* **Note 9**). By default, five pairs of primers will be presented (*see* **Note 10**).
4. Given that primer prediction programs such as Primer3 do not always generate functioning primers, ordering of at least two forward and two reverse oligonucleotides for testing is recommended. Follow manufacturer's instructions to resuspend primers to make a stock solution (20 μM).

Table 2
A typical PCR protocol for generating templates for in vitro transcription

Description	Cycles	Time	Temperature (°C)
Initial denaturation	1	5 min	94
Denaturation Annealing Elongation	40	1 min 1 min 2 min	94 55 72
Final elongation	1	7 min	72
Storage	1	Indefinite	4

3.3 Generating In Vitro Transcription Templates by RT-PCR or Genomic PCR

1. To extract total RNA from embryos for RT-PCR, refer to the instruction manual of the Ambion mirVana miRNA Isolation Kit, with phenol. To synthesize cDNA, refer to the instruction manual of the Invitrogen SuperScript III First-Strand Synthesis System Kit (*see* **Note 11**). To extract genomic DNA from embryos, refer to the instruction manual for the DNeasy Blood and Tissue Kit.
2. To increase the sensitivity of the RT-PCR or genomic PCR reaction, use the Roche GC-RICH PCR System Kit, following the instruction manual. All reagents can be found in the kit except for the dNTP mix. A conservative PCR protocol is shown below (Table 2), but annealing and elongation times can be optimized (*see* **Note 12**). Store PCR products at −20 °C since they will be used as template for a second round of PCR, for synthesis of the in vitro transcription reaction template.
3. Run a 1% agarose gel to analyze PCR amplified products.
4. Order a T7-reverse primer based on the best primer pair with the lowest background.

3.4 Probe Synthesis and Purification

1. To generate a PCR template for in vitro transcription reaction, most common PCR kits can be used. In order to add the necessary T7 promoter site to the template sequence, use a chimeric T7-reverse primer (*see* **Note 13**) instead of the regular reverse primer.
2. For in vitro transcription, follow the nonisotopic label instruction manual of the MAXIscript® T7 In Vitro Transcription Kit provided by Ambion. Use digoxigenin-11-UTP as the nonisotopic label. After incubation, treat with TURBO DNase 1 (supplied by kit).
3. Run part of the in vitro transcription reaction product (typically 1 μL of a 10 μL reaction) on a 1% agarose gel to assess efficiency of synthesis (*see* **Note 14**).

4. For purification of the RNA probe away from digoxigenin-11-UTP-labeled nucleotide, refer to the instruction manual of the RNeasy MinElute Cleanup Kit provided by Qiagen.
5. Optional: Run a 1% agarose gel to quantify the amount of product that has been purified.

3.5 Whole Mount In Situ Hybridization

This protocol was used in [9] and is based on mouse protocols published previously by [64].To facilitate the testing of multiple embryos and probes simultaneously (e.g., 6–48 conditions), Netwells permit rapid washes of multiple samples. Washes are carried out in 12-well cell culture plates that have been fitted with a net-bottom well used to transfer embryos between solutions. For Netwells designed for 12-well plates, the optimal volume per well is 2.5 mL of solution.All washes are carried out on a shaker or rocking platform for 5 min at room temperature unless otherwise specified. For hybridization and pre-hybridization, embryos are transferred into round-bottomed microcentrifuge tubes to reduce volume and amount of probe needed, but hybridizations can also be carried out in cell culture plates.

Day 1

1. Rehydrate embryos in a series of washes (75% MeOH:25% PBT, 50% MeOH:50% PBT, 25% MeOH:75% PBT) (*see* **Note 15**). Wash 2× in PBT.
2. To clear embryos, wash in 6% hydrogen peroxide diluted in PBT for 1 h. Wash 3× in PBT to remove residual hydrogen peroxide.
3. To permeabilize embryos for hybridization with riboprobes, treat embryos with 4.5 μg/mL of proteinase K in PBT for 5–10 min. The length of treatment depends on embryonic size and proteinase K stock batch. For each species, the length of proteinase K treatment should be optimized, but 15 min for embryos approximately 3.2 mm in length, equivalent to a 9.5-day mouse embryo, can be expected.
4. After digestion, block proteinase activity in 2 mg/mL of glycine in PBT 2×.
5. Fix the embryos in 0.2% glutaraldehyde in 4% paraformaldehyde for 20 min (*see* **Note 16**). Wash 2× in PBT.
6. Using a wide-bore Pasteur pipet, gently transfer the embryos from Netwells to 2 mL tubes pre-filled with pre-hybridization buffer (*see* **Notes 17** and **18**). The amount of buffer should be sufficient to completely cover the embryos, ideally at least ten times the estimated volume of the embryos. Pre-hybridize at 70 °C for at least 1 h in the heat block.
7. Before transferring embryos to hybridization buffer, prepare solution by adding torula yeast (10 μL/ 1 mL buffer) and

predetermined amount of riboprobe (*see* **Note 19**) to pre-hybridization buffer. Warm to 70 °C.

8. Carefully remove pre-hybridization solution from tube using a pipet. Note that embryos in pre-hybridization buffer are very translucent and difficult to see, so examination of the tube on the stereodissecting microscope may be needed to prevent accidental pipetting of the embryos. Rapidly add pre-heated hybridization solution and avoid having embryos cool down to room temperature (*see* **Note 20**). Hybridize overnight at 70 °C with gentle shaking using a thermomixer, or if using a standard heat block, manually shake the tubes a few times during the incubation to mix solution. Aliquot and pre-warm solution I and I:II at 70 °C for day 2.

Day 2

9. Using a wide-bore Pasteur pipet, transfer embryos out of 2 mL tubes back into the cell culture plates with Netwells. Wash 2× in Solution I for 30 min at 70 °C in a hybridization oven with a rocking platform.
10. Wash 1× in Solution I:II for 10 min at 70 °C.
11. Wash 3× in Solution II for 5 min at room temperature.
12. To reduce unhybridized probe, incubate embryos in RNAse A (100 μg/mL) diluted in Solution II for 30 min at 37 °C. Embryos can be left in RNAse A up to 50 min.
13. Wash in 1× in Solution II.
14. Wash 1× in Solution III.
15. Wash 2× in Solution III for 30 min at 65 °C.
16. Wash 3× in TBST.
17. Pre-block embryos in 10% heat-inactivated fetal bovine serum diluted in TBST for at least 1 h.
18. Incubate embryos with anti-DIG-AP antibodies (1:2000) diluted in 1% FBS in TBST at 4 °C overnight in the dark (*see* **Note 21**).

Day 3

19. Wash embryos in TBST for 8–10 times, changing the solution every hour. For the final wash, place at 4 °C overnight with rocking.

Day 4

20. Wash 3× in NTMT for 10 min.
21. Prepare cold developer solution on ice and in the dark. Add 4.5 μL NBT/mL and 3.5 μL BCP/mL of NTMT. Place embryos in developer solution and wrap in aluminum foil to protect from light (*see* **Note 22**).

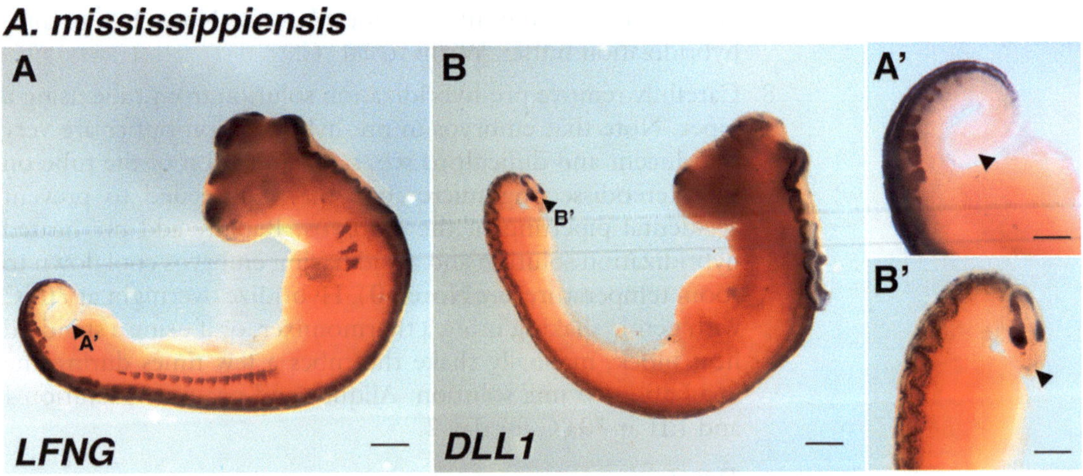

Fig. 2 Expression of developmental genes in *Alligator mississippiensis*. (**a**) Expression of *LFNG* is present in the neural tube and somites but is absent in the presomitic mesoderm (*black arrowhead*). (**b**) *DLL1* expression is localized to the dorsal edges of the closing neural tube and presomitic mesoderm (*black arrowhead*). (**a′**) *LFNG* expression absent in the presomitic mesoderm (*black arrowhead*). (**b′**) *DLL1* is expressed throughout the presomitic mesoderm with increased expression in a cyclical band localized toward the determination front (*black arrowhead*). Figures (**a**) and (**b**), scale bar = 500 μm. Figures (**a′**) and (**b′**), scale bar = 250 μm

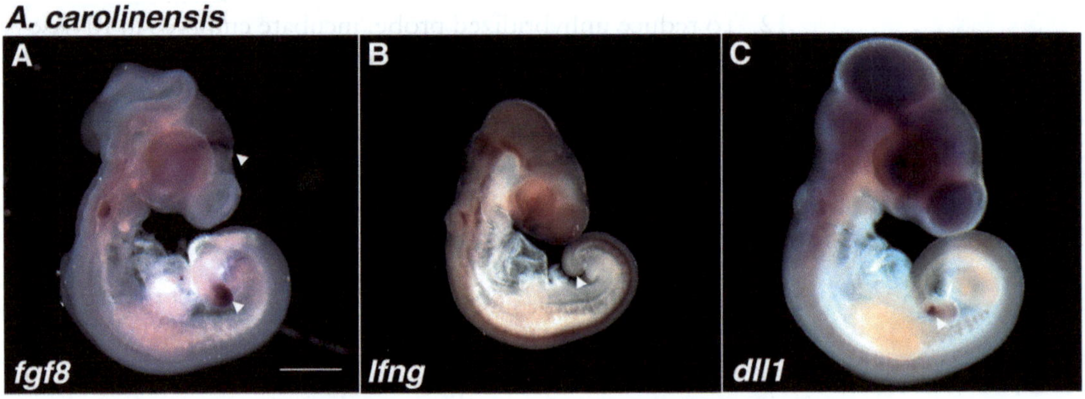

Fig. 3 Expression of developmental genes in the green anole, *Anolis carolinensis*. (**a**) *fgf8* is expressed as a gradient within the caudal presomitic mesoderm as well as at the boundary between the midbrain and hindbrain (*white arrowheads*). (**b**) Expression of *lfng* is absent from the presomitic mesoderm (*white arrowhead*) but is expressed in the developing neural tube. (**c**) *dll1* is expressed in band within the rostral end of the presomitic mesoderm (*white arrowhead*). Scale bar = 500 μm. Images (**b**) and (**c**) courtesy of Walter Eckalbar

22. Stop reaction by washing 2× in NTMT, followed by washing 1× in PBT (pH 5.5). Clear and refix embryos in 0.1% glutaraldehyde diluted in 4% paraformaldehyde for 20 min at 4 °C.
23. Wash 2× in PBT. Either transfer embryos to a 2 mL tube or cover wells with parafilm to reduce evaporation. Store in 50% glycerol diluted in PBS at 4 °C. Gene expression associated with somitogenesis are shown in *Alligator mississippiensis* (Fig. 2) and *Anolis carolinensis* (Fig. 3).

4 Notes

1. Paraformaldehyde is a known formaldehyde-releasing agent and a suspected carcinogen, so follow instructions from your institution's environmental health and safety office for preparation. Protocols for preparing 4% PFA can be found from many sources. Filter 4% PFA solution to remove particulates that may bind to embryos. Aliquot and store at −20 °C. For use, thaw and resolubilize at 37 °C.
2. To design T7 promoter-chimeric primers, add the T7 RNA polymerase promoter sequence to the 5′ end of the reverse primer sequence (relative to the protein coding strand). The T7 sequence is preceded by a 5′ end three nucleotide clamp sequence (GCG) is as follows: 5″- GCG TAA TAC GAC TCA CTA TAG GGA GA-3′.
3. SDS will precipitate over time at room temperature but can be resolubilized when incubated at 37° C.
4. Tween-20 detergent will precipitate at room temperature after a few days. Concentrated stock solution can be prepared ahead of time omitting the Tween-20. Before use, aliquot Tween-20 to the working solution.
5. If dissecting an egg that has hard shell and that is not too small in size, place a piece of cellophane adhesive tape on top of the egg. On the opposite side of the egg, gently insert a needle with syringe and remove a small amount of albumin to depressurize the egg. Then use scissors to poke a hole on the region of shell with tape and cut an opening. This will help keep shell in one piece and avoid the embryo being ejected due to pressure.
6. For larger embryos, cut the tip of the Pasteur pipet with clean scissors to create a wider opening.
7. To generate highly specific probes, we recommend designing primer pairs that incorporate the 3′ untranslated region (3′ UTR). Unlike coding sequences, the 3′ UTR sequence diverges more rapidly dependent upon the species. This can reduce the potential cross-hybridization with paralogous genes, a concern especially when targeting in highly conserved functional domains.
8. Alignment tools are highly sensitive to strand orientation. The directionality of each sequence should always be noted when downloading from its original source.
9. Long riboprobes do not penetrate tissue as deeply but do stain more intensely than short riboprobes. However, long riboprobes can cross hybridize with closely related members of the same gene family making them less specific.

10. In this protocol, "forward" primers correspond to sense strand sequence and "reverse" primers to antisense sequence.
11. cDNA is most commonly used as template for gene amplification because the sequence is free of introns. If using genomic DNA, primer pairs should be designed to amplify within a single large exon and avoid including any intronic sequence.
12. The default parameters for Primer3 include an annealing temperature of approximately 60 °C. The annealing temperature used in the PCR cycles should be reduced to 55 °C to allow for slight differences from the theoretical annealing temperature.
13. To ensure sufficient riboprobe, synthesize multiple replicates. Replicates can be pooled after amplification.
14. Gel electrophoresis is an efficient and well-established method to separate and analyze DNA, but RNA can also be visualized. If staining with ethidium bromide, RNA yields lower intensity bands compared to DNA. A sharp band is typically not observed due to folding and formation of secondary structures.
15. For optimal results, transfer embryos directly into Netwells filled with cold 100% MeOH before starting the graded methanol wash series. When transferring Netwells between wells on a plate, gently tap the bottom of the Netwell against the plate to remove excess solution.
16. Embryos can become overfixed and brittle if left in fixative for more than 20 min.
17. Pre-hybridization solution should be pre-warmed to 70 °C.
18. Remove as much fixative as possible when transferring embryos from wells to tubes with pre-hybridization solution. This can be achieved by placing the pipet tip lightly against the wall of the well and gently squeezing out excess fixative. Then pipet embryos in into the microcentrifuge tube filled with pre-warmed pre-hybridization buffer, avoiding creating bubbles.
19. The amount of probe depends on its labeling efficiency. The recommended amount for initial experiments is 1 μg per mL of hybridization solution, but titration can be used in subsequent experiments to identify the minimal amounts necessary for the embryonic structures being analyzed.
20. First remove pre-hybridization buffer with a narrow-bore micropipet to avoid suctioning out the embryos. Remove pre-hybridization solution starting at the top surface and be careful to not disturb the embryo. Remove as much liquid as possible. Replace with pre-warmed hybridization buffer.
21. If necessary, this protocol can be halted at the antibody incubation step for up to 2 days.

22. Alkaline phosphatase colorimetric reactions proceed at different rates at room temperature vs. 4 °C, affecting development times. At 4 °C, embryos will develop slower, but the background noise may be minimized due to not overdeveloping. When first testing a probe, we recommend developing embryos at room temperature and checking every 30 min to establish the rate at which the colorimetric reaction is observed.

Acknowledgments

We would like to thank Walter Eckalbar for the images of green anole embryonic gene expression of *dll1* and *lfng*. We also thank Catherine May and Ruth Elsey for the alligator embryo collection and dissection and Jon Grove for information on Burmese python reproduction.

References

1. Galis F, Carrier D, van Alphen J, van der Mije SD, Van Dooren TJM, Metz JAJ et al (2014) Fast running restricts evolutionary change of the vertebral column in mammals. Proc Natl Acad Sci U S A 111:11401–11406
2. Müller J, Scheyer TM, Head JJ, Barrett PM, Werneburg I, Ericson PGP et al (2010) Homeotic effects, somitogenesis and the evolution of vertebral numbers in recent and fossil amniotes. Proc Natl Acad Sci U S A 107:2118–2123
3. Mansfield JH, Abzhanov A (2010) Hox expression in the American alligator and evolution of Archosaurian axial patterning. J Exp Zool Part B Mol Dev Evol 314:629–644
4. Gilbert SF, Loredo GA, Brukman A, Burke AC (2001) Morphogenesis of the turtle shell: the development of a novel structure in tetrapod evolution. Evol Dev 3:47–58
5. Hirasawa T, Pascual-Anaya J, Kamezaki N, Taniguchi M, Mine K, Kuratani S (2014) The evolutionary origin of the turtle shell and its dependence on the axial arrest of the embryonic rib cage. J Exp Zool Part B Mol Dev Evol 324:194–207
6. Wiens JJ, Brandley MC, Reeder TW (2006) Why does a trait evolve multiple times within a clade? Repeated evolution of snakelike body form in squamate reptiles. Evolution 60:123–141
7. Romer AS (1956) Osteology of the reptiles. University of Chicago Press, Chicago
8. Polly PD, Head JJ, Cohn MJ (2001) Testing modularity and dissociation: the evolution of regional proportions in snakes. In: Zelditch ML (ed) Beyond Heterochrony: the evolution of development. Wiley-Liss, New York
9. Eckalbar WL, Lasku E, Infante CR, Elsey RM, Markov GJ, Allen AN et al (2012) Somitogenesis in the anole lizard and alligator reveals evolutionary convergence and divergence in the amniote segmentation clock. Dev Biol 363:308–319
10. Christ B, Huang R, Scaal M (2007) Amniote somite derivatives. Dev Dyn 236:2382–2396
11. Brent AE, Tabin CJ (2002) Developmental regulation of somite derivatives: muscle, cartilage and tendon. Curr Opin Genet Dev 12:548–557
12. Beck CW (2015) Development of the vertebrate tailbud. Wiley Interdiscip Rev Dev Biol 4:33–44
13. Ohta S, Suzuki K, Tachibana K, Tanaka H, Yamada G (2007) Cessation of gastrulation is mediated by suppression of epithelial-mesenchymal transition at the ventral ectodermal ridge. Development 134:4315–4324
14. Kusumi K, May CM, Eckalbar WL (2013) A large-scale view of the evolution of amniote development: insights from somitogenesis in reptiles. Curr Opin Genet Dev 23:491–497
15. Koyano-Nakagawa N, Kim J, Anderson D, Kintner C (2000) Hes6 acts in a positive feedback loop with the neurogenins to promote

neuronal differentiation. Development 127:4203–4216

16. Aulehla A, Pourquié O (2010) Signaling gradients during paraxial mesoderm development. Cold Spring Harb Perspect Biol 2:a000869
17. Burke A, Nelson C, Morgan B, Tabin C (1995) Hox genes and the evolution of vertebrate axial morphology. Development 121:333–346
18. Ohya YK, Kuraku S, Kuratani S (2005) Hox code in embryos of Chinese soft-shelled turtle *Pelodiscus sinensis* correlates with the evolutionary innovation in the turtle. J Exp Zool Part B Mol Dev Evol 304:107–118
19. Cohn M, Tickle C (1999) Developmental basis of limblessness and axial patterning in snakes. Nature 399:474–479
20. Di-Poï N, Montoya-Burgos J, Miller H, Pourquié O, Milinkovitch MC (2010) Changes in *Hox* genes′ structure and function during the evolution of the squamate body plan. Nature 464:99–103
21. Woltering JM, Vonk FJ, Müller H, Bardine N, Tuduce IL, de Bakker MAG et al (2009) Axial patterning in snakes and caecilians: evidence for an alternative interpretation of the Hox code. Dev Biol 332:82–89
22. Guerreiro I, Nunes A, Woltering JM, Casaca A, Nóvoa A, Vinagre T et al (2013) Role of a polymorphism in a Hox/Pax-responsive enhancer in the evolution of the vertebrate spine. Proc Natl Acad Sci U S A 110:10682–10686
23. Head JJ, Polly PD (2015) Evolution of the snake body form reveals homoplasy in amniote Hox gene function. Nature 520:86–91
24. Licht P (1973) Influence of temperature and photoperiod on the annual ovarian cycle in the lizard *Anolis carolinensis*. Copeia 1973:465–472
25. Lovern MB, Wade J (2003) Yolk testosterone varies with sex in eggs of the lizard, *Anolis carolinensis*. J Exp Zool A Comp Exp Biol 295:206–210
26. Alföldi J, Di Palma F, Grabherr M, Williams C, Kong L, Mauceli E et al (2011) The genome of the green anole lizard and a comparative analysis with birds and mammals. Nature 477:587–591
27. Eckalbar WL, Hutchins ED, Markov GJ, Allen AN, Corneveaux JJ, Lindblad-Toh K et al (2013) Genome reannotation of the lizard *Anolis carolinensis* based on 14 adult and embryonic deep transcriptomes. BMC Genomics 14:49
28. Ikeuchi I (2004) Male and female reproductive cycles of the Japanese Gecko, *Gekko japonicus*, in Kyoto, Japan. J Herpetol 38:269–274
29. Zhang YP, WG D, Zhu LJ (2009) Differences in body size and female reproductive traits between two sympatric geckos, *Gekko japonicus* and *Gekko hokouensis*. Folia Zool 58:113–122
30. Ji X, Wang PC (1991) Incubation character of eggs of the gecko *Gekko japonicus*. Zool Res 12:28–78
31. Liu Y, Zhou Q, Wang Y, Luo L, Yang J, Yang L et al (2015) *Gekko japonicus* genome reveals evolution of adhesive toe pads and tail regeneration. Nat Commun 6:1–11
32. Cannon MJ (2003) Husbandry and veterinary aspects of the bearded dragon (*Pogona* spp.) in Australia. Semin Avian Exot Pet Med 12:205–214
33. Holleley CE, O'Meally D, Sarre SD, Marshall-Graves JA, Ezaz T, Matsubara K et al (2015) Sex reversal triggers the rapid transition from genetic to temperature-dependent sex. Nature 523:79–82
34. Quinn AE, Georges A, Sarre SD, Guarino F, Ezaz T, Graves JAM (2007) Temperature sex reversal implies sex gene dosage in a reptile. Science 316:411
35. Georges A, Li Q, Lian J, O'Meally D, Deakin J, Wang Z et al (2015) High-coverage sequencing and annotated assembly of the genome of the Australian dragon lizard *Pogona vitticeps*. Gigascience 4:45
36. Wall F (1908) Notes on the incubation and brood of the Indo-Burmese snake lizard or slow worm (*Ophisaurus gracilis*). J Bombay Nat Hist Soc 18:503–504
37. Vitt LJ, Caldwell JP (2009) Herpetology: an introductory biology of amphibians and reptiles, 3rd edn. Academic Press, Massachusetts
38. Song B, Cheng S, Sun Y, Zhong X, Jin J, Guan R et al (2015) A genome draft of the legless anguid lizard, *Ophisaurus gracilis*. Gigascience 4:15–17
39. Mierop LHS V, Barnard SM (1976) Observations on the reproduction of *Python molurus bivittatus* (Reptilia, Serpentes, Boidae). J Herpetol 10:333–340
40. Castoe TA, de Koning APJ, Hall KT, Card DC, Schield DR, Fujita MK et al (2013) The Burmese python genome reveals the molecular basis for extreme adaptation in snakes. Proc Natl Acad Sci U S A 110:20645–20650
41. Chanhome L, Jintakune P, Wilde H, Cox MJ (2001) Venomous snake husbandry in Thailand. Wilderness Environ Med 12:17–23
42. Vonk FJ, Casewell NR, Henkel CV, Heimberg AM, Jansen HJ, McCleary RJR et al (2013) The king cobra genome reveals dynamic gene evolution and adaptation in the snake venom

system. Proc Natl Acad Sci U S A 110:20651–20656
43. Bird WM, Peak P, Baxley DL (2015) Natural history and meristics of an allopatric population of red cornsnakes, *Pantherophis guttatus* (Linnaeus, 1766) in Central Kentucky, USA. J North Am Herpetol 1:6–11
44. Gomez C, Ozbudak EM, Wunderlich J, Baumann D, Lewis J, Pourquié O (2008) Control of segment number in vertebrate embryos. Nature 454:335–339
45. Ullate-Agote A, Milinkovitch MC, Tzika AC (2014) The genome sequence of the corn snake (*Pantherophis guttatus*), a valuable resource for EvoDevo studies in squamates. Int J Dev Biol 58:881–888
46. Krawchuk MA, Brooks RJ (1998) Basking behavior as a measure of reproductive cost and energy allocations in the painted turtles, *Chrysemys picta*. Herpetologica 5:112–121
47. Cordero GA, Janzen FJ (2014) An enhanced developmental staging table for the painted turtle, *Chrysemys picta* (Testudines: Emydidae). J Morphol 275:442–455
48. Rowe JW (1994) Reproductive variation and the egg size-clutch size trade-off within and among populations of painted turtles (*Chrysemys picta bellii*). Oecologia 99:35–44
49. Shaffer HB, Minx P, Warren DE, Shedlock AM, Thomson RC, Valenzuela N et al (2013) The western painted turtle genome, a model for the evolution of extreme physiological adaptations in a slowly evolving lineage. Genome Biol 14:R28
50. Gibbons JW, Greene JL (1990) Reproduction in the slider and other species of turtles. In: Whitfield JW (ed) Life history and ecology of the slider turtle. Smithsonian Institution Press, Washington, D.C
51. Filoramo NI, Janzen FJ (1999) Effects of hydric conditions during incubation on overwintering hatchlings of the red eared slider turtle (*Trachemys scripta elegans*). J Herpetol 33:29–35
52. Spotila LD, Spotila JR, Hall SE (1998) Sequence and expression analysis of *Wt1* and *Sox9* in the red-eared slider turtle, *Trachemys scripta*. J Exp Zool 281:417–427
53. Tucker JK, Paukstis GL, Janzen FJ (1998) Annual and local variation in reproduction in the red-eared slider, *Trachemys scripta elegans*. J Herpetol 32:515–526
54. Kaplinsky NJ, Gilbert SF, Cebra-Thomas J, Lilleväli K, Saare M, Chang EY et al (2013) The embryonic transcriptome of the red-eared slider turtle (*Trachemys scripta*). PLoS One 8: e66357
55. Ji X, Chen F, Du W, Chen H (2003) Incubation temperature affects hatchling growth but not sexual phenotype in the Chinese soft-shelled turtle, *Pelodiscus sinensis* (Trionychidae). J Zool 261:409–416
56. Tokita M, Kuratani S (2001) Normal embryonic stages of the Chinese softshelled turtle *Pelodiscus sinensis* (Trionychidae). Zool Sci 18:705–715
57. Wang Z, Pascual-Anaya J, Zadissa A, Li W, Niimura Y, Huang Z et al (2013) The draft genomes of soft-shell turtle and green sea turtle yield insights into the development and evolution of the turtle-specific body plan. Nat Genet 45:701–706
58. Godley BJ, Broderick AC, Frauenstein R, Glen F, Hays GC (2002) Reproductive seasonality and sexual dimorphism in green turtles. Mar Ecol Prog Ser 226:125–133
59. Miller JD (1985) Embryology of marine turtles. In: Gans C, Billet F, Maderson PFA (eds) Biology of the reptilia, Development A, vol 14. Wiley, New York
60. Ferguson MW (1995) Reproductive biology and embryology of the crocodilians. In: Gans C, Billet F, Maderson PFA (eds) Biology of the reptilia, Development A, vol 14. Wiley, New York
61. St John JA, Braun EL, Isberg SR, Miles LG, Chong AY, Gongora J et al (2012) Sequencing three crocodilian genomes to illuminate the evolution of archosaurs and amniotes. Genome Biol 13:415
62. Hua T, Wang C, Chen B (2004) Stages of embryonic development for *Alligator sinensis*. Zool Res 25:263–271
63. Wan Q, Pan S, Hu L, Zhu Y, Xu P, Xia J et al (2013) Genome analysis and signature discovery for diving and sensory properties of the endangered Chinese alligator. Cell Res 23:1091–1105
64. Xu Q, Wilkinson D (1992) *In situ* hybridization of mRNA with hapten labelled probes. In: Wilkinson D (ed) In situ hybridization, a practical approach. Oxford University Press, New York

Chapter 24

MicroCT Imaging on Living Alligator Teeth Reveals Natural Tooth Cycling

Randall B. Widelitz, Alaa Abdelhamid, M. Khalil Khan, Amr Elkarargy, Cheng-Ming Chuong, and Ping Wu

Abstract

To study tooth cycling in polyphyodont animals, we chose to work on alligators. Alligators have teeth in three phases of development at each tooth location. This assembly of three teeth is called a tooth family unit. As part of the study, in order to study tooth cycling in alligators, we wanted to know the configuration of the tooth family unit in every tooth position. From the surface of the mouth, this is difficult to assess. Therefore, we decided to use MicroCT which can image X-ray dense materials providing a three-dimensional view. MicroCT provided us with valuable information for this study. The method described below can be applied to study tooth cycling in other vertebrate species.

Key words Tooth cycling, Monophyodont, Polyphyodont, Tooth family unit, Functional tooth, Replacement tooth

1 Introduction

Our interest in stem cells led us to wonder why humans can only replace their teeth once (diphyodont), yet other animals, especially reptiles, can replace their teeth multiple times (polyphyodont). We sought to explore tooth renewal in a species whose tooth structure and attachment to the jaw was similar to that found in humans. We chose to investigate alligator tooth renewal because they can replace their teeth annually for approximately 50 years [1], their teeth are anchored by sockets implanted in the maxillary/dental bones [2], and their secondary palate resembles that found in humans [3].

Adult alligators have 20 teeth in each quadrant of their mouths [4, 5] (Fig. 1). In alligators, three teeth at different stages of development are present at each tooth position which comprise a tooth family unit. Teeth cycle through phases of initiation, growth, and resting (Fig. 2). These include the functional tooth, the replacement tooth, and the latent dental lamina [4, 6, 7]. In

Guojun Sheng (ed.), *Avian and Reptilian Developmental Biology: Methods and Protocols*, Methods in Molecular Biology, vol. 1650, DOI 10.1007/978-1-4939-7216-6_24, © Springer Science+Business Media LLC 2017

Fig. 1 Alligator skull showing a lateral view of the teeth in the mandible and maxilla (From [11])

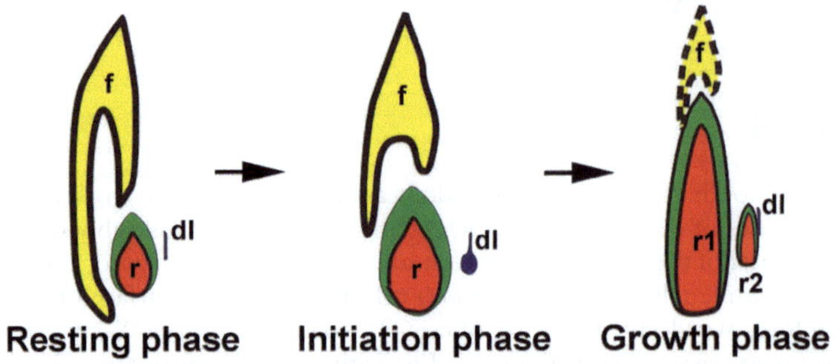

Fig. 2 Diagram showing the three stages (resting, initiation, and growth stages) of tooth development at each tooth position. *f* functional tooth (*yellow*), *r* replacement tooth (*red*), *r1* replacement tooth 1, *r2* replacement tooth 2, *dl* dental lamina

order to begin to understand how alligators have retained their tooth-regenerating capabilities, we performed a number of studies to characterize the organization of the tooth structure. However, we wanted to be sure to collect teeth at appropriate stages. Since we can't determine the state of the tooth family unit from surface observations, we turned to microCT to image tooth family unit development so we could examine the appropriate stages for our studies based on tooth family configurations. Here we present our method for microCT analysis of tooth family unit development.

2 Materials

2.1 Alligator Egg Incubation

Fertilized alligator eggs.

GQF Hova-Bator incubator, Model 1583:

1 group set at 30 °C.

1 group set at 32–33 °C.

Vermiculite.

2.2 Alligator Housing

378 L Rubbermaid bins.

Warm room set at 27 °C with 50% humidity.

2.3 Anesthesia Administration

Isoflurane vaporizer.

Oxygen gas supply.

Flowmeter (0–1000 mL/min).

Induction chamber.

Connection tubing and valves.

Face mask.

Heated chamber to keep anesthetized alligators warm.

2.4 MicroCT

Micro-computed tomography (microCT; MicroCAT II; Siemens).

2.5 Viewing MicroCT Data

ImageJ software.

2.6 3D Reconstruction and Image Analysis

Amira software.

3 Methods

3.1 Alligator Egg Collection

Fertilized alligator eggs were collected from the Rockefeller Wildlife Refuge in Louisiana (*see* **Note 1**). Eggs were incubated at 30–33 °C in a GQF Hova-Bator model 1583 containing vermiculite (Fig. 3). Water was sprinkled on the vermiculite every 48 h to maintain proper humidity. Embryonic alligators were staged according to the work by Ferguson [8]. Alligators hatch after approximately 65 days of incubation. If the eggs are incubated at 30 °C, female offspring are favored, whereas, when incubated at 32–33 °C, males preferentially hatch [9].

3.2 Alligator Housing

Hatchling to 1-year-old juvenile male and female alligators up to 22 in. in length were kept at the University of Southern California vivaria. The alligators are housed in 378 L Rubbermaid bins at 27 °C with 50% humidity. The tanks were placed at an angle on the ground so the alligators could take advantage of dry spots as well as water as deep as 15 cm. All procedures were approved by the local Institutional Animal Care and Use Committee (*see* **Note 2**).

3.3 Micro-computed Tomography

Micro-computed tomography (microCT; MicroCAT II; Siemens) was performed at the University of Southern California's Molecular Imaging Center. To image live animals, alligators were placed in a plexiglass induction chamber in which they are anesthetized with isoflurane/oxygen (2–4% vol/vol). This generally takes about 10 min. Anesthesia was confirmed by pinching the toe to ensure

Fig. 3 Picture of hatching alligator

there was no reaction (*see* **Note 3**). The alligator was then placed on the microCT bed and held in place with tape. A maintenance dose of anesthesia was supplied during microCT imaging via isoflurane/oxygen (2% vol/vol) administered by inhalation through a nose cone. The bed then is drawn in the inner chamber of the MicroCT scanner. The animal is exposed to an X-ray energy level of 250 μA current and 80 kVp with a scan time of about 10–20 min to image the entire anterior head region. The exposure dose was ~10 cGy/scan (*see* **Note 4**). The teeth within the jaw comprise the region of interest. Alligators were then kept warm in a heated chamber and allowed to fully recover from the anesthesia before they were returned to the animal facility. This took from 30 min to several hours depending on how long they were maintained under anesthesia.

3.4 Viewing Sequential X-ray Sections Obtained by MicroCT

MicroCT of the head was performed on three alligators at hatching, 4 months, and 1 year. Fiji software (ImageJ 2.0) was used to view sequential sections that were obtained by microCT. This enabled us to visualize individual planes within the 3D space of the jaw (Fig. 4a) so we can choose appropriate digital images corresponding to the middle section of each tooth family. This view allows us to determine the developmental stage of different tooth family units and judge whether both sides of the jaw develop in parallel or if the two sides differ spatiotemporally. These data also permit comparisons between X-ray density and H&E stained sections showing which regions are epithelial or mesenchymal. H&E staining helps us to clearly see the state of the dental lamina that is not visible in MicroCT.

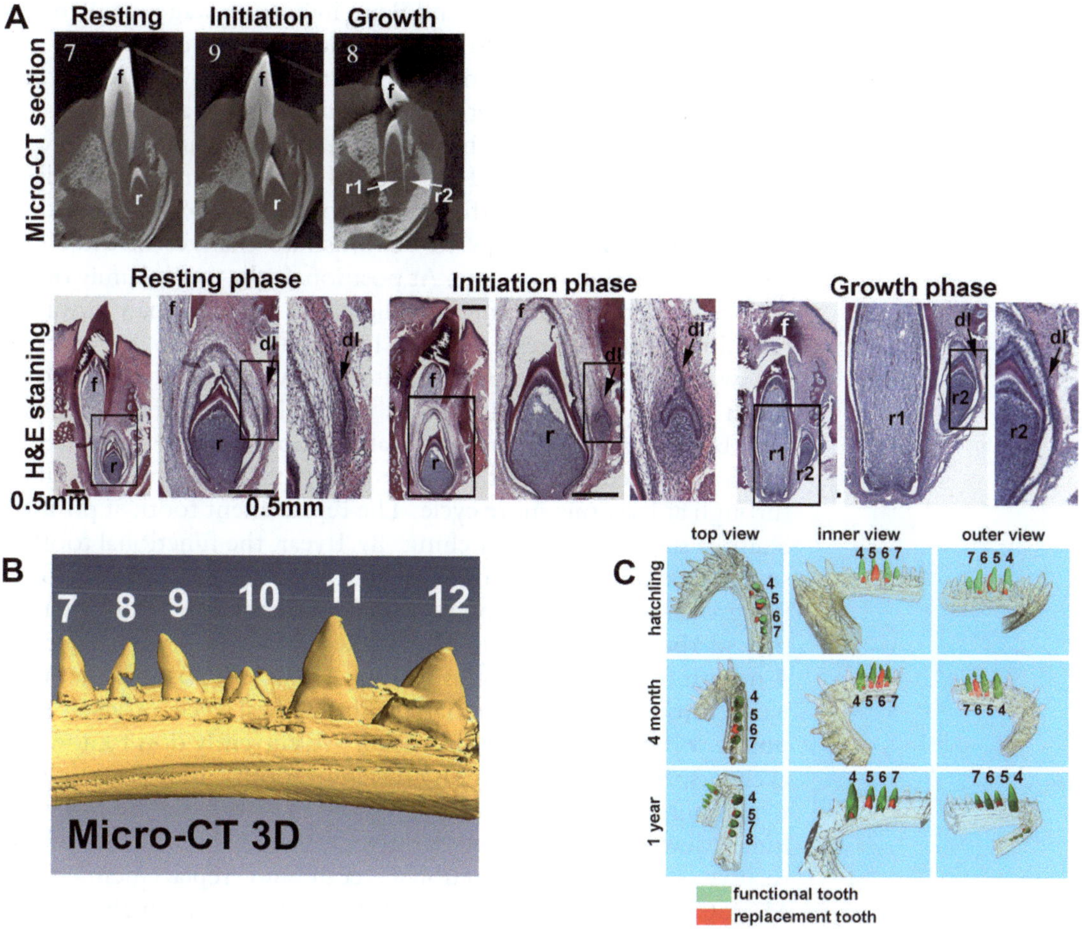

Fig. 4 (**a**) Top panel. MicroCT views using Fiji software show tooth family unit development at resting (tooth family unit 7), initiation (tooth family unit 9), and growth (tooth family unit 8) phases by viewing a single plane of the 3D dataset. Bottom panel. These same stages are shown from H&E staining of sectioned materials. Higher magnification views of the indicated areas are shown. (**b**) 3D visualization of teeth 7–12. (**c**) Tracing of tooth development at different hatching, 4 months and 1 year. *f* functional tooth, *r* replacement tooth, *r1* replacement tooth 1, *r2* replacement tooth 2, *dl* dental lamina. Panels A and C are from [11]

3.5 3D Reconstruction

Amira software was used to make 3D reconstructions in order to evaluate which tooth family to study. 3D reconstructions were used to assess tooth family unit status at each tooth position within the alligator jaw. Regions of interest can be selected to remove extraneous zones from the reconstructed view. Density segmentation can be used to render materials within a select density range to observe an unobstructed view. Choosing appropriate regions of interest allows us to remove extraneous fields that would obscure our view of the tooth families.

3.6 Use of MicroCT to Evaluate Tooth Family Staging

This software enables us to see the relative sizes of the functional tooth and replacement tooth which is necessary to evaluate different tooth family stages (Fig. 4b). From the data presented, we have determined that teeth 7, 8, and 9 are in growth stage, while tooth 10 is in initiation stage, and teeth 11 and 12 are in resting stage. We can also use microCT to measure the rates of growth and cycling during the tooth cycle. Here we imaged the jaw of a single alligator from three time points (hatchling, 4 months, and 1 year old) (Fig. 4c). We highlighted position 4–7 in the right mandible. Here we highlighted the functional tooth (red) and replacement tooth (green) which helps to demonstrate changes in the tooth family progression over time. At position 4, the tooth family of the replacement tooth becomes smaller at 4 months than it was at hatching. This suggests that it has gone through a complete renewal cycle during this time interval. At 1 year the functional tooth is bigger than at 4 months. Since the functional tooth can only become shorter due to root absorption after the replacement tooth is formed, we infer that the tooth at this position has gone through at least one more cycle. The replacement tooth at position 5 also is smaller than at hatching. By 1 year, the functional tooth is smaller (possibly due to root absorption) as the replacement tooth has grown larger. This observation suggests that the tooth family may still be within the same cycle at these two time points. Since the 4-month to 1-year interval is too large, it is still possible that they went through one additional tooth renewal cycle. Closer time points will help to resolve this issue. We currently don't know the cycling speed difference among different tooth family units. This is worth pursuing in the future. We show the jaw from the top view, inner view, and outer view to give a better appreciation of the relative sizes of the functional teeth and replacement teeth. MicroCT also allows us to monitor tooth regeneration after plucking of the functional tooth.

MicroCT can be applied to explore several issues regarding tooth development and renewal in reptiles:

1. *Sequence of alligator tooth formation and renewal.* Alligator tooth family units are not at the same stage. The sequence of their formation can be more precisely determined using microCT. This approach can be used to test existing theories of how teeth are positioned along the anterior-posterior axis of the jaw. These include the Zahreihen theory in which teeth erupt in sequence along this axis versus the concept that two waves progress from the back to the front of the jaw where new teeth are inserted between existing teeth during the second wave of initial tooth family formation [10].
2. *Root formation/decay.* Whether reptile teeth have mammalian-like roots and how the roots get resorbed as the replacement tooth grows bigger.
3. *Tracking the rate of replacement tooth growth after functional tooth extraction.* MicroCT can be used to monitor the state of tooth family unit regeneration after plucking of the functional tooth.

4 Notes

1. *Alligator eggs*: Alligator eggs can be transported at room temperature but should arrive at their destination within 24 h to maintain their survival. The eggs are transported by plane in boxes containing moistened hay or vermiculite. To keep the fertility rate high, alligator eggs should not be exposed to X-ray examination at the luggage check. We have found it is helpful to have letters of approval from the Rockefeller Wildlife Refuge, from the local IACUC, and from the state fish and game department.
2. *Alligator housing*: We have found it to be beneficial to wear thick gloves when handling the alligators. Even young alligators can be aggressive. We recommend that two people assist one another on all procedures involving alligators. The door to the animal room should always be kept closed to prevent escapes.
3. *MicroCT imaging*: During anesthesia, ensure that the nose cone is securely fastened to the alligator. You want to be sure that the alligator is fully anesthetized so it will not awaken during the procedure. Movement will produce imaging artifacts. At the same time, we want to minimize the length of time that the alligator is anesthetized. They may take a long time or even not recover from extended anesthesia.
4. *MicroCT* can only image the hard tissue in the tooth family unit. The dental lamina and its niche are not visible with this method. We are developing ways to improve this shortfall for future studies.

Acknowledgment

The authors thank Ruth Elsey at Rockefeller animal refuge in Louisiana for providing fertilized alligator eggs and Roger Sawyer (University of South Carolina) and Travis Glenn (University of Georgia) for support. Research reported in this publication was supported by the National Institute of Arthritis and Musculoskeletal and Skin Diseases of the National Institutes of Health under Award Numbers AR 47364 and AR 60306 of NIH in the USA and the National Plan for Science, Technology and Innovation (MAARIFAH), King Abdulaziz City for Science and Technology, Kingdom of Saudi Arabia, Award Number 11-BIO 2216-09. The authors declare no potential conflicts of interest with respect to the authorship and/ or publication of this article.

References

1. Poole DFG (1961) Notes on tooth replacement in the Nile crocodile, Crocodilus niloticus. J Zool 136:131–140
2. McIntosh JE et al (2002) Caiman periodontium as an intermediate between basal vertebrate ankylosis-type attachment and mammalian "true" periodontium. Microsc Res Tech 59(5):449–459
3. Ferguson MW (1981) Developmental mechanisms in normal and abnormal palate formation with particular reference to the aetiology, pathogenesis and prevention of cleft palate. Br J Orthod 8(3):115–137
4. Ferguson MW (1981) Review: the value of the American alligator (*Alligator mississippiensis*) as a model for research in craniofacial development. J Craniofac Genet Dev Biol 1 (1):123–144
5. Osborn JW (1975) Tooth replacement: efficiency, patterns and evolution. Evolution 29:180–186
6. Westergaard B, Ferguson MWJ (1986) Development of the dentition in Alligator mississippiensis: early development in the lower jaw. J Zool 210:575–597
7. Westergaard B, Ferguson MWJ (1987) Development of the dentition of *Alligator mississippiensis*: later development in the lower jaws of embryos, hatchlings and young juveniles. J Zool 212:191–222
8. Ferguson M (1985) In: Gans C, Billet F, Maderson P (eds) Biology of the reptilia. Wiley, New York
9. Lang JW, Andrews HV (1994) Temperature-dependent sex determination in crocodilians. J Exp Zool 270(1):28–44
10. Westergaard B, Ferguson MW (1990) Development of the dentition in *Alligator mississippiensis*: upper jaw dental and craniofacial development in embryos, hatchlings, and young juveniles, with a comparison to lower jaw development. Am J Anat 187(4):393–421
11. Wu P et al (2013) Specialized stem cell niche enables repetitive renewal of alligator teeth. Proc Natl Acad Sci U S A 110(22): E2009–E2018

Index

Guojun Sheng (ed.), *Avian and Reptilian Developmental Biology: Methods and Protocols*, Methods in Molecular Biology, vol. 1650, DOI 10.1007/978-1-4939-7216-6, © Springer Science+Business Media LLC 2017

F

G

I

J

L

M

N

O

P

MIX
Papier aus verantwortungsvollen Quellen
Paper from responsible sources
FSC® C105338

If you have any concerns about our products,
you can contact us on
ProductSafety@springernature.com

In case Publisher is established outside the EU,
the EU authorized representative is:
Springer Nature Customer Service Center GmbH
Europaplatz 3, 69115 Heidelberg, Germany

Printed by Libri Plureos GmbH
in Hamburg, Germany